Practical Temperature Measurement

This book is dedicated to Fiona

Practical Temperature Measurement

Peter R.N. Childs B.Sc. (Hons), D.Phil., C.Eng., F.I.Mech.E.
Reader, University of Sussex

OXFORD   AUCKLAND   BOSTON   JOHANNESBURG   MELBOURNE   NEW DELHI

Butterworth-Heinemann
Linacre House, Jordan Hill, Oxford OX2 8DP
225 Wildwood Avenue, Woburn, MA 01801-2041
A division of Reed Educational and Professional Publishing Ltd

℞ A member of the Reed Elsevier plc group

First published 2001

British Library Cataloguing in Publication Data
Childs, Peter R.N.
  Practical temperature measurement
  1. Temperature measurements
  I. Title
  536.5'027

Library of Congress Cataloguing in Publication Data
Childs, Peter R.N.
  Practical temperature measurement/Peter R.N. Childs.
  p. cm.
  Includes bibliographical references and index.
  ISBN 0 7506 5080 X
  1. Temperature measurements.   2. Temperature measuring
  instruments.   I. Title
  QC271.C45 2001
  536'.5'0287–dc21   2001037460

ISBN 0 7506 5080 X

For information on all Butterworth-Heinemann publications visit our
website at www.bh.com

Composition by Genesis Typesetting, Rochester, Kent
Printed and bound in Great Britain by Biddles Ltd
www.biddles.co.uk

FOR EVERY TITLE THAT WE PUBLISH, BUTTERWORTH-HEINEMANN
WILL PAY FOR BTCV TO PLANT AND CARE FOR A TREE.

Contents

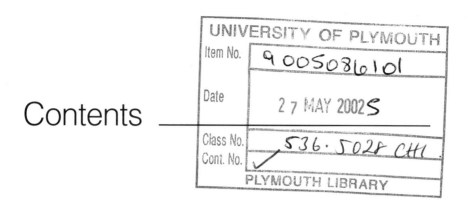

Preface _____

Temperature is both a thermodynamic property and a fundamental unit of measurement. Its measurement is critical to many aspects of human activity from the thermodynamic improvement of heat engines to process control and health applications. Current estimates of the value of the temperature measurement market run at approximately 80% of the sensor market. The range of methods and devices available for temperature measurement are extensive. Options include invasive or contact methods such as thermocouples and resistance thermometers to non-invasive techniques using, for example, infrared detectors. In addition, recent developments in optical methods and micro-manufacturing have resulted in the wider spread availability and use of advanced techniques such as coherent anti-Stokes Raman scattering and thin-film transducers for temperature measurement. The aims of this text are to introduce the concepts of temperature and its measurement, to describe the range of techniques and specific devices available for temperature measurement and to provide guidance for the selection of a particular method for a given application.

The concept of temperature, its definition and practical modelling are described in Chapter 1. Both the thermodynamic temperature scale and the International Temperature Scale of 1990 are considered. General considerations of temperature measurement are explored in Chapter 2 including thermal disturbance effects for both solids and fluids. Consideration is given to steady-state measurements and transient measurements with quantification of time response and phase lag. Critical to measurement of temperature is calibration as this provides the quantitative validation for the uncertainty of a measurement. An indication of a temperature can be worthless without information on the calibration. Methods of calibration are introduced in Chapter 2 and also in subsequent chapters where specific methods and sensors are described. Related to calibration and quantification of uncertainty is the concept of traceability, which describes the management of undertaking temperature measurement, and this is also introduced in Chapter 2.

Practical methods of temperature measurement are introduced in Chapters 3–10. For convenience the methods are categorized according to the degree of contact between the medium of interest and the measurement device. In Chapters 3–7 details of invasive measurement methods where the transducer is in direct contact with the medium such as a thermocouple embedded in a surface are given. In Chapter 8 methods where, say, a surface is treated to facilitate the temperature measurement but observed remotely are considered. An example is the use of thermochromic liquid crystals that change colour with temperature. Methods where the undisturbed medium is observed remotely are described in Chapters 9 and 10.

The range of techniques and sensors available for temperature measurement is extensive. Developments in the areas of micro-manufacture, laser technology and data processing have resulted in an increase and wider availability of measurement techniques. Consequently, where measurements might once have been made with one technology another may now be more appropriate. Chapter 11 provides a guide for the appropriate selection of measurement technique based on the demands of range, uncertainty, sensitivity, life, size, cost, manufacturing constraints, dynamic response, temperature of operation and robustness.

Related to the measurement of temperature is the measurement of heat flux. Heat flux measurement is used in the field of fluid mechanics and heat transfer to quantify the transfer of heat within systems. Several techniques are in common use, including: differential temperature sensors such as thermopile, layered resistance temperature devices or thermocouples and Gardon gauges; calorimetric methods involving a heat balance analysis and transient monitoring of a representative temperature, using, for example, thin-film temperature transducers or temperature-sensitive liquid crystals; energy supply or removal methods using a heater to generate a thermal balance; and finally by measurement of mass transfer which can be linked to heat transfer using the analogy between heat and mass transfer. The various types of heat flux sensors available as well as unique designs for specific applications are described in Chapter 12.

The framework adopted for this text involves description and definition of the physical phenomena involved prior to descriptions of temperature measurement methods and specific sensors. This allows a meaningful appreciation of the method of measurement to be developed and as a result a deeper understanding of its strengths and weaknesses. Descriptions of sensors are accompanied by schematic diagrams, photographs and circuit diagrams thereby facilitating visualisation and practical usage. Nomenclature has been defined both within the text and at the end of each chapter.

This book will be of value to engineering and physics undergraduates studying modules on instrumentation and process control and for practical project work requiring an understanding of temperature measurement methods. Specific undergraduate modules for which this book has applications include Measurement and Instrumentation, Sensors, Mechanical Measurement Technology, Testing and Instrumentation and Process Control. For postgraduates and industrialists faced with the task of selecting a particular measurement method or sensor for an experiment, product or process this text provides both thorough descriptions of the various techniques and guidance for their selection.

In writing this book advice and assistance from a number of sources has been given. I would like to express my gratitude to a former doctorate student Joanne Greenwood for her diligence and enthusiasm in her work on thin-film sensors. The background reading for this research resulted in a number of review papers, which were useful in the preparation of this book. I would also like to thank my colleagues Christopher Long, for providing encouragement at the right time, Alan Turner for his continuing inspiration as a practical engineer and Val Orringe for typing some of the tables. Finally I would like to thank the Editor, Matthew Flynn, for knowing when to accept excuses on delays and when to push!

Peter R.N. Childs

1

Temperature

The aims of this chapter are to introduce the subject of temperature and its measurement. Qualitative and quantitative definitions of temperature are given in Section 1.1 prior to the development of temperature scales in Section 1.2. An overview of measurement considerations is provided in Section 1.3 along with a brief introduction to the techniques available for the measurement of temperature.

1.1 Definition of temperature

Temperature is one of the seven base SI (Le Système International d'Unités, Bureau International des Poids et Mesures (1991)) units, the others being the mole, kilogram, second, ampere, candela and metre. Many physical processes and properties are dependent on temperature and its measurement is crucial in industry and science with applications ranging from process control to the improvement of internal combustion engines.

The concept of temperature is familiar to us from our day-to-day experience of hot and cold objects. Indeed temperature can be defined qualitatively as a measure of hotness. Systems or objects can be ranked in a sequence according to their hotness and each system assigned a number, its temperature. Linked to the concept of hotness is heat transfer, the flow of thermal energy. It is a common experience that heat transfer will occur between a hot and a cold object. Temperature can be viewed accordingly as a potential, and temperature difference as the force that impels heat transfer from one object or system to another at a lower temperature.

The term 'system' is used to define a macroscopic entity, that is, one consisting of a statistically meaningful number of particles, that extends in space and time. An example of a system is the content of an internal combustion engine cylinder with the valves closed. Such a system can be described by specifying the composition of the substance, volume, pressure and temperature. A system is affected in two ways when its temperature rises. There is an increase in the disordered thermal motion of the constituents. The hotter a system is, the faster its particles move or vibrate. Similarly, the colder a system is, the slower the particles will move or vibrate, with a limit occurring when the particles can be considered to be in their most ordered state. As an example consider heating a solid whose atoms are initially vibrating in a lattice. As the temperature rises the atoms will vibrate more vigorously until a point is reached where the atoms can slide past one another and are not held in a fixed position and the substance is classed as a liquid. If more heating is provided raising the temperature further, then atoms will break away from each other breaking any bonds and become a gas. Further heating again will raise the temperature higher causing the molecules to increase their speed to a point where violent collisions between molecules ionizes them turning the gas into a plasma, containing ions and electrons. In addition to the disordered thermal motion of molecules, raising the temperature also excites higher energy states within the constituents of the system.

Although temperature as a concept is very familiar to us, its detailed definition has occupied the attention of many scientists for centuries. Much of our current understanding has come from the science of thermodynamics, the study of heat and work and the relevant physical properties. There are two approaches to thermodynamics: classical and statistical. Classical thermodynamics is predominantly concerned with the use of heating processes to do work, whilst statistical thermodynamics is concerned with linking quantum theory with the properties of matter in equilibrium. Either approach yields the same result for the definition of temperature. A summary of some of the definitions developed to describe temperature is provided in Table 1.1.

Quantitatively, temperature can be defined from the second law of thermodynamics as the rate of change of entropy with energy at constant volume, equation (1.1) (see, for example, Baierlein, 1999):

$$T = \frac{1}{(\partial S/\partial E)_v} \qquad (1.1)$$

where: T = absolute temperature (K)
S = entropy (J/K)
E = energy (J)
V = volume (m^3) or some other fixed external parameter.

Table 1.1 Some descriptions of temperature

Temperature is defined as the degree of hotness or coldness of a body.	M. Planck
The temperature of a system is a property that determines whether a system is in thermal equilibrium with other systems.	Zemansky and Dittman (1981)
Temperature is the parameter of state that is inversely proportional to the rate of change of the log of the number of accessible states as a function of energy for a system in thermal equilibrium.	Quinn (1990)

The definition given in equation (1.1) can appear abstract and for many applications the notion of temperature as a measure of hotness and temperature difference as a potential for the transfer of heat energy from one region to another is quite adequate. Nevertheless a quantitative definition of temperature is the basis of a substantial proportion of science, and it is a necessity for some temperature measurement applications, particularly those where temperatures are low or varying rapidly.

1.2 Temperature scales

In order to provide meaningful comparisons of temperature measurements made by different people it is useful to define a common scale. The definition of temperature scales is an arbitrary undertaking. One approach, referred to as the thermodynamic temperature scale, provides a linear scale that is valid for any substance and temperature range. The thermodynamic temperature scale is based on the ideal reversible Carnot engine cycle. The Carnot cycle consists of isothermal heat transfer at a high temperature, adiabatic expansion, isothermal heat transfer at a low temperature and adiabatic compression back to the high temperature. The Carnot engine efficiency is given by

$$\eta = 1 - \frac{T_2}{T_1} \tag{1.2}$$

where: η = efficiency,
T_1 = the high value of temperature at which isothermal heat transfer takes place in the cycle,
T_2 = the low value of temperature at which isothermal heat transfer takes place in the cycle.

Examination of equation (1.2) provides some insight into the concept of temperature and defines some bounds. Efficiency can never be greater than unity otherwise one would obtain more work out of a system than put in; a gross violation of the laws of physics. Efficiency equal to unity can only be achieved theoretically if T_2 is equal to zero or as T_1 approaches infinity. T_2 equal to zero therefore defines the lowest possible theoretical limit for temperature. As efficiency cannot be greater than unity, T_2 cannot normally be negative although negative absolute temperatures are possible and an example of this are temperatures experienced in a laser (see Purcell and Pound, 1951; Ramsey, 1962; Baierlein, 1999 for further insight into this subject).

The thermodynamic temperature scale provides a means of defining temperature in terms of equal increments of work outputs from ideal Carnot engines operating between two temperatures. In order to define a temperature scale all that is necessary is to decide the size of the increment. The two fixed temperatures used for the thermodynamic temperature scale are zero and the triple point of water. The triple point for a substance is the condition where solid, liquid and vapour phases co-exist simultaneously and this occurs at a unique pressure. The numerical value assigned to the triple point of water is 273.16. The SI unit of temperature is the kelvin, symbol K, and is defined as the fraction 1/273.16 of the temperature of the triple point of water.

The Celsius scale is also used to express temperature. The unit of Celsius temperature is the degree Celsius, symbol °C, and the magnitude of one degree Celsius is numerically equal to one kelvin. Temperature in degrees Celsius is related to that in kelvin by the equation

$$t = T - 273.15 \ (°C) \tag{1.3}$$

where: t = temperature in degrees Celsius (°C),
T = temperature in kelvin (K).

Other scales in use include the Fahrenheit and Rankine scales with symbols °F and R respectively. The conversion relationships between these are given in equations (1.4) and (1.5):

$$T|_{°F} = 1.8t + 32 \ (°F) \tag{1.4}$$

where: t = temperature in degrees Celsius (°C),

$$T|_{R} = T|_{°F} + 459.67 \ (R) \tag{1.5}$$

Whilst useful as an ultimate baseline, the thermodynamic temperature scale is not particularly practical. It is not actually possible to manufacture engines that operate on the Carnot cycle as minor inefficiencies in practical devices cause departures from the ideals demanded. Neither would it be desirable to have set up an elaborate configuration of thermodynamic cycles just to measure temperature. As a result, more practical methods have been proposed to define the temperature scale. The current internationally agreed scale is the International Temperature Scale of 1990 (ITS-90), which is described in Section 1.2.1.

1.2.1 The International Temperature Scale of 1990

The International Temperature Scale of 1990 (Preston-Thomas, 1990) is intended to be a practical internationally agreed best approximation to the thermodynamic temperature scale. It extends from 0.65 K up to the highest temperature practically measurable using the Planck radiation law (see Chapter 9, Section 9.2.1) and is believed to represent thermodynamic temperature within about ±2 mK from 1 to 273 K, ±3.5 mK (one standard deviation limits) at 730 K and ±7 mK at 900 K (Mangum and Furukawa, 1990). The ITS-90 is constructed using a number of overlapping temperature ranges. The ranges are defined between repeatable conditions using so-called fixed points such as the melting, freezing and triple points of a variety of materials. Fixed points are convenient as the conditions at which, say, the freezing of aluminium occurs can be set up in a highly reproducible fashion. The temperatures assigned to these are provided from the best estimates, using thermometers of an approved type, at the time of formulation of the ITS-90. Intermediate temperatures between the fixed points are determined by interpolation using specified equations. The five temperature ranges used in the ITS-90 are:

- 0.65 to 5 K defined in terms of the vapour pressures of helium 3 and helium 4
- 3 to 24.5561 K using a constant volume gas thermometer
- 13.8033 to 273.16 K using a platinum resistance thermometer
- 273.15 to 1234.93 K using a platinum resistance thermometer
- 1234.94 K and above using the Planck law of radiation.

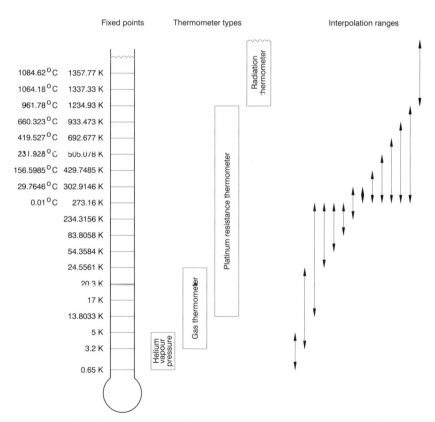

Figure 1.1 Simplified overview of fixed points and devices used for the ITS-90. (Data from Preston-Thomas, 1990.) Figure after Nicholas and White (1994)

Table 1.2 Defining points of the ITS-90 (Preston-Thomas, 1990)

Number	Temperature (K)	Temperature (°C)	Substance	State
1	3 to 5	−270.15 to −268.15	He	Vapour pressure
2	13.8033	−259.3467	H_2	Triple point
3	≈17	≈−256.15	H_2 or He	Vapour pressure
4	≈20.3	≈−252.85	H_2 or He	Vapour pressure
5	24.5561	−248.5939	Ne	Triple point
6	54.3584	−218.7916	O_2	Triple point
7	83.8058	−189.3442	Ar	Triple point
8	234.3156	−38.8344	Hg	Triple point
9	273.16	0.01	H_2O	Triple point
10	302.9146	29.7646	Ga	Melting point
11	429.7485	156.5985	In	Freezing point
12	505.078	231.928	Sn	Freezing point
13	692.677	419.527	Zn	Freezing point
14	933.473	660.323	Al	Freezing point
15	1234.93	961.78	Ag	Freezing point
16	1337.33	1064.18	Au	Freezing point
17	1357.77	1084.62	Cu	Freezing point

An overview of the fixed points, ranges and devices used in defining the ITS-90 is given in Figure 1.1. Table 1.2 provides details of the fixed points used and the temperatures assigned to these.

The significance of the ITS-90 is its value in providing a means of traceability between measurements taken in practical applications and a close approximation to thermodynamic temperature. This is illustrated schematically in Figure 1.2. In order for measurements to be traceable back to the ITS-90, sensors used for a given application should be calibrated against sensors that have a direct link to the ITS-90 (e.g. see UKAS, 2000). This can be achieved in a number of ways. A typical chain for an industrial measurement sensor might be calibration against another sensor that has itself been calibrated by a standards laboratory against a sensor that has been calibrated by a national standards laboratory. In this way the ITS-90 acts as a transfer standard of near thermodynamic temperatures to popular usage.

1.3 An overview of temperature measurement techniques

Temperature cannot be measured directly. Instead the effects on some other physical phenomena must be observed and related to temperature. There are many physical phenomena that are dependent on temperature such as

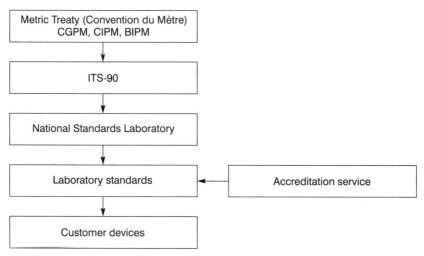

Figure 1.2 Recommended practice to ensure definition and traceability of temperature measurements. (CGPM = Conférence Générale des Poids et Mesures, CIPM = Comité International des Poids et Mesures, BIPM = Bureau International des Poids et Mesures)

resistance, volumetric expansion, vapour pressure and spectral characteristics. Many such phenomena have been exploited to produce devices to measure temperature. For convenience the various temperature measurement techniques can be classified according to the nature of contact between the medium of interest and the device. The categories used here are invasive, semi-invasive and non-invasive.

- Invasive techniques are those where the transducer is in direct contact with the medium of interest. An example is a liquid in glass thermometer immersed in a liquid.
- Semi-invasive techniques are those where the surface of interest is treated in some way to enable remote observation. An example is the use of thermochromic liquid crystals, which change colour with temperature. A surface can be sprayed with these and then observed remotely with, say, a CCD (charged coupled device) camera.
- Non-invasive techniques are those where the medium of interest is observed remotely. An example is the use of infrared thermometry where the sensor is located some distance away from the target material.

The terms 'sensor' and 'transducer' are commonly used in discussions on instrumentation. 'Sensor' is used here to describe the temperature-measuring device as a whole, while the term 'transducer' is used to define the part of the sensor that converts changes in temperature to a measurable quantity.

Temperature measurement sensors associated with invasive instrumentation include liquid-in-glass thermometers, bimetallic strips, thermocouples, resistance temperature detectors and gas thermometers. The traditional liquid-in-glass thermometer comprises a reservoir and capillary tube supported in a stem (Figure 1.3). The reservoir contains a liquid, which expands with temperature forcing a column of liquid to rise in the capillary tube. Liquid-in-glass thermometers permit a quick visual indication of temperature and can with care produce very low uncertainty results. Bimetallic thermometers use two strips of material with differing coefficients of thermal expansion that are bonded together. When the temperature of the assembly is changed, then in the absence of external forces, the strip will bend due to the difference in expansion between the two materials. The extent of bending can be utilized as a measure of temperature. Bimetallic strips are robust and relatively cheap. They are widely used as temperature control devices. Thermocouples exploit the generation of an electromotive force by a conductor when it experiences a thermal gradient. In its simplest practical form a thermocouple can consist of two dissimilar wires connected together at one end with a voltage-measuring device connected across the free ends (a commercial thermocouple and indicator are shown in Figure 1.4). Thermocouples are a common choice in industry and research because of their wide temperature range, rugged nature and low cost. The variation of resistance with temperature is utilized by a number of measuring devices such as platinum resistance thermometers, thermistors and semiconductor-based transducers. A typical resistance temperature detector might comprise the transducer consisting of a portion of material whose resistance is sensitive to temperature mounted in a housing and connected to an electronic circuit that measures the resistance and converts this value to a meaningful measure of temperature for display and data recording. A commercial platinum resistance probe and associated measurement and display system is illustrated in Figure 1.5. The behaviour of many gases is well represented by the ideal gas law. Gas thermometry exploits this relationship by measurement of pressure and volume to determine an unknown temperature.

One of the major drawbacks associated with invasive instrumentation is the distortion of the temperature distribution in the application. The insertion of a temperature probe will inevitably alter the temperature of the application. An

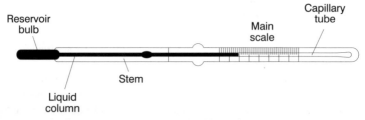

Figure 1.3 Principal features of a liquid-in-glass thermometer

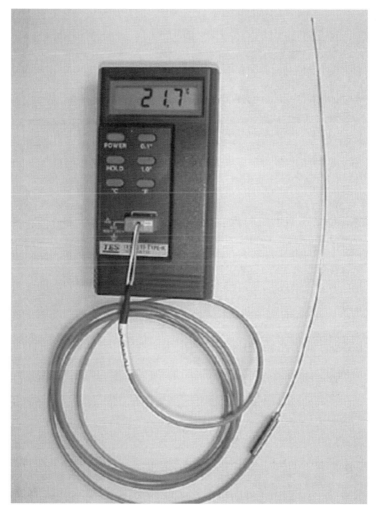

Figure 1.4 A stainless steel sheathed Type K thermocouple and associated handheld display

example is the use of a thermocouple embedded on the surface of a turbine disc. The materials used for the thermocouple and any cement used to hold it in place will have different thermal conductivities from those of the disc and the temperature distribution will be altered as a result. The level of disturbance caused for a given application must be assessed and if significant accounted for or an alternative method considered. The issue of thermal disturbance is considered in Chapter 2.

Semi-invasive techniques where the surface of interest is treated in some manner to facilitate observation at a distance include the use of liquid crystals,

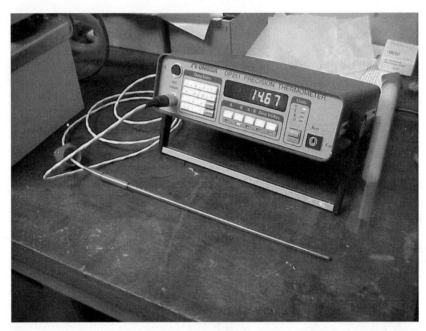

Figure 1.5 A laboratory-use platinum resistance thermometer and associated measurement system and display

heat-sensitive crystalline solids and paints, and thermographic phosphors. Cholesteric liquid crystals are optically active and react to changes in temperature and shear stress by reversible changes in colour and are popularly known as thermochromic liquid crystals. They are available in the form of a water-based slurry and can be applied to a surface by means of painting or spraying. The colours displayed can be recorded by means of a video and subsequently analysed to give an indication of temperature for each pixel location. The fluorescence of certain phosphors is temperature dependent and these materials can also be applied to a surface and observed remotely by an optical system to provide a measure of temperature. These techniques have the merit that they provide a measurement of the variation in temperature over the surface. In order to provide an indication that a certain temperature has been reached, a number of devices in the form of crayons, pellets and paints are available that melt above a certain limit.

Non-invasive temperature measurement, where the medium of interest is observed remotely, include techniques such as infrared thermometry, refractive index methods, absorption and emission spectroscopy, line reversal, spontaneous Rayleigh and Raman scattering, coherent anti-Stokes Raman scattering, laser-induced fluorescence, speckle and acoustic methods. Of these, infrared techniques are the most popular and are based on the measurement of electromagnetic radiation. A typical infrared temperature measurement system may comprise the target, the medium through which the radiant energy is transmitted, an optical system to gather the electromagnetic radiation, a transducer to convert the radiation into a signal related to temperature and amplification and interface circuitry to control, display and record the temperature measurement. Using techniques based on infrared thermometry, thermal imaging provides measurements of the variation of temperature over a region. A commercial handheld thermal imager, designed to be aimed at the target of interest, is illustrated in Figure 1.6. Non-invasive temperature techniques provide a number of significant advantages over invasive instrumentation. Invasive instrumentation must be capable of surviving and operating at the temperatures concerned. In high-temperature or chemically reactive applications such as flames and plasmas the performance of invasive instrumentation can degrade significantly with time if it can

Figure 1.6 A handheld thermal imager designed to be aimed at the target of interest. Photograph courtesy of FLIR Systems Ltd

Table 1.3 Considerations in the selection of a technique for the measurement of temperature

Criteria	Considerations
Temperature range	The sensor selected should be capable of monitoring the minimum and maximum expected temperatures.
Robustness	The sensor selected should be mechanically strong enough to withstand the environment and conditions to which it is subjected such as vibration, mechanical shock and high pressures. In addition consideration must also be given to the physical compatibility of the device to the environment to ensure that the sensor does not chemically or physically degrade.
Disturbance	The insertion of a measurement probe onto or into the medium of interest will cause distortion of the temperature distribution in comparison to the undisturbed case. This must be assessed and if significant then compensated for, or if this is not possible an alternative method of temperature measurement should be considered.
Signal	Many transducers produce an output in the form of an electrical signal. The size and form of the signal is dependent on the type of transducer. Some transducers produce a relatively small signal that is sensitive to electromagnetic interference (electrical noise). Some transducers have high sensitivity where there is a significant change in the signal for a given change in temperature. Devices should be selected that provide characteristics compatible with the purpose of the measurement.
Response	The speed of response of different sensors to a change in temperature varies according to local levels of heat transfer and the thermal properties of the sensor. The speed of response is often characterized as a time constant and for rapid variations in temperature a device with a short time constant compatible with that of the transient is necessary in order for the sensor to register the time-varying temperature.
Uncertainty	Uncertainty defines the closeness of measurement to the true thermodynamic temperature. The uncertainty varies for different devices and can alter over the lifetime of an individual sensor. Uncertainty can be quantified within specified limits by regular calibration. A compounding issue in an application is that the temperature indicated by a sensor is that for the disturbed case where the installation of the sensor may have distorted the temperature distribution.

Criteria	Considerations
Calibration	The relationship between the temperature indicated by a sensor and the corresponding known values should be checked periodically. This task can be undertaken, for some sensors, by the user or in others should be performed in a specialist laboratory. The complexity and cost associated with calibration can affect the choice of sensor.
Availability	It is often important to ensure ready supply of a temperature sensor. The majority of the techniques described in this book are commercially available. Some methods, however, require a substantial skill level in their application and cannot be regarded as an off the shelf method. Other sensors, such as some thermocouples, are available from a large number of commercial suppliers.
Cost	The variation in cost between different temperature measurement techniques is substantial. For example, the cost of a thermal imaging sensor can be of the order of 1000 times more expensive than a thermocouple and associated panel display. In addition the variation in cost between different sensors within the same general category is extensive with, for example, a platinum–rhodium thermocouple costing ten times as much as a constantan thermocouple.
Size	The size of temperature measurement transducers and associated processing and display equipment varies considerably. Generally the smaller the measurement probe, the smaller the thermal disturbance but the less robust the device. It is also necessary to consider the size of the processing equipment as this will have to be installed as well.
Legal	An increasing number of applications must be considered with reference to the relevant national or international standard and temperature measurements undertaken accordingly. Examples include the monitoring of foodstuffs and the measurement of clinical temperature.

function at all. In addition non-invasive instrumentation does not normally involve the level of disruption to the temperature distribution of the application associated with invasive instrumentation but it can require an extensive skills base and be very expensive.

Each method of temperature measurement has different merits and requires particular considerations in an application. For example, liquid-in-glass thermometers are relatively fragile and bulky but can provide very low uncertainty temperature measurements and are self-contained giving a direct visual indication of temperature. If an application required a small compact sensor that could withstand high levels of mechanical shock or vibration then a liquid-in-glass thermometer would obviously be inappropriate. The selection of a particular type of temperature measurement technique therefore requires careful consideration to ensure that the measurement will function and give worthwhile information. In addition consideration must be given to a whole series of factors such as economic and legal aspects. A measurement system will often require a transducer and some form of sensing electronics to convert the signal into a meaningful measure of temperature that can be recorded and displayed. The transducer, such as a thermocouple, may be relatively cheap in comparison to, say, a liquid-in-glass thermometer but the associated sensing electronics and display for the thermocouple could elevate the cost beyond that of a competing method. As a result an overall or holistic approach to temperature measurement technique selection considering all relevant criteria must be undertaken. Some of the relevant aspects are listed in Table 1.3. The various criteria are considered in the introduction of relevant temperature measuring techniques in Chapters 3–10 and are described further in Chapter 11 which concentrates on the selection process.

1.4 Conclusions

Temperature can be defined qualitatively as a measure of hotness and temperature difference between two locations can be viewed as the force that impels the flow of thermal energy from one location at a higher temperature to another at a lower value. In order to ensure that measurements of temperature are made on a consistent basis, it is necessary to use a common system of units. The SI unit of temperature is the kelvin, symbol K, which is defined as the fraction 1/273.16 of the temperature of the triple point of water. Quantitative values can be assigned to temperature using a thermodynamic temperature scale. This was originally developed based on the Carnot cycle and provides a definition that is independent of the properties of any one material. Thermodynamic temperature scales are not, however, particularly practical and in their place the International Temperature Scale of 1990 is currently used. The ITS-90 matches the thermodynamic temperature scale closely but uses practical sensors such as platinum resistance thermometers to provide a measure of temperature. Temperatures measured in practical applications can then be related to temperatures undertaken in accordance with the ITS-90 thereby providing traceability and conformance to an internationally agreed temperature scale.

Temperature cannot be measured directly. Instead the effects of temperature on some other medium must be observed. In many applications this involves the insertion of a temperature sensor into direct thermal contact with the medium of interest and subsequent distortion of the temperature.

The implications of the choice of temperature measurement technique on the medium of interest are described in Chapter 2. Temperature measurement techniques can be classified according to the level of contact between the medium of interest and the temperature sensor as invasive, semi-invasive or non-invasive. These techniques are described in Chapters 3–7, 8 and 9–10 respectively. The number of temperature measuring sensors is extensive and the selection of an appropriate method for a given application can make the difference between worthwhile and wasted effort. Chapter 11 provides an overview of the capabilities of the various techniques described and a guide for selection. Related to the measurement of temperature is the measurement of heat flux and a review of appropriate methods is provided in Chapter 12.

References

Baierlein, R. *Thermal Physics*. Cambridge University Press, 1999.

Bureau International des Poids et Mesures. *Le Système International d'Unités*, 6th edition, 1991.

Mangum, B.W. and Furukawa, G.T. Guidelines for realizing the ITS-90. NIST Technical Note 1265, 1990.

Nicholas, J.V. and White, D.R. *Traceable Temperatures*. Wiley, 1994.

Preston-Thomas, H. The international temperature scale of 1990 (ITS-90). *Metrologia*, **27**, 3–10, 1990.

Purcell, E.M. and Pound, R.V. A nuclear spin system at negative temperature. *Phys. Rev.*, **81**, 279–280, 1951.

Quinn, T.J. *Temperature*, 2nd edition. Academic Press, 1990.

Ramsey, N.F. Thermodynamics and statistical mechanics at negative absolute temperatures. In Herzfeld, C.M. (Editor), *Temperature. Its Measurement and Control in Science and Industry*, Vol. 3, Part 1, pp. 15–20, Reinhold, 1962.

UKAS (United Kingdom Accreditation Service). Traceability of temperature measurement. Platinum resistance thermometers, thermocouples, liquid-in-glass thermometers and radiation thermometers. UKAS Publication LAB 11, 2000.

Zemansky, M.W. and Dittman, R.H. *Heat and Thermodynamics*. 6th edition. McGraw-Hill, 1981.

Nomenclature

E = energy (J)
S = entropy (J/K)
t = temperature in degrees Celsius (°C)
T = absolute temperature in kelvin (K)
V = volume (m^3) or some other fixed external parameter
η = efficiency

2
General temperature measurement considerations

The measurement of temperature involves complex interactions both between the sensor and the medium of interest and within the measurement system itself. The aims of this chapter are to introduce the measurement process and to provide a guide to the manipulation and assessment of the resulting data.

2.1 The measurement process

A measurement informs us about a property of something and gives a magnitude to that property. Temperature can be measured by observing a physical phenomena that is temperature dependent. This may involve inserting a probe containing a transducer into the medium of interest and relating the observed effect on the transducer to temperature. A general arrangement for the measurement of temperature is illustrated in Figure 2.1.

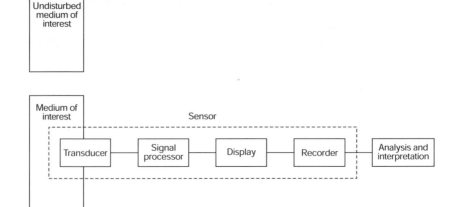

Figure 2.1 A general arrangement for the process of temperature measurement

Here the measurement system comprises a transducer to convert a temperature-dependent phenomena into a signal that is a function of that temperature, a method to transmit the signal from the transducer, some form of signal processing, a display and method of recording the data. In a given application some of these functions such as data recording may be undertaken by the sensor system itself or by a human operator. One example of the latter is the use of a liquid-in-glass thermometer. In this case the transducer is the liquid contained within the bulb which expands with temperature. The observed expansion along the capillary can be related to temperature by use of a scale on the thermometer that has been engraved on the stem following a calibration process relating the meniscus location to the bulb temperature. The recording of data, in this case, is undertaken by a human observer.

The temperature measurement process can be broadly considered to comprise the following discrete processes:

1 It starts with the undisturbed object or medium whose temperature is to be determined.
2 The transducer or sensor is brought into contact with the medium.
3 The transducer converts changes in temperature into a measurable quantity.
4 A calibration is used to convert the measured quantity into a value of temperature.
5 The temperature can then be recorded and its significance subsequently evaluated.

Whenever a measurement is made it is unlikely that the measured value will equal the true value; the actual or exact value of the measured variable. The difference between the measured value and the true value is the error. There are a number of reasons why a measured value will not equal the true value and many of these can be inferred by considering the measurement process illustrated in Figure 2.1. The insertion of a transducer, or sometimes even thermal interactions between a remote sensor and an application, will result in disturbance to the temperature distribution in an application. The magnitude of this disturbance will depend on the heat transfer processes involved and these are introduced in Section 2.2. Natural instabilities associated with the transducer and signal-measuring devices also contribute to a deviation between the true and measured value and these effects are considered with respect to the specific temperature measurement devices considered in Chapters 3–10. Further deviations between the measured and true values are due to the processing of the signal and data and uncertainties arising from the calibration process. Calibration involves relating the output value from a measurement system to a known input value and the subject is introduced in Section 2.3 prior to more detailed considerations in the 'device-specific' chapters.

The measurement process is subject to the same criteria and principles as any scientific activity. It can be approached casually and the significance of a measurement assessed after the data has been collected. Alternatively, and in most cases preferably, the measurement of temperature can be considered in depth before a sensor is installed and data taken. This latter approach falls within the scope of experimental design (see Clarke and Kempson, 1998). Experiment design involves evaluating what quantities need to be measured, where and when they need to be measured, and what level of uncertainty is necessary for these measurements. The uncertainty of a measurement is an expression of the doubt associated with the measurement. Do the measured values need to be 100% certain or can a proportion of measurements be tolerated that are less than 100% certain? Experimental design procedures are useful in the assessment of these decisions and the reader is referred to texts on experiment design for further information on this subject. An introduction to the manipulation of data and assessment of uncertainty is given in Sections 2.4 and 2.5.

2.2 Heat transfer

As indicated in Figure 2.1, a temperature-measurement sensor will interact with the application and almost inevitably disturb the temperature distribution. The magnitude of this disturbance will depend on the heat transfer processes involved and an introduction to these is given here in Section 2.2.1 prior to the consideration of applications. Temperature measurements are typically undertaken in a solid, on a surface, in a liquid or gas and the principal considerations for each of these are described in Sections 2.2.2 and 2.2.3. The subject of unsteady heat transfer, where temperatures at a given location are not constant with time, is described in Section 2.2.4.

2.2.1 Modes of heat transfer

Heat transfer occurs due to temperature differences. It is convenient to distinguish heat transfer according to the dominant physical mechanism. If a temperature gradient exists in a stationary solid or fluid then the term used to describe the mode of heat transfer is conduction. Heat conduction can be determined using Fourier's law, which for a one-dimensional slab of material as illustrated in Figure 2.2 under steady-state conditions is given by

$$q = -k \frac{dT}{dx} \tag{2.1}$$

where: q = heat flux (W/m^2)
k = thermal conductivity (W/m K)
T = temperature (K)
x = location (m)

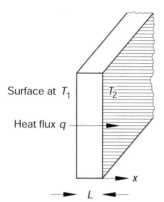

Figure 2.2 Heat conduction through a solid or stationary fluid

Heat flux is the heat transfer rate in a given direction per unit area perpendicular to the direction. The negative sign in equation (2.1) is used to signify that heat is transferred in the direction of decreasing temperature. The thermal conductivity is a transport property of the material concerned. In the form given in equation (2.1) it is assumed that the thermal conductivity is independent of temperature. This is a valid assumption for most materials provided the temperature range is not very large and that a value for the thermal conductivity at an average temperature for the application is used. Data for the thermal conductivity can be obtained in summary form from textbooks on heat transfer such as Incropera and Dewitt (1996), or from data books such as Touloukian *et al.* (1970a,b) and Kaye and Laby (1986).

If a temperature gradient exists in a fluid then it is possible for heat transfer to occur due to both random molecular motion and bulk macroscopic motion of the fluid and this mode of heat transfer is described as convection. Convection from a surface under steady-state conditions as illustrated in Figure 2.3 can be described by Newton's law of cooling:

$$q = h(T_s - T_f) \tag{2.2}$$

where: q = heat flux normal to the surface (W/m²)
h = the heat transfer coefficient (W/m² K)
T_s = surface temperature (K or °C)
T_f = bulk fluid temperature (K or °C)

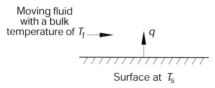

Figure 2.3 Convection from a surface to a moving fluid

The value of the heat transfer coefficient depends principally on conditions in the boundary layer which are influenced by the surface geometry, flow condition and thermodynamic and transport properties.

All matter will emit electromagnetic radiation due to its temperature. Heat transfer due to radiation between a small object and its surroundings, as illustrated in Figure 2.4, can be modelled by

$$q = \varepsilon\sigma(T_s^4 - T_\infty^4) \tag{2.3}$$

where:
q = heat flux (W/m^2)
ε = total surface emissivity
σ = the Stefan–Boltzmann constant (5.67051×10^{-8} Wm^{-2}K^{-4} (Cohen and Taylor, 1999)
T_s = temperature of the surface (K)
T_∞ = temperature of surroundings (K).

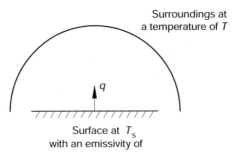

Figure 2.4 Heat transfer by radiation exchange between a surface and its surroundings

The modelling of heat transfer can involve considerably more complex analysis than that indicated by equations (2.1) to (2.3). The temperature distribution may involve gradients in more than one dimension, in which case two- or three-dimensional equations are necessary, or the temperature at a given location may not be constant with time and a transient term will need to be included. The reader is referred to the texts by Bejan (1993), Kreith and Bohn (1997), Incropera and Dewitt (1996), Chapman (1987) and Long (1999) for a thorough treatment of the subject of heat transfer. Nevertheless equations (2.1) to (2.3) can be highly useful in identifying the characteristics of a heat transfer application and, for example, the magnitude of the disturbance caused by the use of a particular temperature-measurement technique.

2.2.2 Solid and surface temperature measurement

The measurement of the temperature within a solid body of material can be achieved in a number of ways. A transducer can be embedded in the material by, for instance, drilling a hole, inserting the device and, in the case of an electrically based transducer, feeding the wires out through the drill hole. This method can also be used for liquid in glass or bimetallic thermometers where the stem of the device is inserted to a specific depth bringing the transducer into thermal contact with the region of interest. In the case of a liquid-in-glass thermometer it may need to be removed from the application in order to reveal the scale to enable the temperature to be read. Care is necessary to ensure that the transducer takes up the local temperature. This means that the thermal contact between the transducer and the local solid must be satisfactory. Thermal contact can be improved in some cases by filling the hole or pocket with a high thermal conductivity oil which will provide a much better conduction path than if air is filling the space between the probe and the hole walls. Errors arising from the effects of unwanted heat conduction along the connecting wires, or the presence of the probe, should be reduced as far as possible. If the temperature of the solid material is acceptably uniform then a non-invasive temperature measurement technique may be suitable.

The measurement of the temperature of a surface which itself is in thermal contact with a gas and also subject to exchange of thermal radiation with its surroundings has a series of possible errors that could contribute to deviations between a measured temperature and that of the undisturbed object. For example a transducer contained within a cylindrical stem, such as bimetallic thermometer, platinum resistance thermometer or thermocouple, can be placed in contact with the surface as illustrated in Figure 2.5. Some of these devices are available with flat bases, spring clips, magnetic mountings or flat

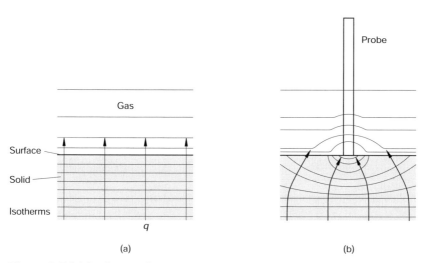

Figure 2.5 (a) Isotherms, lines representing constant temperature, in a solid body with a convective boundary condition at a surface. (b) Distortion caused to the isotherms due to the presence of a surface probe

adhesive pads to facilitate attachment to a surface. If the heat flux from the surface prior to the installation of the probe is uniform then the isotherms, lines representing values of constant temperature, will be horizontal. The direction of the local heat flux is represented by the normal to an isotherm and is shown by the arrows in Figure 2.5(a). Once the probe is brought into contact with the surface, heat transfer will occur between both the sensor and the surface and between the sensor and local environment, in addition to that between the surface and local environment. The likely outcome is a distortion of the temperature distribution in the solid as indicated in Figure 2.5(b). The magnitude of these effects can be identified in many cases using simple one-dimensional analysis techniques. In order to obtain the minimum thermal disturbance sensor probes should be small and possibly recessed in shallow grooves in order to maintain the flow conditions at the surface. An alternative method of surface temperature measurement is by non-invasive techniques such as infrared thermometry or semi-invasive techniques such as the use of thermochromic liquid crystals, thermal paints or thermographic phosphors.

Example 2.1

In order to measure the surface temperature of a steel block it is proposed to coat the entire surface with a 0.5 mm thick layer of polyamide onto which platinum resistance thermometers have been applied. The base temperature of the steel block is 150°C and the block is exposed to a fluid at the opposite boundary with a temperature of 20°C and a heat transfer coefficient of 100 W/m²·K. Determine an analytical value for the temperature indicated by the sensors and compare this value to the surface temperature of the undisturbed metal block. For the purpose of this example, heat transfer by radiation from the surface can be neglected. The thermal conductivity of the steel and polyamide layer can be taken as 40 W/m·K and 0.1 W/m·K respectively. The thickness of the steel layer is 10 mm. The coated and uncoated conditions are illustrated in Figure 2.6.

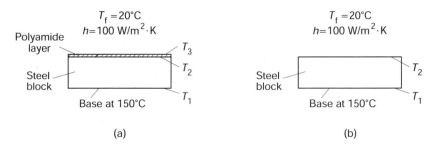

(a)   (b)

Figure 2.6 (a) A surface coating applied to a metal block to facilitate the installation of surface temperature measurement transducers. (b) The uncoated steel block

Solution

The heat flux through the steel layer will be equal to that through the polyamide coating and is dissipated by convection at the surface. Following equation (2.1) the heat flux through the steel is given by

$$q = -k_1 \frac{(T_2 - T_1)}{\Delta x_1}$$

where k_1 and Δx_1 are the thermal conductivity and the thickness of the steel layer respectively. The heat flux through the polyamide layer is given by

$$q = -k_2 \frac{(T_3 - T_2)}{\Delta x_2}$$

where k_2 and Δx_2 are the thermal conductivity and the thickness of the polyamide layer respectively.

From equation (2.2), the heat flux at the surface is given by

$$q = h(T_3 - T_f)$$

Combining the above equations and substituting for the unknowns T_2 and T_3 gives an equation for evaluating the heat flux through the surface:

$$q = \frac{T_1 - T_f}{\Delta x_1/k_1 + \Delta x_2/k_2 + 1/h}$$

Substituting the numerical values gives

$$q = \frac{150 - 20}{0.01/40 + 0.0005/0.1 + 1/100} = 8524.6 \text{ W/m}^2$$

Using Newton's law of cooling, equation (2.2),

$$T_3 = \frac{q}{h} + T_f = 105.3°C$$

For the undisturbed steel block, the heat flux through the steel is given by

$$q = -k_1 \frac{(T_2 - T_1)}{\Delta x_1}$$

From equation (2.2) the heat flux at the surface is given by

$$q = h(T_2 - T_f)$$

Combining the previous two equations and substituting for the unknown, T_2, gives an equation for evaluating the heat flux through the surface:

$$q = \frac{T_1 - T_f}{\Delta x_1/k_1 + 1/h}$$

so

$$q = \frac{150 - 20}{0.01/40 + 1/100} = 12682.9 \, \text{W/m}^2$$

From equation (2.2)

$$T_2 = \frac{q}{h} + T_f = 146.8°\text{C}$$

Comment

The influence of the surface coating is therefore to substantially decrease the surface temperature in comparison to the undisturbed case. A sensor here would therefore indicate a much lower temperature than the undisturbed temperature. Use of instrumentation in this way, however, is quite often acceptable and has a number of advantages, provided due account is made for the thermal disturbance caused.

2.2.3 Fluid temperature measurement

The measurement of liquid temperatures usually involves using an immersed sensor such as a liquid-in-glass or bimetallic thermometer, thermocouple or resistance temperature detector. The actual device will be dependent on the temperature range and the level of uncertainty required for the temperature measurement. Sometimes the sensor will require protection from the local environment, which may be corrosive, at a high or very low pressure. In these cases use can be made of protection tubes or thermowells. A protection tube usually takes the form of a stainless steel cylinder, closed at one end, within which the transducer and connection wires are contained. Thermowells consist of constant section or tapered tubes into which a blind hole has been machined. A thermowell can be fitted into a hole in a pipe or enclosure of interest. A sensor probe can then be inserted into the thermowell. It is also possible in some cases to measure liquid temperatures by use of non-invasive instrumentation such as infrared or acoustic thermometry. These methods are used to determine the temperature in the processing of molten metal and glass. Uncertainty about values for the surface emissivity and transparency make these methods less straightforward.

Many of the considerations that influence the measurement of liquid temperatures also apply to gases. However, the heat transfer for similar velocities is generally much less with gases than for liquids. Thus the effects of unwanted heat conduction along the sensor, or its connecting wires, and radiation exchange are likely to be more significant. Also, high gas velocities can give rise to dynamic heating with an immersed probe and due allowance must be made for this effect. For simple, isolated measurements, liquid-in-glass or bimetallic thermometers may well be adequate. If electrical signals are required, the choice again lies between thermocouples, platinum resistance thermometers, thermistors or transistor instruments. If protection has to be added, the increase in thermal mass and thermal resistance will tend to degrade the response of the sensor more seriously in a gas than in a liquid. The use of invasive instrumentation involves a disturbance, which manifests itself as a difference between the temperature being measured and that, which would exist in the absence of the instrumentation.

The temperature reached, for example by an invasive instrument in contact with a gas, is determined by the balance of convective heat transfer from the gas to the sensor surface, conduction in the sensor itself and its supports and connections, and radiative heat transfer between the probe and its surroundings. All these considerations must be taken into account when undertaking a measurement of the temperature of a flowing fluid. However, in order to develop the concepts, the error due to conduction along the probe supports is described first in isolation, followed by consideration of both conductive and radiative effects, followed by consideration of how the temperature indicated by the probe can be related to the temperature of the fluid.

For some gas temperature measurements, generally associated with combustion processes, it may be undesirable or impossible to use conventional sensors and the use of a non-invasive technique such as one based on refractive index or spectroscopy may be suitable. The methods are described in Chapter 10 but tend to be complex and expensive and may require the expenditure of a considerable amount of time and effort to produce useful results.

As mentioned in this section, transducers are commonly enclosed in a cylindrical casing in order to protect the device from both mechanical and chemical damage. This approach is used for some resistance temperature

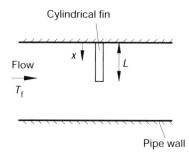

Figure 2.7 A cylindrical fin protruding from a surface, representative of some temperature transducer protection tubes

detectors, thermocouples and bimetallic thermometers. A cylindrical or tapered tube, however, provides a conduction path between the support and the transducer, which can distort the temperature indicated. The resulting geometry is familiar in heat transfer studies and can be modelled as a fin of constant cross-section (Figure 2.7). The equation representing the heat transfer process, assuming steady-state conditions, is

$$\frac{d^2 T}{dx^2} - \frac{hP}{kA_c}(T - T_f) = 0 \tag{2.4}$$

where: T = local fin temperature (K)
          x = location away from the fin base (m)
          A_c = the cross-sectional area (m²)
          P = the fin perimeter (m)
          k = the thermal conductivity of the fin (W/m·K)
          h = heat transfer coefficient (W/m²·K)
          T_f = bulk temperature of the surrounding fluid (K)

Solution of equation (2.4), assuming negligible loss from the tip of the fin, gives

$$\frac{T - T_f}{T_b - T_f} = \frac{\cosh m(L - x)}{\cosh mL} \tag{2.5}$$

where L = length of the fin (m), T_b = temperature at the base of the fin (K) and

$$m = \sqrt{\frac{hP}{kA_c}} \tag{2.6}$$

For the case of a cylindrical fin, $P = 2\pi r$ and $A_c = \pi r^2$, so

$$m = \sqrt{\frac{2h}{kr}} \tag{2.7}$$

Equation (2.5) can be used to evaluate the deviation of the temperature of a transducer, located part-way along a cylindrical sheath, from that of the local fluid temperature.

Example 2.2

A 10 mm external diameter stainless steel sheathed thermocouple is inserted into a pipe containing a fluid (Figure 2.8). If the heat transfer coefficient is 140 W/m²·K and the fluid temperature is 90°C and the length of the sensor protruding from the wall is 60 mm, determine analytically the temperature indicated by the transducer. The transducer is located 5 mm from the tip of the

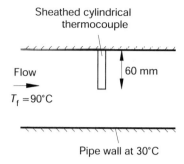

Figure 2.8 A sheathed thermocouple protruding from the pipe wall

sheath and the pipe wall temperature can be taken as 30°C. The thermal conductivity of the sensor can be taken as 16 W/m·K and the effects of heat transfer by radiation can be neglected.

Solution

From equation (2.7), taking the radius of the probe as 0.005 m,

$$m = \sqrt{\frac{2 \times 140}{16 \times 0.005}} = 59.2$$

Using equation (2.5), taking $x = 60 - 5 = 55$ mm and recalling that

$$\cosh n = \frac{e^n + e^{-n}}{2}$$

$$\frac{T - 90}{30 - 90} = \frac{\cosh(59.2(0.06 - 0.055))}{\cosh (59.2 \times 0.06)} = 0.06$$

So the temperature indicated by the transducer will be $T = 86.4°C$. This is 3.6°C below the true temperature of the fluid.

Comment

At the tip, $x = L$, the error can be defined as $= T_f - T_L$

$$\text{error} = T_f - T_L = \frac{T_f - T_b}{\cosh mL} \qquad (2.8)$$

The error is proportional to $T_f - T_b$ so a pragmatic approach to minimize the error is to insulate the wall or pipe therefore elevating T_b. If control is available over the choice of sensor, then one with a small radius, low thermal conductivity and large immersion length is desirable.

It should be noted that the analysis presented in equations (2.4)–(2.6) and in the previous example does not take into account the detailed assembly of the sensor. In practice some sensors consist of a multi-core assembly with the transducer isolated electrically from the sheath by a ceramic insulator.

Example 2.3

A 2 mm external diameter thermocouple is used to measure the temperature of a gas flowing in a pipe, (Figure 2.9). If the pipe wall temperature is 300°C and the gas temperature is 650°C, determine whether the thermocouple will take up the temperature of the gas stream. The immersed length of the probe is 50 mm and the thermal conductivity of the probe can be taken as 6 W/m·K. The heat transfer coefficient is 250 W/m²·K and the total emissivity of probe can be taken as 0.7.

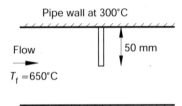

Pipe wall at 300°C

Flow

50 mm

$T_f = 650°C$

Figure 2.9 A thermocouple used to measure the temperature of a gas flowing in a pipe

Solution

The thermocouple here will be subject to a number of phenomena which could cause errors between the indicated temperature and the true temperature of the gas including conduction along the support and thermal radiation exchange between the sensor and the pipe walls.

Using the analysis of the previous example the effect of conduction along the probe length will cause an error of

$$T_f - T_L = \frac{T_f - T_b}{\cosh mL} = \frac{650 - 300}{\cosh \left[\left(\sqrt{\dfrac{2 \times 250}{6 \times 0.001}} \right) 0.05 \right]}$$

$$= \frac{350}{\cosh 14.43} = 3.8 \times 10^{-4}°C$$

This value is very small and can effectively be ignored in this case in comparison to error due to radiation heat transfer.

The thermocouple can be modelled as a small object within a large enclosure and the heat balance for the thermocouple between heat transfer by convection and that by radiation is given by

$$A\varepsilon\sigma(T_{ic}^4 - T_w^4) = Ah(T_f - T_{tc})$$ (2.9)

where: T_f = the bulk temperature of the fluid (K)
T_{tc} = the temperature of the thermocouple (K)
T_w = the temperature of the walls of the surrounding enclosure (K).

Equation (2.9) is a transcendental equation and must be solved for T_{tc} by using an iterative method. Guessing an initial value $T_{tc} = 650°C$, then

$$T_{tc} = T_f - \frac{\varepsilon\sigma(T_{tc}^4 - T_w^4)}{h} = 650 - \frac{0.7 \times 5.67 \times 10^{-8}(650^4 - 300^4)}{250}$$

$$= 622.9°C$$

Using the above value as an updated estimate for T_{tc} gives

$$T_{tc} = 650 - \frac{0.7 \times 5.67 \times 10^{-8}(622.9^4 - 300^4)}{250} = 627.4°C$$

Using the above value as an updated estimate for T_{tc} gives

$$T_{tc} = 650 - \frac{0.7 \times 5.67 \times 10^{-8}(627.4^4 - 300^4)}{250} = 626.7°C$$

Using the above value as an updated estimate for T_{tc} gives

$$T_{tc} = 650 - \frac{0.7 \times 5.67 \times 10^{-8}(626.7^4 - 300^4)}{250} = 626.8°C$$

This is similar to the previous value and the solution can be assumed to be converged.

The difference between T_f and T_{tc} is 23.2°C and the thermocouple does not take up the true temperature of the gas stream.

Comment

In this case the difference between the gas stream and that indicated by the thermocouple is about 23°C. This difference can be reduced by use of a radiation shield and by increasing the temperature of the wall. Radiation shields are discussed in the next example. The temperature of the wall could be elevated by use of thermal insulation, which would serve to reduce the heat transfer due to radiation between the transducer and the wall.

If the heat transfer coefficient is unknown it can usually be estimated using one of the many correlations that are available for standard geometries. In this case for a cylinder in cross flow the correlation provided by Churchill and Bernstein (1977) would be suitable.

Example 2.4

For the data of the previous example, determine the effect of the inclusion of a cylindrical radiation shield around the sensor as illustrated schematically in Figure 2.10. The total emissivity of the shield can be taken as 0.26 and the surface areas of the thermocouple and shield can be taken as $2.2 \times 10^{-5}\,\mathrm{m^2}$ and $1.885 \times 10^{-4}\,\mathrm{m^2}$ respectively.

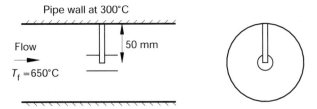

Figure 2.10 A radiation shield used to limit the exchange of radiation between a transducer and the surrounding environment

Solution

Ignoring the effects of heat conduction, the thermal equilibrium balance for the shielded thermocouple is given by considering the radiative heat exchange between the thermocouple and the shield and the convective heat exchange between the gas and the thermocouple:

$$A_{tc}\varepsilon_{tc}\,\sigma(T_{tc}^4 - T_{sh}^4) = A_{tc}h(T_f - T_{tc}) \tag{2.10}$$

where ε_{tc} is the total emissivity of the thermocouple. The radiative heat exchange between the thermocouple, shield and the wall is given by

$$A_{tc}\varepsilon_{tc}\,\sigma(T_{tc}^4 - T_{sh}^4) = A_{sh}\varepsilon_{sh}\,\sigma(T_{sh}^4 - T_w^4) \tag{2.11}$$

where ε_{sh} is the total emissivity of the shield.
This equation can be rearranged to give the temperature of the shield:

$$T_{sh}^4 = \frac{A_{tc}\,\varepsilon_{tc}\,T_{tc}^4 + A_{sh}\,\varepsilon_{sh}\,T_w^4}{A_{sh}\,\varepsilon_{sh} + A_{tc}\,\varepsilon_{tc}} \tag{2.12}$$

Substitution of equation (2.12) in (2.10) gives

$$h(T_f - T_{tc}) = \varepsilon_{tc}\,\sigma\left(T_{tc} - \frac{A_{tc}\,\varepsilon_{tc}\,T_{tc}^4 + A_{sh}\,\varepsilon_{sh}\,T_w^4}{A_{sh}\,\varepsilon_{sh} + A_{tc}\,\varepsilon_{tc}}\right)$$

Again a solution for T_{tc} can be found by using an iterative method.

Guessing an initial value $T_{tc} = 870$ K, then

$$T_{tc} = T_f - \frac{\varepsilon_{tc}\,\sigma}{h}\left(T_{tc}^4 - \frac{A_{tc}\,\varepsilon_{tc}\,T_{tc}^4 + A_{sh}\,\varepsilon_{sh}\,T_w^4}{A_{sh}\,\varepsilon_{sh} + A_{tc}\,\varepsilon_{tc}}\right)$$

$$= 923 - \frac{0.7 \times 5.67 \times 10^{-8}}{250}$$

$$\left(870^4 - \frac{2.2 \times 10^{-5} \times 0.7 \times 870^4 + 1.885 \times 10^{-4} \times 0.26 \times 573^4}{1.885 \times 10^{-4} \times 0.26 + 2.2 \times 10^{-5} \times 0.7}\right)$$

$$= 867°C$$

Using the above value as an updated estimate for T_{tc} gives

$$T_{tc} = 923 - \frac{0.7 \times 5.67 \times 10^{-8}}{250}$$

$$\left(867^4 - \frac{2.2 \times 10^{-5} \times 0.7 \times 867^4 + 1.835 \times 10^{-4} \times 0.26 \times 573^4}{1.885 \times 10^{-4} \times 0.26 + 2.2 \times 10^{-5} \times 0.7}\right)$$

$$= 868°C$$

This is similar to the previous value indicating that the solution can be assumed to be converged.

The difference between T_f and T_{tc} is 55°C. The thermocouple does not take up the exact temperature of the gas stream but is within about 6%.

Comment

The inclusion of a single radiation shield has substantially reduced the difference between the indicated and actual fluid temperature in comparison to the previous example, where there was no radiation shield. The difference could be further reduced by the addition of extra shields around the first.

The temperature of a moving fluid can be characterized by the following equation:

$$T_{total} = T_{static} + \frac{U^2}{2c_p} \tag{2.14}$$

where: T_{total} = the total temperature (°C)
  T_{static} = the static temperature (°C)
  U = the bulk free stream velocity (m/s)
  c_p = specific heat capacity (J/kg·K).

The total temperature represents the temperature that an idealized thermometer, inserted in an ideal gas stream, would measure at the point at which the fluid is brought to rest or stagnated. The total temperature is also sometimes known as the stagnation temperature. It is made up of two parts, the static temperature, T_{static}, and the dynamic temperature, $U^2/2c_p$. The static temperature is the actual temperature of the gas. It is the temperature that would be sensed by an adiabatic probe in thermal equilibrium and at zero relative velocity with respect to the probe. In other words a sensor travelling at the same speed as the gas and one experiencing no heat transfer exchanges with the temperature probe would indicate the static temperature of the gas. The static temperature is sometimes associated with the random translational kinetic energy of the gas molecules. The dynamic temperature is the thermal equivalent of the directed kinetic energy at that part in the gas continuum.

Some typical configurations of probes used for measuring fluid temperatures are illustrated in Figure 2.11. If a temperature probe is placed in a flow then the fluid will be brought to rest on the surface of the probe as a result of stagnation and viscosity, as described by the no-slip condition. However, this process will not be achieved in a thermodynamically reversible fashion and the gas concerned may not behave exactly according to the ideal gas law. The deceleration of the flow on a probe will convert a proportion of the directed kinetic energy of the gas into thermal energy. Furthermore a real probe will transfer heat by radiation to the surroundings and by conduction along supports and lead wires. The fraction of the kinetic energy recovered in practice by a diabatic non-ideal probe as thermal energy can be modelled by a dynamic correction factor, which is defined by

$$K = \frac{T_{probe} - T_{static}}{U^2/2c_p} \tag{2.15}$$

where: K = the dynamic correction factor
$\quad\quad\quad T_{probe}$ = the equilibrium temperature indicated by a stationary probe (K).

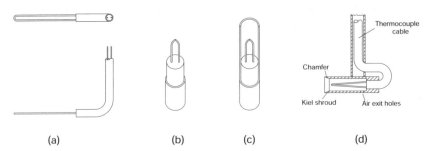

(a)   (b)   (c)   (d)

Figure 2.11 Dynamic correction factors for thermocouples. (a) Probe at 90° to the flow. (b) Probe parallel to the flow. (c) A half-shielded transducer normal to the flow. (d) A Kiel probe

Table 2.1 Dynamic correction factors for thermocouples

Wire configuration	Dynamic correction factor	Source of information
Wires normal to the flow	0.68±0.07 (95%)	Moffat (1962)
Wires parallel to the flow	0.86±0.09 (95%)	Moffat (1962)
Half shielded thermocouple normal to the flow	0.96	Benedict (1959)

Details of dynamic correction factors for thermocouple wires are presented in Table 2.1.

The total and static temperatures can be evaluated from the following equations as

$$T_{\text{total}} = T_{\text{probe}} + (1 + K) \frac{U^2}{2c_{\text{p}}} \tag{2.16}$$

$$T_{\text{static}} = T_{\text{probe}} - K \frac{U^2}{2c_{\text{p}}} \tag{2.17}$$

Example 2.5

A bare thermocouple is mounted at right angles to a gas stream flowing at 130 m/s. Assuming that the dynamic correction factor for a bare thermocouple mounted at right angles to the flow direction is 0.7, determine the static temperature of the gas if the probe indicates a temperature of 215°C. The specific heat capacity of the gas can be taken as 1033 J/kg·K.

Solution

From equation (2.17),

$$T = T_{\text{probe}} - K \frac{U^2}{2c_{\text{p}}} = 215 - 0.7 \frac{130^2}{2 \times 1033} = 209.3°C$$

2.2.4 Transient heat transfer

In many practical applications where the temperature of an object is being measured, its temperature is not constant but varying with time. Even if the temperature of an object was constant, if a transducer is brought into thermal contact with it, then unless the temperatures of the two are equal then the temperature of the object will alter as heat is transferred between the two and this process will occur over a certain period of time until thermal equilibrium is reached. No heat transfer process produces instantaneous changes in bulk temperature and the time taken for a temperature sensor to change its temperature to a value approaching that of the application is called the response time. The response time of a sensor is dependent on the thermal properties of the sensor, the heat transfer coefficient, its size and the response time of any signal processing associated with it. Because a sensor takes a certain time to change its temperature and indicate this change there is always a time lag between the sensor and the temperature of the object being measured. Transient heat transfer can be due to conduction, convection or radiation although transient conduction tends to be the most common mode of interest. In temperature measurement applications this is because in many of the common sensors such as liquid-in-glass thermometers, platinum resistance thermometers and thermocouples heat is conducted through the solid sensor or protective enclosure around the transducer.

Transient conduction can be modelled in a number of ways. The simplest is to assume that temperature gradients within a body are negligible and performing an energy balance for the system. This method is sometimes referred to as the 'lumped capacitance method'. Using this approach the temperature variation of a body exposed to a convective boundary condition can be evaluated with respect to time by

$$\frac{T - T_{\infty}}{T_{i} - T_{\infty}} = \exp\left[-\left(\frac{hA_{s}}{\rho V c_{p}}\right)t\right] \qquad (2.18)$$

where: ρ = density (kg/m^3)
A_{s} = surface area of the body (m^2)
c_{p} = specific heat at constant pressure (J/kg·K)
V = volume of the body (m^3)
h = heat transfer coefficient (W/m^2·K)
t = time (s)
T_{∞} = bulk fluid temperature (K)
T_{i} = initial temperature (K).

The quantity $\rho V c_{p}/hA_{s}$ defines the characteristic speed of response of the system to an input and is known as the time constant, τ:

$$\tau = \frac{\rho V c_{p}}{hA_{s}} \qquad (2.19)$$

Equation (2.18) is valid when a dimensionless quantity called the Biot number satisfies the following inequality:

$$Bi = \frac{hL_{c}}{k} < 0.1 \qquad (2.20)$$

where L_{c} is a characteristic length and can be taken for a general object as the ratio of V/A_{s}.

Example 2.6

A thermocouple junction can be formed by welding the two ends of wire together to form a bead which can be spherical in shape. If the diameter of the junction bead can be taken as 0.6 mm, then by modelling a thermocouple junction as a sphere determine the length of time required for the junction to reach 195°C when it is suddenly exposed to a gas stream at a temperature of 200°C. The initial temperature of the thermocouple is 20°C and the heat transfer coefficient is 350 W/m²·K. The thermal conductivity, specific heat capacity and density of the thermocouple material can be taken as 18 W/m·K, 384 J/kg·K and 8920 kg/m³ respectively.

Solution

In order to identify whether the lumped capacitance method is valid, it is necessary to determine the Biot number, which, in this case is

$$Bi = \frac{hL}{k} = \frac{350 \times 0.0006}{18} = 0.01167$$

As the Biot number is much less than 0.1 the error associated with the assumption that thermal gradients within the bead are negligible is small. Equation (2.18) can therefore be used to determine the time taken for the thermocouple, initially at 20°C, to reach 195°C.

The surface area of a sphere is given by $A_s = \pi D^2 = 1.13 \times 10^{-6}$ m² and the volume by $V = \pi D^3/6 = 1.131 \times 10^{-10}$ m³:

$$\frac{195 - 200}{20 - 200} = \exp\left[-\left(\frac{350 \times 1.13 \times 10^{-6}}{8920 \times 1.131 \times 10^{-10} \times 384}\right)t\right]$$

$$0.027 = \exp(-1.0209t)$$

Taking natural logs, $-3.612 = -1.0209t$, $t = 3.5$ s. So the time required for the thermocouple to reach 19.5°C is approximately 3.5 s. The time constant is given by

$$\tau = \frac{\rho V c_p}{hA_s} = \frac{8920 \times 1.131 \times 10^{-10} \times 384}{350 \times 1.13 \times 10^{-6}} = 0.98\,\text{s}$$

Comment

The time constant given here refers to the response to an instantaneous step change in conditions. It is useful in that it provides an indication of the speed of response of a sensor. For a time constant evaluated in this way, at $t = \tau$ the output of the sensor will have reached 63.2% of the input condition, at $t = 3\tau$ the output will have reached 95% of the input conditions and at $t = 5\tau$ the response will be 99%.

Example 2.7

A 0.5 mm external diameter sheathed thermocouple is used in a workshop to measure the temperatures of various different mediums such as oven temperatures and the temperatures of vats of water. The temperature of the workshop is generally about 25°C. Determine the variation of the sensor temperature with time, the time constant and the time it is necessary to leave the thermocouple before taking a reading if the sensor is suddenly immersed in the following applications:

(a) air in an oven at 80°C
(b) water in a vat at 80°C.

The heat transfer coefficients for the air and water applications can be taken as 12 W/m²·K and 100 W/m²·K respectively. The thermal conductivity, specific heat capacity and density of the thermocouple material can be taken as 19 W/m·K, 420 J/kg·K and 8500 kg/m³ respectively.

Solution

The surface area of the cylindrical probe is given by $A_s = 2\pi rL = 1.571 \times 10^{-3}L$ m² and the volume by $V = \pi D^2 L/4 = 1.963 \times 10^{-7}L$ m³.

(a)
$$\frac{T - 80}{25 - 80} = \exp\left[-\left(\frac{12 \times 1.571 \times 10^{-3}\, L}{8500 \times 1.963 \times 10^{-7}\, L \times 420}\right) t\right]$$

So

$$T = 80 - 55\,\exp(-0.0269t)$$

$$\tau = \frac{\rho V c_p}{h A_s} = \frac{8500 \times 1.963 \times 10^{-7}\, L \times 420}{12 \times 1.571 \times 10^{-3}\, L} = 37.2\,\text{s}$$

(b)
$$\frac{T - 80}{25 - 80} = \exp\left[-\left(\frac{100 \times 1.571 \times 10^{-3}\, L}{8500 \times 1.963 \times 10^{-7}\, L \times 420}\right) t\right]$$

So

$$T = 80 - 55\,\exp(-0.224t)$$

$$\tau = \frac{\rho V c_p}{h A_s} = \frac{8500 \times 1.963 \times 10^{-7}\, L \times 420}{100 \times 1.571 \times 10^{-3}\, L} = 4.5\,\text{s}$$

The results of the temperature variation with time for the thermocouple in the two different applications are shown in Figure 2.12. The question asked

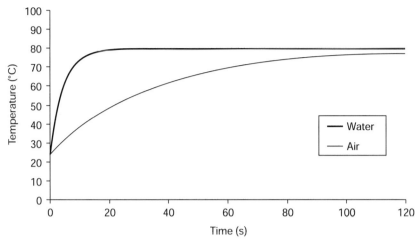

Figure 2.12 Temperature response of a 0.5 mm diameter sheathed thermocouple immersed in air and water with heat transfer coefficients of 12 and 100 W/m^2·K respectively

what length of time should be left before taking a reading. This is subjective. As can be seen in the graph, the sensor takes much longer to respond in the case of air, the low convective heat transfer application. If a 99% response is desired then from the time constant the answers are $t = 5\tau = 186$ s in the case of the air and $t = 5\tau = 22.5$ s in the case of the water application.

Comment

Many transducers such as bare thermocouple wires, thermistor beads and other temperature probes that are physically small and have a high thermal conductivity are used in a high heat transfer coefficient environment can be reasonably well modelled in this way. Sensors that have a response of the form indicated by equation (2.8) are known as first-order systems.

Time constants are also useful in determining the relationship between a continuously varying process temperature and that indicated by a sensor. If the temperature of an application is rising and falling with time then the temperature indicated by a sensor in thermal contact with the application will also rise and fall. However, due to the thermal characteristics of the sensor, there will be a time lapse and a temperature difference between the temperature of the application and that indicated by the sensor. For example, if the temperature of an application is varying sinusoidally, then as the temperature reaches a maximum the sensor will still be responding to the temperature rise, due to the time it takes for heat to be transferred from one location to another, as the temperature of the application then begins to reduce. Modelling a thermometer as a first-order system, then if the input takes the form of a sinusoid, the output can be modelled by

$$\text{output amplitude} = \frac{\text{input amplitude}}{\sqrt{\omega^2 \tau^2 + 1}} \qquad (2.21)$$

and

$$\phi = \tan^{-1} \omega\tau \qquad (2.22)$$

where: ϕ = phase lag (rad)
ω = angular frequency of the sine wave (rad/s).

Example 2.8

A transducer is used to measure the temperature of a liquid in a pipe. The indicated temperature varies approximately sinusoidally between 176°C and 209°C with a time period of 22 s. The transducer has a time constant of 3.8 s. Determine the temperature variation of the application and the delay between corresponding points on the cycle of the actual and indicated temperatures.

Solution

The frequency of the cycle is given by

$$\text{frequency} = \frac{1}{\text{time period}} = \frac{1}{22} = 0.0455\,\text{s}$$

The angular frequency, $\omega = 2\pi f = 2\pi \times 0.0455 = 0.286\,\text{rad/s}$. The amplitude of the recorded temperatures is given by

$$\text{amplitude} = \frac{T_{max} - T_{min}}{2} = \frac{209 - 176}{2} = 16.5°C$$

$$\text{output amplitude} = \frac{\text{input amplitude}}{\sqrt{\omega^2 \tau^2 + 1}}$$

so

$$\text{input amplitude} = 16.5\,\sqrt{(0.286 \times 3.8)^2 + 1} = 24.4°C$$

The temperature of the application is therefore varying between 216.9°C and 168.1°C.

The phase lag between the sensor and the application is

$$\phi = \tan^{-1} \omega\tau = \tan^{-1}(0.286 \times 3.8) = 0.827\,\text{rad} = 47.4°$$

The delay is given by

$$\text{Delay} = \frac{0.827}{2\pi} \times 22 = 2.9\,\text{s}$$

Many practical sensors cannot be modelled by the lumped capacitance method as thermal gradients are significant within the sensor. In these cases the concept of the thermal time constant is inappropriate and instead the response time is more suitable as it refers to way a sensor changes its indicated value with time. This data can be obtained from experiments undertaken by the sensor manufacturer or by solution of the three-dimensional transient heat conduction equation.

2.3 Calibration and traceability

In order to ensure that a measurement taken by one person can be communicated with confidence to another person, it is necessary to ensure that a common scale is being used. A temperature scale on its own, however, is of little use as it is also necessary to ensure that the numerical value indicated by a sensor is a valid indication of that value. If a sensor shows a particular value how is it is known that it is correct and with what confidence and tolerance is it known? These concepts fall within the categories of traceability and uncertainty analysis. Traceability involves relating a measurement to a scale and the physical phenomena the scale represents. In the case of temperature this is currently accepted to mean relating a measurement to the International Temperature Scale of 1990 which is a close approximation to true thermodynamic temperature, which is itself a measure of the parameter of state that is inversely proportional to the rate of change of the log of the number of accessible states as a function of energy for a system in thermal equilibrium. This can be achieved by the process of calibration and management of the measurement process.

Like many words, the term 'calibration' takes different meanings depending on the context and the user. Calibration is often used to refer to the process of checking that a sensor is indicating a value within some acceptable tolerance and adjusting the scale if it is not. This process is sometimes associated with sending a sensor back to a manufacturer or to a laboratory. During the process the scale may have been physically adjusted or new constants programmed into the sensor in order to generate an output indication within a particular tolerance that matches known conditions, imposed on the sensor in the laboratory. This interpretation of the word 'calibration' can, however, be associated with checking and adjusting the sensor. A more robust definition, used by the International Standards Organization, defines calibration as the set of operations that establish the relationship between values indicated by a sensor and corresponding known values of the measurand.

Calibration can be achieved in a number of ways. The sensor to be calibrated can be placed in an environment at a known temperature such as one of the fixed points listed in Table 1.2, for example the triple point of water. The output from the sensor can be recorded and, within the uncertainties associated with the temperature of the fixed point, taken as an indication of that temperature. The sensor can then be exposed to a different temperature using a different fixed point, for example the melting point of gallium, and the process repeated. For intermediate values of temperature between the fixed point, the special interpolation equations documented in the ITS-90 can be used. The process described so far is appropriate for producing a sensor with very low uncertainty capability. This, however, is an expensive undertaking and is rarely necessary for the majority of scientific and industrial applications. Instead a sensor that has been calibrated in this way and is stable is used to calibrate another sensor by comparison. A calibration and uncalibrated sensor are placed in thermal contact in an enclosure such as a bath of liquid or in a specially constructed heating block or furnace. The medium is heated and the output of both the calibrated and the uncalibrated sensor recorded. The output from the uncalibrated sensor can then be related to the temperature indicated by the calibrated sensor. In this way a transfer between values of temperatures set up in the ITS-90 and sensors used in practice can be achieved and is illustrated schematically in Figure 2.13. It is normally recommended that there should be no more than three or four links between a practical measurement and the ITS-90. The details of the calibration process depend in part on the type of sensor concerned typically involving established procedures for record keeping, inspection and conditioning of the sensor, generic checks, intercomparison either with another sensor or with a medium at a known temperature, analysis of the data, quantification of uncertainties and completion of records. Additional details on calibration are included in the device-specific chapters.

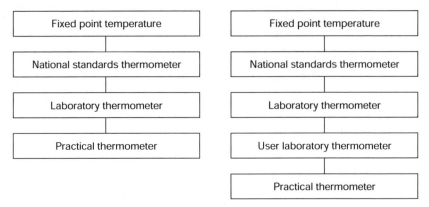

Figure 2.13 There should normally be no more than three or four links between a practical measurement and the ITS-90

Traceability involves ensuring that appropriate procedures have been followed, and in the current industrial setting, this is typically associated with quality standards. Two approaches can be adopted. The first involves following a set of standards and obtaining the appropriate approval and registration certificates for that quality system. The second approach is to accept the principles of the process commended within quality standards and to follow them closely enough to be able to state that the spirit of the standard has been followed. This latter approach negates the need for registration to a standards system but does require a level of trust.

Formal adherence to quality standards typically involves procedures and documentation to support:

- the process of selection and purchase of a sensor
- calibration of the sensor by an approved laboratory
- transportation of the sensor
- appropriate storage
- use by trained personnel
- establishment of calibration at regular intervals
- checking of the sensor
- data recording and handling
- mistake reporting and remedy
- data storage.

These aspects are common sense and their merit is that they provide confidence that a measurement has been sensibly undertaken and that the level of uncertainty stated with the measurement has been assessed properly.

2.4 Data manipulation

The measurement of temperature usually results in a series of values. It is rare that only one temperature measurement will be made. General measurement wisdom dictates that a measurement should be made a minimum of three times. If just one measurement is made then a mistake could go unnoticed. If two measurements are made and they differ then how does one know which is correct? If three measurements are made then if two agree, suspicion can be cast on the third. If repeated measurements give different answers it does not necessarily mean that anything wrong is being done. The differences in the readings may be due to natural variations in the process. For example, if a measurement is made of the temperature of an air stream, the indicated temperature may vary because the temperature of the air stream itself is varying. Broadly speaking, the more measurements that are made, the better the estimate of the true value. The ideal number of measurements to obtain the true value is infinite. This would, however, be far from ideal in terms of effort, expense and practicality! A rule of thumb concerning this dictates that a number of between four and ten measurements is usually suitable unless the statistical tools introduced here suggest that more results are necessary in order to provide better confidence in the result.

2.4.1 Average values

An average provides an estimate of the true value. An average or arithmetic mean is usually given by a symbol with an over-bar. For example, $\bar{x}$ is the mean value of x. The average of a set of data $x_1, x_2, x_3, \ldots, x_\infty$ is given by

$$\bar{x} = \frac{1}{n} \sum_{i=1}^{n} x_i \qquad (2.23)$$

where: $\bar{x}$ = mean value of x
  n = total number of measurements
  x_i = individual measurements.

Example 2.9

Determine the average temperature for the ten temperature measurement readings given in Table 2.2.

Table 2.2 Example temperature readings

Measurement number	1	2	3	4	5	6	7	8	9	10
Temperature reading (°C)	58.6	58.2	57.7	57.8	58.0	58.3	58.9	58.6	57.9	57.6

Solution

The average temperature can be determined using equation (2.23):

$$\bar{T} = \frac{1}{10} \sum_{1}^{10} T_i$$

$$= \frac{1}{10} \; (58.6 + 58.2 + 57.7 + 57.8 + 58.0 + 58.3 + 58.9 + 58.6 + 57.9 + 57.6)$$

$$= 58.16°C$$

2.4.2 Quantifying the spread of data

It is useful to know how widely spread repeated readings are. One measure of the spread of data is the range, which is the difference between the highest and lowest values. For the data of the previous example, the range is 58.9–57.6 = 1.3°C. Information about the range on its own can, however, be misleading, especially if a single reading is markedly different from the rest. A useful way of quantifying the spread of data is to use the variance and the standard deviation, which is the square of the variance. The standard deviation of a set of numbers provides an indication of how the readings typically differ from the average of the set and, for most distributions, is directly proportional to the width of the distribution.

The standard deviation of a series of measurements can be estimated by

$$s = \sqrt{\frac{\sum\limits_{i=1}^{n} (x_i - \bar{x})^2}{n - 1}} \qquad (2.24)$$

where: s = estimated standard deviation
  n = total number of measurements
  x_i = individual measurements
  $\bar{x}$ = estimated mean value of x.

Note that the value given by equation (2.24) is an estimate of the standard deviation. A true value of the standard deviation could only be given by analysis of all the possible data points. In normal circumstances ten readings are usually sufficient in order to obtain an acceptable estimate for the standard deviation.

Example 2.10

Determine an estimate for the standard deviation for the temperature measurements presented in the previous example.

Solution

It is convenient when doing such a calculation to determine the mean, the differences between the individual readings and the mean and the squares of these values as indicated in Table 2.3. The estimate of the standard deviation can then be readily determined using equation (2.24):

$$s = \sqrt{\frac{0.16 + 0 + 0.25 + 0.16 + 0.04 + 0.01 + 0.49 + 0.16 + 0.09 + 0.36}{10 - 1}}$$

$$= 0.44°C$$

Table 2.3 $T_i - \overline{T}$ and $(T_i - \overline{T})^2$

Measurement number	1	2	3	4	5	6	7	8	9	10
T_i (°C)	58.6	58.2	57.7	57.8	58.0	58.3	58.9	58.6	57.9	57.6
$T_i - \overline{T}$ (°C)	0.4	0	−0.5	−0.4	−0.2	0.1	0.7	0.4	−0.3	−0.6
$(T_i - \overline{T})^2$ (°C)²	0.16	0	0.25	0.16	0.04	0.01	0.49	0.16	0.09	0.36

Comment

The value of the estimated standard deviation, 0.44°C in this case, can be used to provide an indication of the spread of the data. If it is assumed that the data is modelled by the normal distribution, which is introduced in the next section, then it can be said that 99.73% of the data falls within a bound of ±3 s, 95% within ±2 s and 68% within 1± s. It is this facility for bounding data that makes the standard deviation a useful concept. The calculation performed in this example can also be undertaken far less laboriously using the function keys on a scientific or programmable calculator or in spreadsheet software.

2.4.3 Data distributions

For a set of data it is possible for the spread of data to take many forms. For example, the data listed in Table 2.4 and illustrated in Figure 2.14, in the form of a blob plot, is regularly distributed. This data can be plotted in the form of a frequency distribution curve, similar to a histogram, which indicates the number of occurrences for which data with the same value occurs. The frequency distribution curve for the data of Table 2.4 is shown in Figure 2.15. The form of this particular distribution is rectangular.

Table 2.4 Example of regularly distributed data

Measurement number	1	2	3	4	5	6	7	8	9	10
Temperature reading (°C)	58.3	58.4	58.5	58.6	58.7	58.8	58.9	59.0	59.1	59.2

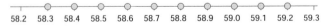

58.2 58.3 58.4 58.5 58.6 58.7 58.8 58.9 59.0 59.1 59.2 59.3

Figure 2.14 Blob plot for the data of Table 2.4

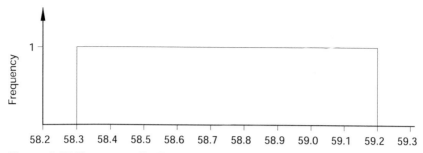

Figure 2.15 Frequency distribution curve for the data of Table 2.4

It is possible for the majority of values in a set of data to be nearer the average than further away. An example of such data is given in Table 2.5 and is also plotted in Figures 2.16 and 2.17 in the form of a blob plot and frequency distribution curve respectively.

The form of the data shown in Figure 2.17 is found to occur commonly in practice. The normal or Gaussian distribution is a mathematical model that

Table 2.5 Example of data distributed with the majority of values nearer to the mean than further away. Here, $\overline{T} = 703.4°C$, $s = 4.9°C$

Measurement number	T_i (°C)	$T_i - \overline{T}$ (°C)	$(T_i - \overline{T})^2$ (°C)^2
1	699.2	−4.17	17.3889
2	701.1	−2.27	5.1529
3	702.4	−0.97	0.9409
4	703.6	0.23	0.0529
5	704.7	1.33	1.7689
6	703.2	−0.17	0.0289
7	702.6	−0.77	0.5929
8	698.9	−4.47	19.9809
9	705.3	1.93	3.7249
10	707.5	4.13	17.0569
11	706.8	3.43	11.7649
12	705.3	1.93	3.7249
13	712	8.63	74.4769
14	706.1	2.73	7.4529
15	696.3	−7.07	49.9849
16	695.5	−7.87	61.9369
17	701.3	−2.07	4.2849
18	707.2	3.83	14.6689
19	713.4	10.03	100.6009
20	695.2	−8.17	66.7489

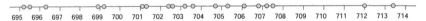

Figure 2.16 Blob plot for the data of Table 2.5

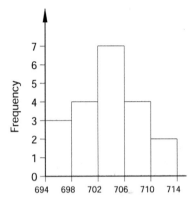

Figure 2.17 Frequency distribution curve for the data of Table 2.5

was developed to represent the spread of data in this form. For data that follows the normal distribution, the frequency of occurrence of a data of value x is defined by

$$f(x) = \frac{1}{\sigma\sqrt{2\pi}} \exp\left[\frac{1}{2}\left(\frac{x-\mu}{\sigma}\right)^2\right] \quad -\infty < x < \infty \quad (2.25)$$

where: $f(x)$ = the probability density function of the continuously random variable x

μ = the mean

σ = the standard deviation.

The 'normal' or Gaussian distribution is illustrated in Figure 2.18. The normal distribution is defined such that 68.26% of all the output is within

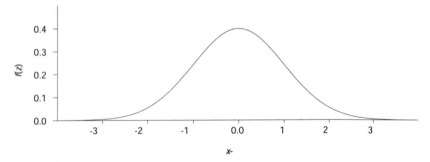

Figure 2.18 The normal distribution

one standard deviation of the mean, 95.44% of all the output is within two standard deviations of the mean and 99.73% of all the output is within three standard deviations of the mean.

The usefulness of the normal distribution stems from its ability to model sets of data arising from a large number of applications. If a set of data follows the normal distribution then equation (2.25) can be used to determine the frequency with which a particular range of data values will occur. An alternative way of saying this is that it will predict the probability with which data will occur.

Table 2.6 The normal distribution. Data reproduced from Chatfield (1983)

					Areas under the normal curve F(z)					
	0.00	0.01	0.02	0.03	0.04	0.05	0.06	0.07	0.08	0.09
0.0	0.5000	0.5040	0.5080	0.5120	0.5160	0.5199	0.5239	0.5279	0.5319	0.5359
0.1	0.5398	0.5438	0.5478	0.5517	0.5557	0.5596	0.5636	0.5675	0.5714	0.5753
0.2	0.5793	0.5832	0.5871	0.5910	0.5948	0.5987	0.6026	0.6064	0.6103	0.6141
0.3	0.6179	0.6217	0.6255	0.6293	0.6331	0.6368	0.6406	0.6443	0.6480	0.6517
0.4	0.6554	0.6591	0.6628	0.6664	0.6700	0.6736	0.6772	0.6808	0.6844	0.6879
0.5	0.6915	0.6950	0.6985	0.7019	0.7054	0.7088	0.7123	0.7157	0.7190	0.7224
0.6	0.7257	0.7291	0.7324	0.7357	0.7389	0.7422	0.7454	0.7486	0.7517	0.7549
0.7	0.7580	0.7611	0.7642	0.7673	0.7703	0.7734	0.7764	0.7793	0.7823	0.7852
0.8	0.7881	0.7910	0.7939	0.7967	0.7995	0.8023	0.8051	0.8078	0.8106	0.8133
0.9	0.8159	0.8186	0.8212	0.8238	0.8264	0.8289	0.8315	0.8340	0.8365	0.8389
1.0	0.8413	0.8438	0.8461	0.8485	0.8508	0.8531	0.8554	0.8577	0.8599	0.8621
1.1	0.8643	0.8665	0.8686	0.8708	0.8729	0.8749	0.8770	0.8790	0.8810	0.8830
1.2	0.8849	0.8869	0.8888	0.8906	0.8925	0.8943	0.8962	0.8980	0.8997	0.9015
1.3	0.9032	0.9049	0.9066	0.9082	0.9099	0.9115	0.9131	0.9147	0.9162	0.9177
1.4	0.9192	0.9207	0.9222	0.9236	0.9251	0.9265	0.9279	0.9292	0.9306	0.9319
1.5	0.9332	0.9345	0.9357	0.9370	0.9382	0.9394	0.9406	0.9418	0.9429	0.9441
1.6	0.9452	0.9463	0.9474	0.9484	0.9495	0.9505	0.9515	0.9525	0.9535	0.9545
1.7	0.9554	0.9564	0.9573	0.9582	0.9591	0.9599	0.9608	0.9616	0.9625	0.9633
1.8	0.9641	0.9648	0.9656	0.9664	0.9671	0.9678	0.9686	0.9693	0.9699	0.9706
1.9	0.9713	0.9719	0.9726	0.9732	0.9738	0.9744	0.9750	0.9756	0.9761	0.9767
2.0	0.9772	0.9778	0.9783	0.9788	0.9793	0.9798	0.9803	0.9808	0.9812	0.9817
2.1	0.9821	0.9826	0.9830	0.9834	0.9838	0.9842	0.9846	0.9850	0.9854	0.9857
2.2	0.9861	0.9864	0.9868	0.9871	0.9875	0.9878	0.9881	0.9884	0.9887	0.9890
2.3	0.9893	0.9896	0.9898	0.9901	0.9904	0.9906	0.9909	0.9911	0.9913	0.9916
2.4	0.9918	0.9920	0.9922	0.9924	0.9927	0.9929	0.9930	0.9932	0.9934	0.9936
2.5	0.9938	0.9940	0.9941	0.9043	0.9045	0.9940	0.9948	0.9949	0.9951	0.9952
2.6	0.9953	0.9955	0.9956	0.9957	0.9959	0.9960	0.9961	0.9962	0.9963	0.9964
2.7	0.9965	0.9966	0.9967	0.9968	0.9969	0.9970	0.9971	0.9972	0.9973	0.9974
2.8	0.9974	0.9975	0.9976	0.9977	0.9977	0.9978	0.9979	0.9979	0.9980	0.9981
2.9	0.9981	0.9982	0.9982	0.9983	0.9984	0.9984	0.9985	0.9985	0.9986	0.9986
3.0	0.9986	0.9987	0.9987	0.9988	0.9988	0.9989	0.9989	0.9989	0.9990	0.9990
3.1	0.9990	0.9991	0.9991	0.9991	0.9992	0.9992	0.9992	0.9992	0.9993	0.9993
3.2	0.9993	0.9993	0.9994	0.9994	0.9994	0.9994	0.9994	0.9995	0.9995	0.9995
3.3	0.9995	0.9995	0.9995	0.9996	0.9996	0.9996	0.9996	0.9996	0.9996	0.9996
3.4	0.9997	0.9997	0.9997	0.9997	0.9997	0.9997	0.9997	0.9997	0.9997	0.9998
3.5	0.9998	0.9998	0.9998	0.9998	0.9998	0.9998	0.9998	0.9998	0.9998	0.9998
3.6	0.9998	0.9998	0.9998	0.9999	0.9999	0.9999	0.9999	0.9999	0.9999	0.9999

Equation (2.25) can be solved numerically and the resulting values are given in Table 2.6. The table provides values that have been generalized for any mean and standard deviation using the dimensionless group

$$z = \frac{x - \mu}{\sigma} \qquad (2.26)$$

The second to the eleventh columns give values for $F(z)$ which is the proportion of data to the left-hand side of z. This represents the probability with which z is likely to occur. For example, if $F(z) = 0.81594$ then from Table 2.6, 81.594% of data will have a value of z of less than 0.9.

In the normal distribution data is assumed to be equally distributed about the mean and to lie between values for z from $-\infty$ to $+\infty$. Measurements are also often equally distributed about a mean value and we can choose to state the distribution of the measurements in the form

$$x = \bar{x} \pm \delta x \qquad (2.27)$$

where: δx represents the spread of data above and below the average value.

To identify the proportion of data lying between $\pm z$, in Figure 2.19, where z is some value other than infinity then scan the first column to find the value of z. Read across the row to give the value for $F(z)$ which represents the proportion of data with values less than z. To identify the proportion of data with values larger than z, subtract $F(z)$ from 1. Because the normal curve is symmetrical, the data to the right-hand side of z will be the same proportion as that to the left-hand side of $-z$. So to identify the data not included in the interval $-z$ to $+z$, the value required is $2(1 - F(z))$.

Data can also follow other distributions, e.g. rectangular with the occurrence of data evenly spread, or with the data skewed with a non-symmetry about the mean. The form of some of these is illustrated in Figure 2.20. Data can be analysed in order to identify whether it follows a particular

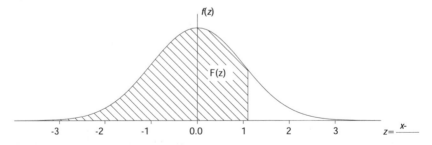

Figure 2.19 The normal distribution

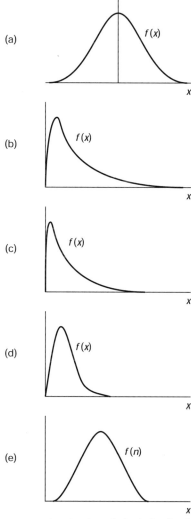

Figure 2.20 Some common standard statistical distributions. (a) The normal distribution. (b) The log normal distribution. (c) The Poisson distribution. (d) The Weibull distribution. (e) The binomial distribution

distribution. One technique for doing this is called the goodness of fit test and this is described in Section 2.4.4 below with particular relevance to identifying whether a data set can be modelled by the normal distribution.

2.4.4 Goodness of fit test

Rather than assuming that data follows, say, a normal distribution based merely on eyeballing the data or viewing a graph, a more rigorous approach is to use the goodness of fit test, which is also known as the chi-squared test. This provides a measure of the discrepancy between the measured variation of a set of data and the variation predicted by the assumed distribution function.

The test statistic for the goodness of fit test can be written as

$$\chi^2 = \sum \frac{(\text{observed} - \text{expected})^2}{\text{expected}} \qquad (2.28)$$

or

$$\chi^2 = \sum_{i=1}^{k} \frac{(n_o - n_e)^2}{n_e} \qquad (2.29)$$

where: k = number of group intervals
n_o = the actual number of observations in the ith group interval
n_e = the expected number of observations in the ith group interval based on the assumed frequency distribution.

In the goodness of fit test, one degree of freedom is deducted for each parameter estimated in the data. The degree of freedom is defined as

$$v = n - m \qquad (2.30)$$

where: n = number of observations or intervals
m = number of conditions imposed on the distribution.

If $\chi^2 = 0$ then the match is perfect. Values of $\chi^2 > 0$ indicate that there is a possibility that the data are not well represented by the proposed distribution.

The procedure to verify whether data follows an assumed distribution is as follows:

1 Divide the data into k convenient intervals. For small data sets k should be chosen such that $n_o > 5$ for at least one interval. For data sets with more than about 40 observations, then k can be given by

$$k = 1.87(n - 0.4)^{0.4} + 1 \qquad (2.31)$$

2 Find the number of occurrences, n_e, that are predicted by the assumed distribution for each interval.
3 Determine the value of $(n_o - n_e)^2/n_e$ for each interval.
4 Calculate χ^2.
5 Calculate the degrees of freedom, $v = k - m$.
6 By examining the corresponding row for v in the χ^2 table, (Table 2.7) identify the column which indicates a value of χ^2_α nearest to that calculated in step 4. If necessary, use interpolation to identify the value of α.

7 The probability of $P(\chi^2) = 1 - \alpha$, where α is called the level of significance.

If $P(\chi^2) = 1 - \alpha \geq 0.95$ then the evidence is that the assumed distribution does not model the data very well.

If $0.95 \geq P(\chi^2) = 1 - \alpha \geq 0.05$ then there is no reason to reject the hypothesis that assumed distribution models the data.

If $P(\chi^2) = 1 - \alpha \leq 0.05$ then the evidence is that the assumed distribution does model the data.

Table 2.7 Percentage points of the χ^2 distribution. Data reproduced from Chatfield (1983)

α ν	0.995	0.99	0.975	0.95	0.50	0.20	0.10	0.05	0.025	0.01	0.005
1	0.000	0.0002	0.001	0.0039	0.45	1.64	2.71	3.84	5.02	6.63	7.88
2	0.010	0.020	0.051	0.103	1.39	3.22	4.61	5.99	7.38	9.21	10.60
3	0.072	0.115	0.216	0.352	2.37	4.64	6.25	7.81	9.35	11.34	12.84
4	0.207	0.30	0.484	0.71	3.36	5.99	7.78	9.49	11.14	13.28	14.86
5	0.412	0.55	0.831	1.15	4.35	7.29	9.24	11.07	12.83	15.09	16.75
6	0.676	0.87	1.24	1.64	5.35	8.56	10.64	12.59	14.45	16.81	18.55
7	0.989	1.24	1.69	2.17	6.35	9.80	12.02	14.07	16.01	18.48	20.28
8	1.34	1.65	2.18	2.73	7.34	11.03	13.36	15.51	17.53	20.09	21.95
9	1.73	2.09	2.70	3.33	8.34	12.24	14.68	16.92	19.02	21.67	23.59
10	2.16	2.56	3.25	3.94	9.34	13.44	15.99	18.31	20.48	23.21	25.19
11	2.60	3.05	3.82	4.57	10.34	14.63	17.28	19.68	21.92	24.72	26.76
12	3.07	3.57	4.40	5.23	11.34	15.81	18.55	21.03	23.34	26.22	28.30
13	3.57	4.11	5.01	5.89	12.34	16.98	19.81	22.36	24.74	27.69	29.82
14	4.07	4.66	5.63	6.57	13.34	18.15	21.06	23.68	26.12	29.14	31.32
15	4.60	5.23	6.26	7.26	14.34	19.31	22.31	25.00	27.49	30.58	32.80
16	5.14	5.81	6.91	7.96	15.34	20.47	23.54	26.30	28.85	32.00	34.27
17	5.70	6.41	7.56	8.67	16.34	21.61	24.77	27.59	30.19	33.41	35.72
18	6.26	7.02	8.23	9.39	17.34	22.76	25.99	28.87	31.53	34.81	37.16
19	6.84	7.63	8.91	10.12	18.34	23.90	27.20	30.14	32.85	36.19	38.58
20	7.43	8.26	9.59	10.85	19.34	25.04	28.41	31.41	34.17	37.57	40.00
21	8.03	8.90	10.28	11.59	20.34	26.17	29.62	32.67	35.48	38.93	41.40
22	8.64	9.54	10.98	12.34	21.34	27.30	30.81	33.92	36.78	40.29	42.80
23	9.26	10.20	11.69	13.09	23.34	28.43	32.01	35.17	38.08	41.64	44.18
24	9.89	10.86	12.40	13.85	23.34	29.55	33.20	36.42	39.36	42.98	45.56
25	10.52	11.52	13.12	14.61	24.34	30.68	34.38	37.65	40.65	44.31	46.93
26	11.16	12.20	13.84	15.38	25.34	31.79	35.56	38.89	41.92	45.64	48.29
27	11.81	12.88	14.57	16.15	26.34	32.91	36.74	40.11	43.19	46.96	49.64
28	12.46	13.57	15.31	16.93	27.34	34.03	37.92	41.34	44.46	48.28	50.99
29	13.12	14.26	16.05	17.71	28.34	35.14	39.09	42.56	45.72	49.59	52.34
30	13.79	14.95	16.79	18.49	29.34	36.25	40.26	43.77	46.98	50.89	53.67
40	20.71	22.16	24.43	26.51	39.34	47.27	51.81	55.76	59.34	63.69	66.77
50	27.99	29.71	32.36	34.76	49.33	58.16	63.17	67.50	71.41	76.15	79.49
60	35.53	37.48	40.48	43.19	59.33	68.97	74.40	79.08	83.30	88.38	91.95
70	43.28	45.44	48.76	51.74	69.33	79.71	85.53	90.53	95.02	100.43	104.2
80	51.17	53.54	57.15	60.39	79.33	90.41	96.58	101.88	106.63	112.33	116.3
90	59.20	61.75	65.65	69.13	89.33	101.05	107.57	113.15	118.14	124.12	128.3
100	67.33	70.06	74.22	77.93	99.33	111.67	118.50	124.34	129.56	135.81	140.2

Table 2.8 Example data

Measurement number	T_i (°C)	$T_i - \bar{T}$ (°C)	$(T_i - \bar{T})^2$ (°C)^2
1	70.2	−0.5	0.25
2	70.4	−0.3	0.09
3	70.4	−0.3	0.09
4	70.6	−0.1	0.01
5	70.6	−0.1	0.01
6	70.5	−0.2	0.04
7	70.5	−0.2	0.04
8	70.6	−0.1	0.01
9	70.7	0	0
10	70.7	0	0
11	70.8	0.1	0.01
12	70.7	0	0
13	70.8	0.1	0.01
14	70.8	0.1	0.01
15	70.7	0	0
16	70.8	0.1	0.01
17	70.9	0.2	0.04
18	71.0	0.3	0.09
19	71.0	0.3	0.09
20	71.2	0.5	0.25

Example 2.11

Determine whether the data listed in Table 2.8 can sensibly be modelled using the normal distribution.

Solution

$$\bar{T} = \sum_{i=1}^{20} T = 70.7°C$$

and

$$s = \frac{1}{20 - 1} \sum_{1}^{20} (T - \bar{T})^2 = 0.24°C$$

The number of intervals can be approximated using equation (2.31),

$$k = 1.87(n - 0.4)^{0.4} + 1 = 1.87(20 - 0.4)^{0.4} - 1 = 7.148$$

Taking the number of intervals as 7 with the interval is $(71.2 - 70.2)/7 = 0.142$. Rounding up, a convenient interval is 0.15. The first interval covers the range $70.2 \leq x_i < 70.35$. We need to calculate the probability $P(70.2 \leq x_i < 70.35)$ which is equal to $P(70.35 \leq x_i < \mu) - P(70.2 \leq x_i < \mu) = P(z_2) - P(z_1)$ where, using the estimate for the mean and standard deviation,

$$z_1 = \frac{70.2 - \bar{T}}{s} \quad \text{and} \quad z_2 = \frac{70.35 - \bar{T}}{s}$$

$$P(z_2)-P(z_1) = P(-1.46)-P(-2.08)$$

From Table 2.6, $P(-2.08) = 1 - 0.9812 = 0.0188$ and $P(-1.46) = 1 - 0.9279 = 0.0721$.

$$P(z_2) - P(z_1) = 0.0721 - 0.0188 = 0.0533$$

So for a normal distribution, 5.33% of the measured data should be expected to fall within the interval between 70.2 and 70.35°C. For twenty measurements, the number of expected occurrences is

$$n_e = 20 \times 0.0533 = 1.066$$

From Table 2.8, the number of observed values in this interval was one:

$$\frac{(n_o - n_e)^2}{n_e} = \frac{(1-1.066)^2}{1.066} = 4.086 \times 10^{-3}$$

Values for all the intervals are listed in Table 2.9.

As there are two calculated statistical values, $\bar{T}$ and s, $m = 2$. There are seven intervals, so $n = 7$ and $v = n - m = 7 - 2 = 5$.

From Table 2.7, the nearest value of α for $\chi^2_\alpha(5) = 0.959$ is between $\alpha = 0.975$ and $\alpha = 0.95$.

Table 2.9 Calculation of χ^2

Interval number	Interval	z_1	$P(z_1)$	z_2	$P(z_2)$	$P(z_2)-P(z_1)$	n_e	n_o	$\frac{(n_o - n_e)^2}{n_e}$
1	$70.2 \leq x_i < 70.35$	−2.08	0.0188	−1.46	0.0721	0.0533	1.066	1	0.004
2	$70.35 \leq x_i < 70.5$	−1.46	0.0721	−0.83	0.2033	0.1312	2.624	2	0.148
3	$70.5 \leq x_i < 70.65$	−0.83	0.2033	−0.21	0.4168	0.2135	4.27	5	0.125
4	$70.65 \leq x_i < 70.8$	−0.21	0.4168	0.42	0.6628	0.246	4.92	4	0.172
5	$70.8 \leq x_i < 70.95$	0.42	0.6628	1.04	0.8508	0.188	3.76	5	0.409
6	$70.95 \leq x_i < 71.1$	1.04	0.8508	1.67	0.9525	0.1017	2.034	2	0.0006
7	$71.1 \leq x_i < 71.25$	1.67	0.9525	2.29	0.9890	0.0365	0.73	1	0.0999
									$\chi^2 = 0.959$

By linear interpolation,

$$\alpha = 0.975 + \frac{0.959 - 1.15}{1.15 - 0.831}(0.975 - 0.95) = 0.96$$

$$\chi^2_{0.96}(5) = 0.959 \text{ or } P(\chi^2) = 1 - \alpha = 1 - 0.96 = 0.04$$

As $P(\chi^2) \le 0.05$, there is evidence that the data can be modelled by the normal distribution.

2.5 Uncertainty

Accuracy is a qualitative term (Bell, 1999). A measurement can be accurate or not accurate. Uncertainty is a quantitative term. Whenever bounds are placed on a measurement then these define the uncertainty associated with that measurement. In practice, however, the terms 'accuracy' and 'uncertainty' are used loosely and interchangeably. This is especially the case in technical specifications where a manufacturer may quote the accuracy for a device to be within a ± bound or within a percentage of the output. Strictly speaking, using the above definitions then the manufacturer should be using the term 'uncertainty' for this.

The uncertainty of a measurement expresses the doubt that exists about the result of the measurement. Even with the utmost care there is also a margin of doubt with a measurement. In order to quantify the uncertainty of a measurement it is necessary to identify how large the margin is and how bad the doubt is. The margin can be expressed by means of the width of the margin, or interval. The doubt can be expressed by means of a confidence level, which states the certainty that the true value is within the margin. For example, the measurement of body temperature might be stated as 37°C ± 1.5°C at the 95% confidence level. This statement indicates that we are 95% sure (or 19/20ths sure) that the temperature is between 35.5°C and 38.5°C; the interval here is 3°C. The uncertainty of a measurement is important in defining the calibration of a sensor, to identify the outcome of a test and whether a tolerance has been met.

As indicated by the discussion within this chapter there are a large number of terms associated with measurement data. A summary of some of these is given in Table 2.10.

Errors in temperature measurement can occur due to a number of sources including bias errors, precision errors and measurement blunders. These errors may arise due to inadequacies of the calibration, instabilities of the transducer, limitations of the data-acquisition system or errors or loss of information in the data reduction. In the case of fixed-point calibration, errors occur due to the difference between the temperature taken up by a sensor and the actual temperature of the transfer medium. For comparison calibrations, a tem-

Table 2.10 Definitions of some common terms in measurement

Accuracy	A term used to describe the closeness of a measurement to the true value. Accuracy is a qualitative term.
Bias	The constant offset between the average indicated value and the true value.
Calibration	The act of applying a known input to a system in order to observe the system output.
Confidence level	A number, for example 95% or 0.95, expressing the degree of confidence or certainty in the result.
Error	Error is defined as the difference between the measured value and the true value of the quantity being measured.
Mean	The mean is the average of a set of numbers.
Precision	Precision defines the closeness of data to the average value.
Range	The range defines the difference between the highest and the lowest of a set of values.
Repeatability	Repeatability defines the closeness of the agreement between repeated measurements of the same property under the same conditions.
Resolution	The resolution defines the smallest difference that can be meaningfully distinguished.
Sensitivity	Sensitivity is the ratio of the change in response of an instrument to the change in the stimulus.
True value	The actual value of the measured variable.
Uncertainty	Uncertainty is a quantification of the doubt associated with the measurement result.

perature difference between the test and the standard thermometer is inevitable due to thermal disturbance effects. Installation effects mean that a transducer will not indicate the undisturbed temperature of a medium of interest and in the case of transient applications, this effect is compounded by an amplitude difference and phase lag between the indicated and actual temperatures. The data-acquisition system must read the input signal, manipulate and communicate it in a convenient form. Differences between the true and indicated temperature will be dependent in part on the calibration fit and truncation error associated with the data acquisition system.

Bias error represents a constant offset between the average indicated value and the true value as illustrated in Figure 2.21. Bias error remains constant during a series of measurements. Repetition of more and more measurements,

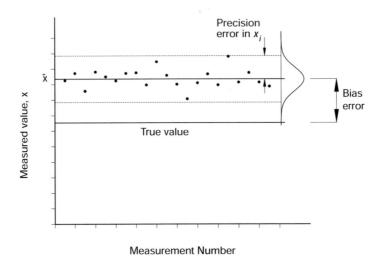

Figure 2.21 Illustration of bias and precision error

therefore, does not assist in either eliminating or reducing the bias error. The level of bias error can only be estimated by comparison. This can be achieved by calibration, by concomitant methodology, which is the use of different methods to estimate the same thing and comparison of the results (see Nicholas and White, 1994), or by inter-laboratory comparisons of measurements of the same quantity. Corrections identified due to a bias error should be incorporated in to the data. If a calibration reveals a bias error of, say, a 0.2°C underestimate of temperature then all the results should be adjusted by this amount. However, the uncertainty associated with the calibration itself will remain. For example, if the uncertainty of the calibration method is ±0.01°C then this value will propagate into the uncertainty of the result and remains associated with the adjusted data.

Precision defines the closeness of data to the average value as indicated in Figure 2.21. So-called high-precision data will not have significant scatter around the mean while low-precision data will be highly scattered. Precision is affected by repeatability and resolution, temporal and spatial variations and changes in environmental conditions.

Measurement blunders do occur. For example, thermocouples can be wrongly located or even their location unknown. The data arising from measurement blunders should generally be ignored.

Calibration will generally reduce errors and will set bounds on them. For a measurement device that has not been calibrated then an estimate of the difference between the true value and the measurement value can be made. An example of this process in common practice is the use of Class A and Class B thermocouples. Here the individual thermocouples have not been calibrated. However, the wire from which they are made has been used to make a number of thermocouples and these have been tested. As such there is a level of confidence associated with the thermocouples and their output can be assumed to be within bounds.

The uncertainty of the result of a measurement usually consists of several components which, following the CIPM approach, can be grouped into two categories according to the method used to estimate their values. These are:

(a) those evaluated by statistical methods
(b) those evaluated by other means.

For most distributions the square root of the variance, the standard deviation, is directly proportional to the width of variation. It can therefore be used to characterise the uncertainty associated with a value and using a bilateral tolerance the measurement can be stated as:

$$\text{Value} = \text{mean} \pm \text{standard deviation} \qquad (2.32)$$

A measurement stated in this way is known as the standard uncertainty and accounts for 68.26% of measurements provided the data is normally distributed. That is, we could be confident that 68% of measurements would lie within the range $\mu - \sigma$ to $\mu + \sigma$. However, 68% limits are not suitable for all measurements. A convenient form for reporting measurements is given by

$$\text{Value} = \text{mean} \pm k\sigma \qquad (2.33)$$

where $k =$ the coverage factor which determines the range covered. If $k = 2.6$, then $\pm 2.6\sigma$ or 99% of data would be expected to fall within the range of the confidence interval.

The confidence level associated with a bilateral tolerance should always be reported along with the tolerance. The statement $T = \overline{T} \pm 0.4°C$ is not that useful. The statement fails to inform whether 68%, 95%, or even 100% of data can be assumed to fall within these limits. Common values used for the coverage factor, k, are 1, 2 and 3, corresponding to 68.26%, 95.44% and 99.73% of measurements. A value for the coverage factor of 2 seems to be suitable for the majority of temperature measurement applications, unless the measurement is safety-critical, in which case a value of 3 would be more appropriate.

Knowing the value of k, a measurement can be reported in the form:

$$\text{Value} = \text{mean} \pm k\sigma \ (n\%) \qquad (2.34)$$

where n is the corresponding confidence level. For example, if $k = 2$, then the confidence level is 95%, and for data with a mean of 40.8°C and a standard deviation of 0.7°C, the measurement could sensibly be reported as

$$T = 40.8 \pm 1.4°C \ (95\%)$$

Table 2.11 Percentage points of Student's *t* distribution. Data reproduced from Chatfield (1983).

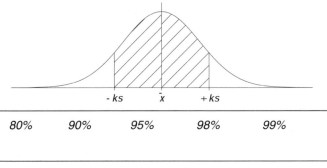

P%	80%	90%	95%	98%	99%	99.8%
v						
1	3.078	6.314	12.706	31.821	63.657	318.310
2	1.886	2.920	4.303	6.965	9.925	22.327
3	1.638	2.353	3.182	4.541	5.841	10.215
4	1.533	2.132	2.776	3.747	4.604	7.173
5	1.476	2.015	2.571	3.365	4.032	5.893
6	1.440	1.943	2.447	3.143	3.707	5.208
7	1.415	1.895	2.365	2.998	3.499	4.786
8	1.397	1.860	2.306	2.896	3.355	4.501
9	1.383	1.833	2.262	2.821	3.250	4.297
10	1.372	1.812	2.228	2.764	3.169	4.144
11	1.363	1.796	2.201	2.718	3.106	4.025
12	1.356	1.782	2.179	2.681	3.055	3.930
13	1.350	1.771	2.160	2.650	3.012	3.852
14	1.345	1.761	2.145	2.624	2.977	3.787
15	1.341	1.753	2.131	2.602	2.947	3.733
16	1.337	1.746	2.120	2.583	2.921	3.686
17	1.333	1.740	2.110	2.567	2.898	3.646
18	1.330	1.734	2.101	2.552	2.878	3.610
19	1.328	1.729	2.093	2.539	2.861	3.579
20	1.325	1.725	2.086	2.528	2.845	3.552
21	1.323	1.721	2.080	2.518	2.831	3.527
22	1.321	1.717	2.074	2.508	2.819	3.505
23	1.319	1.714	2.069	2.500	2.807	3.485
24	1.318	1.711	2.064	2.492	2.797	3.467
25	1.316	1.708	2.060	2.485	2.787	3.450
26	1.315	1.706	2.056	2.479	2.779	3.435
27	1.314	1.703	2.052	2.473	2.771	3.421
28	1.313	1.701	2.048	2.467	2.763	3.408
29	1.311	1.699	2.045	2.462	2.756	3.396
30	1.310	1.697	2.042	2.457	2.750	3.385
40	1.303	1.684	2.021	2.423	2.704	3.307
60	1.296	1.671	2.000	2.390	2.660	3.232
120	1.289	1.658	1.980	2.358	2.617	3.160
∞	1.282	1.645	1.960	2.326	2.576	3.090

The mean and variance of a quantity are, in practice, not known exactly, as we rarely know all the possible values and typically measure only samples of data. Instead it is necessary to account for the decrease in confidence levels due to the uncertainty associated with the mean and variance. The mean of a set of n independent measurements will be random and distributed with a variance

$$\sigma_m^2 = \frac{\sigma^2}{n} \tag{2.35}$$

Similarly, the variance is also distributed with a variance of

$$\sigma_{s^2}^2 = \frac{2\sigma^4}{n} \tag{2.36}$$

There are two approaches to cope with this fuzziness introduced by the uncertainty associated with μ and σ^2. One approach is to take more measurements to increase confidence in the values, and the second is to increase the k factor. The Student's t distribution (Table 2.11) can be used to identify the appropriate inflation of the coverage factor for a specified confidence level to account for the uncertainties in μ and σ^2. In the tables the parameter υ represents the number of degrees of freedom. The number of degrees of freedom in a data set is the number of independent measurements, n. However, for a set of data with a central tendency around a mean, the freedom of data to adopt any value is restricted by the mean and so the number of degrees of freedom is reduced by one. If n measurements have been made then $\upsilon = n - 1$.

Example 2.12

For the data presented in Table 2.8 estimate the interval of values over which 95% of measurements could be assumed to fall within.

Solution

The sample mean and standard deviation are 70.7°C and 0.24°C respectively.

The number of degrees of freedom, $\upsilon = n - 1 = 20 - 1 = 19$.

From Table 2.11, for $\upsilon = 19$ and 95% confidence limits, $t = 2.093$.

So $T = \bar{T} \pm t\sigma = 70.7 \pm 2.093 \times 0.24 = 70.7 \pm 0.5°C$ (95%)

Each measurement error will combine with each other to increase the uncertainty associated with the data. The combination of errors from different variables is considered in Sections 2.5.1 to 2.5.3.

2.5.1 Sure-fit or extreme variability

The sure-fit or extreme variability of a function of several uncorrelated random variables, $y = f(x_1, x_2, \ldots, x_j, \ldots x_\infty)$, represents the bounds for which 100% of data can be assumed to fall within. It could in theory be found by substituting values for the variables to find the maximum and the minimum value of the function. Alternatively, the following expression can be used to approximate the extreme variability:

$$\delta y \cong \sum_{j=1}^{n} \frac{\partial f}{\partial x_j} \delta x_j \tag{2.37}$$

δy represents the width of variability of the function. Bilateral tolerances for y would be $\pm(\delta y/2)$.

2.5.2 Linear functions or tolerance chains

Often it is necessary to know the overall tolerance in a dimension chain. Taking a simple linear chain of variable of overall sum z:

$$z = x_1 + x_2 + x_3 + \ldots + x_n \tag{2.38}$$

The expected value of z is given by:

$$E(z) = \mu_1 + \mu_2 + \ldots + \mu_n = \mu_z \tag{2.39}$$

and the variance by:

$$\text{variance}(z) = \sigma_1^2 + \sigma_2^2 + \sigma_3^2 + \ldots + \sigma_n^2 = \sigma_z^2 \tag{2.40}$$

If the response variable is a linear function of several measured variables, each of which is normally distributed then the response variable will also be normally distributed. In other words, when a number of random variables are added together the result tends to also be normally distributed.

2.5.3 Several independent, uncorrelated random variables

The theoretical preliminaries that enable analysis of functions of several independent, uncorrelated random variables are given by Furman (1981). Let y be a function of several independent, uncorrelated random variables.

$$y = f(x_1, x_2, \ldots, x_j, \ldots x_n) \tag{2.41}$$

The statistical mean and standard deviation of a function of several independent, uncorrelated random variables are given by:

$$\mu_y \cong f(\mu_1, \mu_2, \ldots, \mu_n) + \frac{1}{2} \sum_{j=1}^{n} \frac{\partial^2 f}{\partial x_j^2} \sigma_{xj}^2 \qquad (2.42)$$

$$\sigma_y^2 \cong \sum_{j=1}^{n} \left(\frac{\partial f}{\partial x_j}\right)^2 \sigma_{x_j}^2 + \frac{1}{2} \sum_{j=1}^{n} \left(\frac{\partial^2 f}{\partial x_j^2}\right)^2 \sigma_{x_j}^4 \approx \sum_{j=1}^{n} \left(\frac{\partial f}{\partial x_j}\right)^2 \sigma_{x_j}^2 \qquad (2.43)$$

Consider a measurement E that is subject to, say, three variables e_1, e_2 and e_3 with standard deviations σ_1, σ_2, and σ_3 respectively. If the variables are independent, uncorrelated and random then the overall standard deviation can be modelled, approximately, using equation (2.43) by

$$\sigma_E = \sqrt{\left|\frac{\partial E}{\partial e_1}\right|^2 \sigma_1^2 + \left|\frac{\partial E}{\partial e_2}\right|^2 \sigma_2^2 + \left|\frac{\partial E}{\partial e_3}\right|^2 \sigma_3^2}$$

The overall variability on the measurement could be expressed in the form of a bilateral tolerance, using 95% ($\pm 2\sigma$) limits as

$$E = \bar{E} \pm (2\sigma_E) \ (95\%)$$

Example 2.13

A temperature difference is determined using two different thermocouples and the average measurements of the two sensors are 124.3°C and 55.7°C. The 95% confidence limit uncertainties for each device are ±2.2°C and ±1.7°C respectively. Determine the overall uncertainty for the temperature difference.

Solution

As two different sensors are used the errors can be assumed to be independent and uncorrelated. From equation (2.41),

$$\sigma_{\Delta T}^2 = \sigma_{T1}^2 + \sigma_{T2}^2$$

Assuming 95% limits (k = 2), then the standard deviation, $\sigma_{T1} = 2.2/2 = 1.1$°C and $\sigma_{T2} = 1.7/2 = 0.85$°C;

$$\sigma_{\Delta T} = \sqrt{1.1^2 + 0.85^2} = 1.39°C$$

The bilateral tolerance, taking $k = 2$, 95% confidence limits, is then ±(2 × 1.39) = ±2.78°C. So the temperature difference

$$\Delta T = T_1 - T_2 = 124.3 - 55.7 = 68.6 \pm 2.78°C \ (95\%)$$

Example 2.14

The relationship between a transistor-based temperature sensor output voltage and the temperature is given by the equation $T = kV$, where k is a constant, V is the voltage and T is the temperature in degrees Celsius. If $k = 100.0°C/V$ with an uncertainty of $\pm1.2°C/V$ (95% limits) and the voltage is 0.5 V with an uncertainty of ±0.01 V (95%) determine the overall uncertainty.

Solution

There are two uncorrelated random variables to consider, k and V:

$$\frac{\partial T}{\partial k} = V = 0.5$$

$$\frac{\partial T}{\partial V} = k = 100$$

From equation (2.43),

$$\sigma_T = \sqrt{\left| \frac{\partial T}{\partial k} \right|^2 \sigma_k^2 + \left| \frac{\partial T}{\partial V} \right|^2 \sigma_v^2}$$

$$= \sqrt{0.5^2 \times 0.6^2 + 100^2 \times 0.005^2} = 0.583°C$$

The bilateral tolerance, taking $k = 2$, 95% confidence limits, is then $\pm(2 \times 0.583) \approx \pm1.2°C$:

$$T = 100 \times 0.5 = 50°C$$

So the overall uncertainty is given by

$$T = 50 \pm 1.2°C \ (95\%).$$

Example 2.15

A proposal for a personal computer-based temperature measuring system consists of a Type K thermocouple, a reference junction compensator and a data acquisition board as illustrated in Figure 2.22. The system is designed to measure temperatures nominally at 300°C. The corresponding output voltage for a Type K thermocouple at 300°C is approximately 12.2 mV. The data acquisition board initially chosen has an analogue input range of 0 to 0.1 V, with a 12-bit A/D converter with an uncertainty of 0.01% (95% confidence limits) of the reading. The system makes use of a reference junction

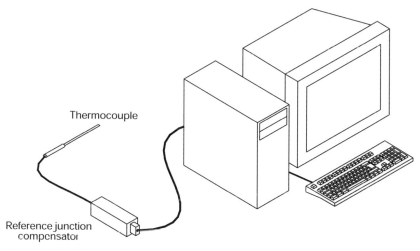

Figure 2.22 PC based temperature measuring system

compensation system, which has an uncertainty of ±0.5°C (95%) over the range 0–35°C. The uncertainty of the thermocouple is ±1.1°C (95%). The market requirement for the system dictates that the overall uncertainty must be less than 1.5°C. Based on the above specifications determine whether the system matches the requirement. (Example after Figliola and Beasley, 2000.)

Solution

The 12-bit A/D converter divides the full scale voltage range into 2^{12} = 4096 intervals. The resolution of the A/D converter is given by dividing the analogue input range, 0.1 V, by the number of intervals:

$$\frac{0.1}{4096} = 0.0000244 \, \text{V} = 0.0244 \, \text{mV}$$

The uncertainty of the A/D converter is 0.01% of the reading. The voltage of the thermocouple at the nominal operating temperature of 300°C is approximately 12.2 mV. The uncertainty in the voltage measurement is therefore

$$\frac{0.01}{100} \times 12.2 = 0.00122 \, \text{mV}$$

The uncertainty of the temperature measurement system associated with the measurement of the voltage consists of contributions from both the minimum resolution capability and the uncertainty of the A/D converter:

$$\sigma_{meas}^2 = \sigma_{res}^2 + \sigma_{A/D}^2$$

$$\sigma_{res} = 0.0244/2 = 0.0122\,mV$$

$$\sigma_{A/D} = 0.00122/2 = 0.00061\,mV$$

$$\sigma_{meas} = t\,\sqrt{0.0122^2 + 0.00061^2} = \pm0.0122\,mV$$

Taking the approximate sensitivity of Type K thermocouples as 12.2 mV/300 = 0.04 mV/°C the corresponding uncertainty in temperature can be evaluated:

$$\frac{0.0122}{0.04} = \pm0.305°C$$

The overall uncertainty arising from the data acquisition system, the reference junction compensator and the uncertainty of the thermocouple can now be evaluated:

$$\sigma_{DAS} = \sqrt{\sigma_{meas}^2 + \sigma_{tc}^2 + \sigma_{comp}^2} = \sqrt{0.305^2 + 0.55^2 + 0.25^2}$$

$$= 0.677°C$$

The overall uncertainty using $k = 2$ (95% confidence limits) $= \pm2\sigma = \pm1.35°C$ (95%). This is within the specification of 1.5°C.

Example 2.16

A temperature difference is determined using the same sensor. If the error associated with the sensor can be considered to comprise two components, a random error of $\pm1.4\,V$ (95%) and a systematic error of $\pm0.8\,V$ (95%), determine the overall uncertainty associated with the measurement. The average values for the two measurements are 124.3°C and 55.7°C.

Solution

The first measurement can be modelled as

$$T_{1m} = T_1 + T_{es1} + T_{er1}$$

where: T_1 = the true value
T_{es} = the systematic error
T_{er1} = the random error.

Similarly

$$T_{2m} = T_2 + T_{es2} + T_{er2}$$

As the same sensor is used for both measurements the systematic error will be the same for both measurements:

$$T_{es1} = T_{es2}$$

The temperature difference is given by

$$\Delta T = (T_1 + T_{es1} + T_{er1}) - (T_2 + T_{es2} + T_{er2}) = T_1 - T_2 + T_{er1} - T_{er2}$$

In determining the temperature difference, the systematic error has cancelled out and so the variability is due entirely to random uncorrelated effects. Using equation (2.41)

$$\sigma_{\Delta T}^2 = \sigma_{er1}^2 + \sigma_{er2}^2$$

So

$$\sigma_{\Delta T} = \sqrt{0.7^2 + 0.7^2} = 0.99°C$$

Therefore the temperature difference

$$\Delta T = T_1 - T_2 = 124.3 - 55.7 = 68.6 \pm 1.98°C \ (95\%)$$

Comment

By using the same sensor for the two measurements the variability in the temperature difference has been reduced, in comparison to that of Example 2.13, because the systematic error has cancelled out. Had the systematic error not cancelled then a mathematical approach taking account of correlated variables would need to be used. The treatment of correlated variables and their combination is covered in texts such as Coleman and Steele (1999).

2.6 Conclusions

The measurement of temperature will typically result in the disturbance of the temperature distribution in the medium of interest. The difference between the measured and the true temperature is called the error. This chapter has introduced methods for quantifying the magnitude of the disturbance caused by the insertion of a measurement probe. In addition, methods for evaluating and communicating data and its associated uncertainty have been developed. While the information provided in this chapter is of use in processing data it does not provide sufficient insight into the assessment of the significance of the data itself. For example, a common goal in the production of data is to compare one set of measurements with another in order to identify whether an improvement in a process has occurred. Values for the average from two sets of data on their own are not sufficient to enable this decision to be made. In addition, information about the spread and reliability of the data is also necessary. There are standard statistical techniques, known as significance tests, which can be applied to data to assist in the decision-making process. The reader is referred to the text by Chatfield (1983) for an introduction to these.

References

Books and papers

Bejan, A. *Heat Transfer*. Wiley, 1993.

Bell, S. *A Beginner's Guide to Uncertainty of Measurement*. Measurement Good Practice Guide, No. 11, NPL, 1999.

Benedict, R.P. Temperature measurement in moving fluids. ASME Paper 59A–257, 1959.

Chapman, A.J. *Fundamentals of Heat Transfer*. Macmillan, 1987.

Chatfield, C. *Statistics for Technology*, 3rd edition. Chapman and Hall, 1983.

Churchill, S.W. and Bernstein, M. A correlating equation for forced convection from gases and liquids to a circular cylinder in crossflow. *Journal of Heat Transfer, Trans. ASME*, **99**, 300–306, 1977.

Clarke, G.M. and Kempson, R.E. *An Introduction to the Design and Analysis of Experiments*. Edward Arnold, 1998.

Cohen, E.R. and Taylor, B.N. The fundamental physical constants. *Physics Today*, BG5–BG9, 1999.

Coleman, H.W. and Steele, W.G. *Experimentation and Uncertainty Analysis for Engineers*. Wiley, 1999.

Figliola, R.S. and Beasley, D.E. *Theory and Design for Mechanical Measurements*, 3rd edition. Wiley, 2000.

Furman, T.T. *Approximate Methods in Engineering Design*. Academic Press, 1981.

Incropera, F.P. and DeWitt, D.P. *Fundamentals of Heat and Mass Transfer*, 4th edition. Wiley, 1996.

Kaye, G.W.C. and Laby, T.H. *Tables of Physical and Chemical Constants*. Longman, 1986.

Kreith, F. and Bohn, M.S. *Principles of Heat Transfer*. PWS Publishing Co., 1997.

Long, C.A. *Essential Heat Transfer*. Longman, 1999.

Moffat, R.J. Gas temperature measurements. In Herzfeld, C.H. (Editor), *Temperature. Its Measurement and Control in Science and Industry*, Vol. 3, Part 2. Reinhold, 1962.

Nicholas, J.V. and White, D.R. *Traceable Temperatures*. Wiley, 1994.

Touloukian, Y.S., Powell, R.W., Ho, C.Y. and Klemens, P.G. *Thermophysical Properties of Matter, Vol. 1. Thermal Conductivity. Metallic Elements and Alloys*. IFI Plenum, 1970a.

Touloukian, Y.S., Powell, R.W., Ho, C.Y. and Klemens, P.G. *Thermophysical Properties of Matter, Vol. 2. Thermal Conductivity. Non-metallic Solids*. IFI Plenum, 1970b.

Web sites

At the time of going to press the world wide web contained useful information relating to this chapter at the following sites.

http://physics.nist.gov/cuu/Uncertainty/index.html
http://www.ukas.com/docs/technical-uncertainty.htm

Nomenclature

A	=	area (m^2)
A_c	=	cross-sectional area (m^2)
A_s	=	surface area (m^2)
Bi	=	Biot number
c_p	=	specific heat capacity (J/kg·K)
h	=	heat transfer coefficient (W/m^2·K)
k	=	thermal conductivity (W/m·K) or coverage factor
K	=	dynamic correction factor
L	=	length (m)
L_c	=	characteristic length (m)
n	=	total number of measurements
n_o	=	number of observations
n_e	=	expected number of observations
P	=	perimeter (m)
q	=	heat flux (W/m^2)
r	–	radius (m)
s	=	estimated standard deviation
t	=	time (s)
T	=	temperature (K or °C)
T_b	=	base temperature (K or °C)
T_f	=	bulk fluid temperature (K or °C)
T_i	=	initial temperature (K)
T_{probe}	=	equilibrium temperature indicated by a stationary probe (K)
T_s	=	surface temperature (K or °C)
T_{static}	=	static temperature (°C or K)
T_{tc}	=	temperature of the thermocouple (°C or K)
T_{total}	=	total temperature (°C or K)
T_w	=	temperature of the walls of the surrounding enclosure (°C or K)
T_∞	=	temperature of surroundings (K)
U	=	bulk free stream velocity (m/s)
V	=	volume (m^3)
x	=	location (m)
α	=	the level of significance
ε	=	total surface emissivity

ε_{sh} = total emissivity of the shield

ε_{tc} = total emissivity of the thermocouple

ϕ = phase lag (rad)

μ = the mean

ρ = density (kg/m^3)

σ = the Stefan–Boltzmann constant (= 5.67051×10^{-8} $W \cdot m^{-2} K^{-4}$ (Cohen and Taylor, 1999) or the standard deviation

τ = time constant (s)

υ = number of degrees of freedom

ω = angular frequency (rad/s)

3

Bimetallic thermometers

Bimetallic thermometers consist of two strips of dissimilar materials bonded together, normally in the form of a coil. On heating the coil will twist and a needle mounted on one end can be used to indicate temperature on a dial. Bimetallic strips are also widely used as control devices to activate machinery at a particular temperature. This chapter introduces the physical phenomena involved, the practical modelling of bimetallic strips and the principal configuration and uses of bimetallic thermometers.

3.1 Introduction

Bimetallic thermometers use two strips of material with different coefficients of thermal expansion that are bonded together. When the temperature of the assembly is changed, in the absence of external forces, the strip will bend or warp due to the difference in expansion between the two materials. In its

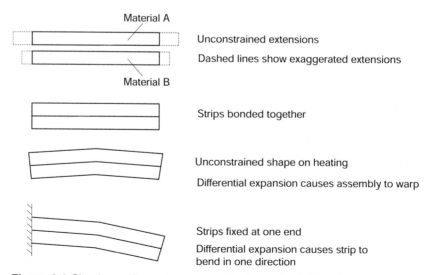

Material A

Unconstrained extensions

Dashed lines show exaggerated extensions

Material B

Strips bonded together

Unconstrained shape on heating

Differential expansion causes assembly to warp

Strips fixed at one end

Differential expansion causes strip to bend in one direction

Figure 3.1 Simple cantilever-type assembly for a bimetallic strip

Table 3.1 Properties of selected materials used in bimetallic elements. After Stephenson et al. (1999), data from Meijer (1994), Goodfellows catalogue (2000)

Material	Density (kg/m^3)	Young's modulus (GPa)	Heat capacity (J/kg·K)	Coefficient of thermal expansion, 10^{-6} K^{-1}	Thermal conductivity (W/m·K)
Al	2700	61–71	896	24	237
Brass				19	
Cu	8954	129.8	383.1	17	386
Cr	7100	279	518	6.5	94
Au	19300	78.5	129	14.1	318
Fe	7870	211.4	444	12.1	80.4
Ni	8906	199.5	446	13.3	90
Ag	10524	82.7	234.0	19.1	419
Sn	7304	49.9	226.5	23.5	64
Ti	4500	120.2	523	8.9	21.9
W	19350	411	134.4	4.5	163
Invar (Fe64/Ni36)	8000	140–150		1.7–2.0	13
Si	2340	113	703	4.7–7.6	80–150
n-Si	2328	130–190	700	2.6	150
p-Si	2300	150–170	770		30
Si_3N_4	3100	304	600–800	3.0	9–30
SiO_2	2200	57–85	730	0.5	1.4

simplest form strips of materials are used supported at one end as illustrated in Figure 3.1. When the assembly is heated it will deform in the shape of an arc. To maximize bending of the assembly, materials with significantly different coefficients of thermal expansion are selected. Table 3.1 lists some of the materials used in bimetallic thermometers. Traditionally, as the name implies, only metals have been used, such as brass and iron or invar, an iron–nickel alloy with a very low coefficient of thermal expansion, and iron. However, any two dissimilar materials could be used in theory and use of ceramics and semiconductors to form the differential expansion strip has been demonstrated (O'Connor, 1995).

3.2 Bimetallic strip modelling

The general equation defining the curvature of a bimetallic strip heated from T_1 to T_2 in the absence of external forces is given by Eskin (1940) and Timoshenko (1953):

$$\frac{1}{R_2} - \frac{1}{R_1} = \frac{6(1 + m)^2 \, (\alpha_2 - \alpha_1)(T_2 - T_1)}{t[3(1 + m)^2 + (1 + mn)(m^2 + 1/mn)]} \qquad (3.1)$$

where: $1/R_1$ = initial curvature of the strip at temperature T_1 (m^{-1})

α_1 = coefficient of thermal expansion of the material with the lower expansivity (K^{-1})

α_2 = coefficient of thermal expansion of the material with the higher expansivity (K^{-1})

n = E_1/E_2 where E_1 and E_2 are the Young's moduli for the high- and low-expansivity materials respectively

m = t_1/t_2 where t_1 and t_2 are the respective thicknesses of the metallic strips

tt = $t_1 + t_2$, the overall thickness of the strip (m)

T_1 = initial temperature (K)

T_2 = final temperature (K)

and the width of the strip is taken as unity.

In the case of simply supported ends and assuming that the deflection and thickness are less than 10% of the length of the strip then the deflection of the middle of the strip is approximated by

$$d = \frac{L^2}{8R} = \frac{3L^2(1 + m)^2 \, (\alpha_2 - \alpha_1)(T_2 - T_1)}{4t[3(1 + m)^2 + (1 + mn)(m^2 + 1/mn)]} \qquad (3.2)$$

where: L = length of the bimetallic strip (m)

d = maximum lateral deflection of the strip (m).

Example 3.1

A 100 mm long bimetallic strip is formed from two layers of material using iron for the high-expansivity layer and Invar for the low-expansivity layer. The thickness of each layer is 0.5 mm and the strip is simply supported at each end. Determine the deflection at the centre of the strip if the strip is heated from 15°C to 100°C.

Solution

The deflection will be a maximum at the middle of the strip as given by equation (3.2). From Table 3.1, Young's modulus for Invar is 140×10^9 Pa and for iron it is 211×10^9 Pa. The coefficient of thermal expansion for Invar is 1.7×10^{-6} K^{-1} and for iron is 12.1×10^{-6} K^{-1}. For the parameters given, $m = t_1/t_2 = 1$, $n = E_1/E_2 = 0.664$, $t = t_1 + t_2 = 0.001$ m. Therefore

$$d = \frac{3(1 + 1)^2 \, (0.1^2)(12.1 \times 10^{-6} - 1.7 \times 10^{-6})(100 - 15)}{4 \times 0.001[3(1 + 1)^2 + (1 + 0.664)(1^2 + 1/0.664)]}$$

$$= 1.64 \times 10^{-3} \, m = 1.64 \, mm$$

In the case of a cantilevered bimetallic strip with one end fixed and the other end free to move then the deflection of the free end is approximated by

$$d = \frac{L^2}{8R} = \frac{3L^2(1 + m)^2 \, (\alpha_2 - \alpha_1)(T_2 - T_1)}{t[3(1 + m)^2 + (1 + mn)(m^2 + 1/mn)]} \tag{3.3}$$

Example 3.2

A 30 mm long bimetallic strip is fixed at one end to form a cantilevered beam. It is formed from two layers of material using iron for the high-expansivity layer and Invar for the low expansivity-layer. The thickness of each layer is 0.6 mm. Determine the deflection at the end of the strip if the strip is heated from 15°C to 40°C.

Solution

The deflection will be a maximum at the end of the strip and is given by equation (3.3). From Table 3.1, Young's modulus for Invar is 140×10^9 Pa and for iron it is 211×10^9 Pa. The coefficient of thermal expansion for Invar is $1.7 \times 10^{-6} \, K^{-1}$ and for iron it is $12.1 \times 10^{-6} \, K^{-1}$. For the parameters given, $m = t_1/t_2 = 1$, $n = E_1/E_2 = 0.664$, $t = t_1 + t_2 = 0.0012 \, m$. Therefore

$$d = \frac{3(0.03^2)(1 + 1)^2(12.1 \times 10^{-6} - 1.7 \times 10^{-6})(40 - 15)}{0.0012[3(1 + 1)^2 + (1 + 0.664)(1^2 + 1/0.664)]}$$

$$= 1.45 \times 10^{-4} \, m = 0.145 \, mm$$

3.3 Standard materials

Standards are available for bimetallic strips, such as DIN 1715, Part 1 (1983) and ASTM B388. ASTM B388 defines a constant called the flexivity, k, which is given by

$$k = \frac{3}{2}(\alpha_2 - \alpha_1) = \frac{t/R}{T_2 - T_1} \tag{3.4}$$

and can be defined as the change of curvature of a bimetallic strip per unit of temperature change times thickness. In Europe the flexivity is known as the specific curvature. In equations (3.2) and (3.3) the magnitude of the ratio n has little effect on the resulting deflection for the range of materials commonly used in bimetallic strips. Assuming a value of unity for n allows equation (3.3) to be simplified to

$$d = \frac{kL^2}{2t}(T_2 - T_1) \tag{3.5}$$

Values for the flexivity can be determined by testing to ASTM B388 and selected values are listed in Table 3.2 for ASTM designation materials and Table 3.3 for DIN standard materials. Information on the composition of various alloys used in bimetallic strips is given in Table 3.4 and considerations in the selection of a bimetallic thermometer are discussed by Divita (1994).

Table 3.2 Selected ASTM bimetallic materials (ASTM B388)

ASTM type	Flexivity $(10^{-6}K^{-1})$	Maximum sensitivity temperature range (°C)	Maximum operating temperature (°C)
TM1	27.0±5%	−18 to 149	538
TM2	38.7±5%	−18 to 204	260
TM5	11.3±6%	149 to 454	538
TM10	23.6±6%	−18 to 149	482
TM15	26.6±5.5%	−18 to 149	482
TM20	25.0±5%	−18 to 149	482

Table 3.3 Selected DIN bimetallic materials (DIN 1715)

DIN type	Flexivity $(10^{-6}K^{-1})$	Linear range (°C)	Maximum operating temperature (°C)
TB0965	18.6±5%	−20 to 425	450
TB1075	20.0±5%	−20 to 200	550
TB1170A	22.0±5%	−20 to 380	450
TB1577A	28.5±5%	−20 to 200	450
TB20110	39.0±5%	−20 to 200	350

Table 3.4 Alloy composition for selected ASTM and DIN bimetallic materials (HE = high coefficient of thermal expansion material, LE = low coefficient of thermal expansion material). (ASTM B 388, DIN 1715 Part 1)

	Ni	Cr	Mn	Cu	Fe	Co
TM1 HE	22	3	–	–	75	–
TM1 LE	36	–	–	–	64	–
TM2 HE	10	–	72	18	–	–
TM2 LE	36	–	–	–	64	–
TM5 HE	25	8.5	–	–	66.5	–
TM5 LE	50	–	–	–	50	–
TM10 HE	22	3	–	–	75	–
TM10 LE	36	–	–	–	64	–
TM15 HE	22	3	–	–	75	–
TM15 LE	36	–	–	–	64	–
TM20 HE	18	11.5	–	–	70.5	–
TM20 LE	36	–	–	–	64	–
TB0965 HE	20	–	6	–	74	–
TB0965 LE	46	–	–	–	54	–
TB1075 HE	16	11	–	–	73	–
TB1075 LE	20	8	–	–	46	26
TB1170A HE	20	–	6	–	74	–
TB1170A LE	42	–	–	–	58	–
TB 1577A HE	20	–	6	–	74	–
TB1577A LE	36	–	–	–	64	–
TB20110 HE	10–16	–	65.5–79.5	18–10	0.5	–
TB20110 LE	36	–	–	–	64	–

3.4 Bimetallic thermometer construction

A spiral or helical configuration of a bimetallic strip is useful for indicating temperature on a dial. The spiral serves to allow a long length of the bimetallic strip within a confined geometry and hence high sensitivity. This form is commonly used in dial thermometers. These tend to be rugged devices and are used in applications where other sensors might fail. They have the advantage of no need for a power supply and do not suffer from the fragility of liquid in glass thermometers, for example. In a dial-type thermometer one end of the bimetallic strip is fixed to the closed end of a stainless steel tube. The bimetallic strip is wound in a helix so that the deflection of the strip with temperature causes rotation of the free end. This is attached to the shaft of a pointer and a scale is provided on a facia fixed to the steel tube to indicate the temperature (Figure 3.2). Diameters for commercial gauges vary from 25 mm to 125 mm with stem lengths up to 610 mm.

Figure 3.2 Dial-type bimetallic thermometer with an adjustable stem and dial to allow flexibility in installation and viewing. Photograph courtesy of WIKA Alexander Wiegand GmbH & Co.

The angular displacement of a bimetallic coil is given by

$$\theta = \left(\frac{1}{R_2} + \frac{1}{R_1} \right) L \qquad (3.6)$$

where: L = length of strip (m)
R_1 = initial radius (assumed constant over the length of the coil) (m)
R_2 = final radius (assumed constant over the length of the coil) (m).

In terms of the specific deflection, a = angular deflection per degree temperature rise ($a = k/2$), and equation (3.6) can be written as

$$\theta = \frac{2aL}{t} (T_2 - T_1) \frac{360}{2\pi} \qquad (3.7)$$

The uncertainty of a typical commercial product is 1–2% of the full-scale deflection with an operating range of −70 to 600°C. However, a large length and thin section can be used to provide high levels of sensitivity and Huston (1962) demonstrated a device with a sensitivity of 0.0035°C/mm and a repeatability of 0.027°C. Whilst bimetallic thermometers have the merits of being easily read, can be utilized to provide both an indication of temperature and to act as an actuator, are relatively inexpensive and do not require an independent power supply they are subject to drift, are relatively inaccurate compared to, say, thermocouples and industrial PRTs and cannot provide a remote indication of temperature.

References

Books and papers

Divita, R.V. Bimetal thermometer selection. *Measurements and Control*, **167**, 93–95, 1994.

Eskin, S.G. Thermostatic bimetals. *Trans. ASME*, 433–442, 1940.

Huston, W.D. The accuracy and reliability of bimetallic temperature measuring elements. In Herzfield, C.M. (Editor), *Temperature. Its Measurement and Control in Science and Industry. Vol. 3*, pp. 949–957, Rheinhold, 1962.

Meijer, G.C.M. and van Herwaarden, A.W. *Thermal Sensors*. IOP, 1994.

O'Connor, L. A bimetallic silicon microvalve. *Mechanical Engineering*, **117**, 1995.

Stevenson, R.J., Moulin, A.M., Welland, M.E. Bimaterials Thermometers. Section 32.1 in Webster, J.G. (Editor), *The Measurement Instrumentation and Sensors Handbook*. CRC Press, 1999.

Timoshenko, S.P. *The Collected Papers*. McGraw-Hill, 1953.

Standards

ASTM B388–96 Standard specification for thermostat metal sheet and strip.

ASTM B106–96 Standard test methods for flexivity of thermostat metals.

ASTM B389–81 (1998) Standard test method for thermal deflection rate of spiral and helical coils of thermostat metal.

BS 1041: Section 2.2: 1989. British Standard. Temperature measurement. Part 2. Expansion thermometers. Section 2.2 Guide to selection and use of dial-type expansion thermometers.

DIN 1715–1: 1983–11 Thermobimetalle; Technische Lieferbedingungen.

DIN 1715–2: 1983–11 Thermobimetalle; Prüfung der spezifischen thermischen Krümmung.

Web sites

At the time of going to press the world wide web contained useful information relating to this chapter at the following sites.

http://www.ametekusg.com/thbimet.htm

http://www.bristolbabcock.com/products/helicoid/thermofeatures.htm

http://www.csiworld.com/moeller/motherm.htm

http://www.dresserinstruments.com/productnav/ashcroft_categ.html#Bimet

http://www.dresserinstruments.com/weksler/weks_catalog/weks_index.pdf
http://www.howstuffworks.com/therm.htm
http://www.kanthal.se/product/bimet/bimet.html
http://www.marshallinstruments.com/
http://www.moellerinstrument.com/motherm.htm
http://www.ptc1.com/therstem.htm
http://www.ptc1.com/thersur.htm
http://www.reotemp.com/temperature.html#bi-industrial
http://www.telatemp.com/Thrshld/Bimetl.htm
http://www.teltru.com/temp/bi-metal/choice.htm
http://www.thomasregister.com/olc/weissinstruments/bimetal.htm
http://www.wekslerglass.com/bimetal.html
http://www.wika.com/web/TempInstru/THERINDX.HTM

Nomenclature

d = maximum lateral deflection of the strip (m)
E = Young's modulus (N/m^2)
k = flexivity (K^{-1})
L = length of the bimetallic strip (m)
m = ratio of thickness t_1/t_2
n = ratio of Young's moduli E_1/E_2
R = radius (m)
t = thickness (m)
T = temperature (K)
α = coefficient of thermal expansion (K^{-1})
θ = angular displacement (rad)

4

Liquid-in-glass thermometers

Liquid-in-glass thermometers remain highly popular. They permit a quick visual indication of temperature and can with care be very accurate. This chapter introduces the physical phenomena involved, common designs, sources of error, calibration and some specialist devices.

4.1 Introduction

The traditional liquid-in-glass thermometer comprising a reservoir and capillary tube supported in a stem follows the designs proposed by Daniel Fahrenheit in 1714. The specifications and design for liquid-in-glass thermometers depend on the required application. Some thermometers can be read to 0.001°C over a limited temperature range. Other designs can be used at temperatures up to 630°C with coarser scales and some spirit-filled thermometers can be used to indicate temperatures down to −196°C. Despite their fragility, liquid-in-glass thermometers remain in popular use due to the chemically inert nature of glass and the lack of external connections and need for a power supply. In addition like-for-like replacement of existing liquid-in-glass thermometers serves to maintain their widespread use.

There is a wide variety of liquid-in-glass thermometer types but there are various common features. The solid-stem form of liquid-in-glass thermometer consists of four main components, the bulb, stem, liquid and markings, and is illustrated in Figure 4.1. The bulb is a thin glass container, typically 0.35–0.45 mm thick, holding a thermometric liquid such as mercury or an organic spirit. The bulb is connected to a glass stem that contains a small-bore capillary tube. The space above the thermometric liquid can be filled with an inert gas or evacuated. Liquid-in-glass thermometers make use of the expansion of liquid with temperature. As the liquid in the bulb gets hotter it expands and is forced up the capillary stem. The temperature of the bulb is then indicated by the position of the top of the liquid column against a scale engraved or printed on the glass stem. The expansion chamber at the top of the capillary allows for gas compression and gives a slight margin of safety in the case of overheating the thermometer. The so-called contraction chamber is a widening of the capillary in the main stem, usually below the main scale. As

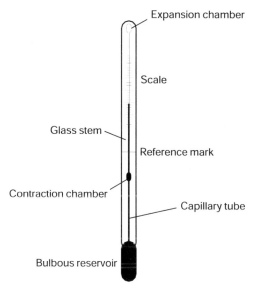

Figure 4.1 The main features of a solid-stem liquid-in-glass thermometer

it is an expansion of the bore, thermometric liquid filling its volume could be equivalent to several centimetres of capillary tube so it shortens the overall length of the thermometer for convenience. In addition, it serves to prevent all the liquid from retreating into the bulb which can cause gas bubbles to become trapped in the bulb. Some thermometers include one or more auxiliary scales marked on the stem as shown in Figure 4.2. An auxiliary scale is included for thermometers when the main scale does not include, say, the ice point and is used for calibration and checking the device. A reference point is sometimes marked on the main or auxiliary scale that is usually at the ice, or less commonly at the steam, point and is used for calibration purposes.

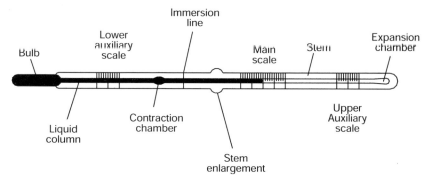

Figure 4.2 The principal features of an ASTM (American Society for Testing and Materials) specification solid stem thermometer

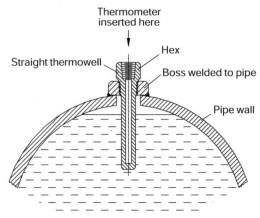

Figure 4.3 Use of a thermowell to provide protection for a measuring device such as a liquid-in-glass thermometer whilst maintaining reasonable thermal contact with the medium of interest

A thermometer should be brought into thermal equilibrium with the medium of interest. That is, it should be in good thermal contact with the application. This can be achieved in the case of some pipe flows by locating the thermometer in a thermowell as illustrated in Figure 4.3. The thermowell can be filled with a high-conductivity liquid to improve heat transfer. Most liquid-in-glass thermometers are normally designed to be used and read in the vertical position, unless otherwise stated on the calibration certificate. The temperature is read by looking at the position of the meniscus against the temperature scale on the

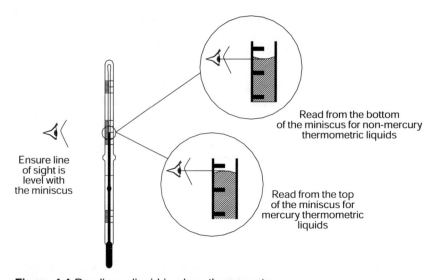

Figure 4.4 Reading a liquid-in-glass thermometer

thermometer. The meniscus is the surface shape taken up by a fluid in a container. The flattest part of the liquid meniscus is used for measurement against the scale. Surface tension effects in mercury cause the meniscus to be predominantly convex and the top of the meniscus should be used to indicate the temperature against the scale as shown in Figure 4.4. In the case of organic thermometric liquids the bottom of the meniscus should be used. Care needs to be taken to avoid parallax errors caused by refraction through the glass. The ideal is to view the meniscus at an angle of 90° to the stem. If a 6 mm diameter thermometer, for example, is viewed at 4° to the normal an error of 0.05°C can occur. For precise work the meniscus should be viewed with a 5–10 times magnification microscope. This can be mounted on the stem with a simple spring mechanism to allow it to slide up and down. It should be noted that the interval between two marks on a scale should be taken between the middle of the markings as shown in Figure 4.5.

Considerations in the design and selection of liquid-in-glass thermometers include temperature range, chemical compatibility, cost, sensitivity, uncertainty and speed of response. The expansion of thermometric liquids with temperature can be modelled by

$$V = V_0 \left(1 + \alpha T + \beta T^2\right) \qquad\qquad (4.1)$$

where: V = volume of liquid at temperature T (m^3)
  V_0 = volume of liquid at 0°C (m^3)
  T = temperature (°C)
  α = first coefficient of thermal expansion (°C^{-1})
  β = second coefficient of thermal expansion (°C^{-2}).

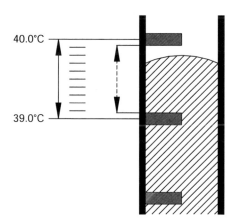

Figure 4.5 Reading a liquid-in-glass thermometer when the meniscus is located between markings. The interval should be divided in the middle of the marking, not between the markings as illustrated by the dotted line. The reading in this case should be taken as 39.8°C. Example after Nicholas and White (1994)

For mercury $\alpha = 1.8 \times 10^{-4}\,°C^{-1}$ and $\beta = 5 \times 10^{-8}\,°C^{-2}$ and as the second coefficient is so small that it is often neglected. Equation (4.1) models the expansion of liquid in an ideal thermometer for which the stem and bulb do not also expand. The glass used in a typical thermometer has a coefficient of thermal expansion of about $2 \times 10^{-5}\,°C^{-1}$ and hence will also expand with temperature. The term 'apparent expansion coefficient' is used to represent the combined effect of expansion of both the thermometric liquid and the glass enclosure. Table 4.1 lists typical values for the apparent expansion coefficients for a variety of thermometric liquids commonly used. The characteristic desired of a thermometric liquid is a high expansion coefficient, chemical stability and compatibility with the container, which is usually a type of glass. Mercury is particularly good in this respect as it does not wet (does not stick to) glass. The use of organic liquids is necessary near or below the freezing point of mercury; mercury freezes at −38.9°C. However, these liquids wet glass, so following a reduction of temperature, the liquid will drain down the surface of the glass over a period of time and as a result the reading given will vary with time. The monitoring of transient temperatures must be undertaken with due consideration of this effect. The possible temperature range for various thermometric liquids is also given in Table 4.1.

Glass is the common choice for thermometers as it is relatively stable in a wide variety of chemical environments. Types of glass used in thermometry include normal glass (soda-lime glass containing zinc oxide and alumina), borosilicate glass, combustion glass, lead glass and silica. Table 4.2 lists the maximum operating temperature for a variety of thermometric glasses.

Variations on the stem thermometer principle include enclosed and the external scale thermometers. In the case of the enclosed scale thermometer, the scale is printed on a separate piece of material to the capillary tube, attached to the capillary or bulb and encapsulated within a tube. An alignment datum mark can be included on the scale strip and the thermometer body to indicate whether the scale has become detached. In addition to reducing

Table 4.1 Apparent thermal expansion coefficients for various thermometric liquids in thermometer glass around room temperature and possible useful temperature ranges. (After Nicholas, 1999)

Liquid	Typical apparent thermal expansion coefficient ($°C^{-1}$)	Possible temperature range (°C)
Mercury	0.00016	−35 to 510
Ethanol	0.00104	−80 to 60
Pentane	0.00145	−200 to 30
Toluene	0.00103	−80 to 100
Xylene		−80 to 50

Table 4.2 Common thermometric glasses and temperature limits of operation. (After Horrigan, 1998)

Glass	Maximum operating temperature ($^{\circ}C$)
Normal lead glass	400
Jena normal	430
Borosilicate	460
Supremax (silica)	595

manufacturing costs, the use of an enclosed scale can reduce parallax errors as the scale and the meniscus are close together. A disadvantage of this form of design is the increased fragility of the device. An alternative and much cheaper option is to attach the scale strip to the external body of the thermometer. Again an alignment mark can be included to indicate that the scale is in the original calibration position. External scale thermometers are popular for domestic applications such as cooking thermometers where low price is an economic driving factor.

The speed of response is governed primarily by the bulb volume, the thickness of glass used for the bulb and the materials. To increase the speed of response the bulb volume and glass thickness should be as small as possible. A thin-walled bulb will, however, be sensitive to pressure and will expand or contract according to external and internal pressures. Thermometer design and selection therefore requires careful compromise between these two conflicting requirements. Time-response characteristics can be determined experimentally or estimated theoretically. A liquid-in-glass thermometer can be modelled as a cylinder experiencing conduction and convection similar to the methods used to determine the conduction error for a sheathed thermocouple, (Chapter 2). Table 4.3 gives the $1/e$ time constants for a 5 mm diameter bulb in a variety of flows.

Table 4.3 Time constants for a 5 mm diameter mercury-in-glass thermometer. (After Nicholas, 1999)

Medium	Still medium	0.5 m/s flow	∞ m/s
Water	10 s	2.4 s	2.2 s
Oil	40 s	4.8 s	2.2 s
Air	190 s	71 s	2.2 s

The sensitivity of a liquid-in-glass thermometer increases as the diameter of the capillary bore is reduced. However, the minimum bore diameter is limited by the effects of surface tension in the thermometric liquid. As the diameter is reduced the surface tension in the liquid becomes significant enough to cause variations in the volume of the bulb. Essentially, as the temperature is increased, the meniscus becomes more convex, increasing the pressure on the bulb and the bulb consequently expands, distorting the temperature indicated by the thermometer. In modern thermometers the capillary is usually circular or elliptical in cross-section with a diameter of between 0.02 and 0.4 mm. A constant-bore section is essential to avoid non-linear response to temperature changes.

4.2 Error sources

A variety of considerations and error sources exist in the use of liquid-in-glass thermometers. These include:

- changes in the glass structure known as secular change
- temporary depression of the temperature reading whilst the glass structure responds to a thermal transient
- immersion effects to account for when not all the thermometric liquid is at the same temperature
- pressure effects expanding or contracting the bulb
- stiction, the tendency of a thermometric liquid to adhere to the capillary and indicate a false reading
- construction faults and malfunctions.

4.2.1 Secular change

Glass is an unusual material with an irregular or amorphous material structure, which allows the molecules some degree of mobility (Liberatore and Whitcomb, 1952). Despite careful annealing during manufacture, long-term drift occurs and continues for the life of the thermometer and is called secular change. Secular change is most significant following manu-facture and the majority of the effect normally occurs in the first year. Secular change manifests itself by a rise in the temperature indicated by the thermometer and is due to an irreversible contraction of the bulb. During the first year the indicated temperature at the ice point can rise by up to 0.04°C with a rise of 0.01°C per year in subsequent years until it stabilizes after four or five years. For accurate results the extent of secular change should be monitored over the lifetime of a thermometer, say every six months, by checking the ice point reading and adjustment of the calibration as appropriate.

4.2.2 Temporary depression

During cooling from a high temperature most materials contract quickly to their original size at that temperature. Glass, however, allows the molecules some degree of mobility without changing its bulk characteristics significantly. A subsequent increase in temperature following cooling not only causes an expansion of the glass but also increases the mobility of the molecules and can result in structural changes. Slow cooling provides time for the structure to respond and the glass can return to its original size. If the cooling occurs rapidly, however, structural changes are temporarily frozen in and the glass can take a long time to return to its original size. Consequently a glass thermometer would be temporarily larger than its equilibrium volume and the reading would show a low temperature. This effect is called, 'temporary depression' and is often noticed at the ice point, when a thermometer is brought from a high temperature for checking purposes, although it can occur at any temperature. It can take days to months for the glass structure to recover and for the reference point reading to return to its original value. The magnitude of error caused by temporary depression depends on the temperature difference experienced by the thermometer and the type of thermometer and can be about 0.05°C after heating to 100°C and rapid cooling to 0°C. If necessary for the achievement of low measurement uncertainty there are two ways of dealing with temporary depression. The first is to measure the ice point of the thermometer immediately after a rapid cooling and adjusting the results for the application measurement as appropriate. The second method is to wait until the ice point has recovered before making a measurement with a thermometer. Van Dijk *et al.* (1958) demonstrated, for example, that if a thermometer is cooled over a period of 15 or more hours no temporary depression is observed.

4.2.3 Immersion effects

In a theoretically ideal thermometer all the thermometric liquid will be at the same temperature. In practice the temperature along the stem and hence along the length of the thermometric liquid may vary somewhat as it is rare that an application is free of thermal gradients and thermometers have to be in the line of sight to be read. The error caused by such thermal gradients can be accounted for either at the design stage or by applying a correction to the reading. In general, three types of thermometers are classified according to the level of contact between the medium of interest and the thermometer. These are partial immersion, total immersion and complete immersion thermometers (Figure 4.6).

Total immersion thermometers are designed to be used so that the bulb and the majority of the liquid column are submerged. The level of submersion of the thermometer is adjusted so that only about 1 mm of the liquid column protrudes at the surface, which is sufficient in most cases to enable the observer to read the

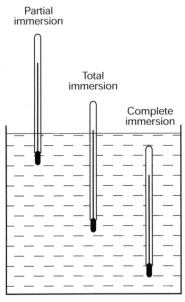

Figure 4.6 Use of partial, total and complete immersion thermometers

temperature. Above 150°C the risk of mercury distillation increases and it is advisable to maintain the thermometer at the total immersion condition for only a short period before the reading is made (BS 1041, Section 2.1). The thermometer should then be withdrawn from the application by a few centimetres to reduce the meniscus temperature. The best measurement uncertainties obtainable for total immersion thermometers compatible with BS 1900 for particular temperature ranges are listed in Table 4.4.

Table 4.4 Measurement uncertainties for total immersion thermometers compatible with BS 1900

Best measurement uncertainty (°C)	Temperature range (°C)
±0.20	−80 to 30
±0.05	−40 to 2
±0.01	0 to 100
±0.05	100 to 200
±0.10	200 to 300
±0.20	200 to 450
±1.00	95 to 500

Partial immersion thermometers require the thermometer to be immersed to a specific depth in the medium of interest. This is usually engraved on the stem by both a ring and a numerical inscription. In use a column of exposed thermometric liquid, called the emergent liquid column or elc, will rise above the application. The temperature of the emergent liquid column is likely to be different from that of the medium of interest and can cause a significant error unless an appropriate correction for the variation of temperature in the column is made. There are two methods for determining the emergent liquid column temperature: one using a series of auxiliary thermometers and the other instruments called Faden thermometers. Figure 4.7(a) illustrates the use of a series of auxiliary thermometers, with the bottom of the bulb of the first at a distance of 10 mm from the point of emergence, and subsequent thermometers evenly spaced at intervals of not more than 100 mm (BS 1041: Section 2.1). A final thermometer is located with its bulb level with the meniscus. The average-elc temperature can be calculated by

$$T_{elc} = \frac{d_1 T_1 + d_2 T_2 + d_3 T_3}{d_1 + d_2 + d_3} \tag{4.2}$$

where d_i represents the number of degrees equivalent length covered by the auxiliary thermometer reading T_i.

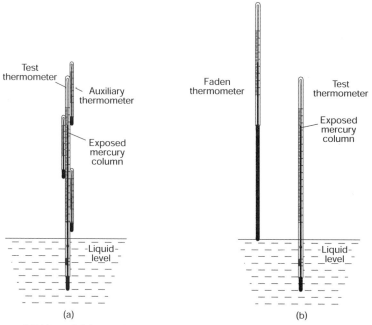

Figure 4.7 Use of (a) a series of auxiliary thermometers and (b) a Faden thermometer to account for the difference in temperature between the emergent liquid column and the medium of interest

If the application temperature exceeds 100°C a more precise method can be applied which makes use of specialized calibration thermometers called Faden thermometers whose bulb length is equal, within about 1 cm, to the emergent liquid column (Figure 4.7(b)). The long bulb length is designed to simulate conditions in the emergent liquid column of the test thermometer. A number of Faden thermometers can be used and immersed in the medium alongside the partial immersion thermometer. The average-elc temperature can be calculated by

$$T_{\text{elc}} = \frac{L_{\text{Faden 1}} \, T_{\text{Faden 1}} + L_{\text{Faden 2}} \, T_{\text{Faden 2}} + \ldots + L_{\text{Faden n}} \, T_{\text{Faden n}}}{L_{\text{elc}}}$$

(4.3)

where: $L_{\text{Faden n}}$ = length of the bulb of Faden thermometer n (m)
  L_{elc} = length of the emergent liquid column (m)
  $T_{\text{Faden n}}$ = temperature reading of Faden thermometer n (°C).

Faden thermometers can be purchased singly or as a set with bulb lengths varying from about 50 mm to 300 mm. For example, if the emergent liquid column is 25 cm long it could be measured by two Faden thermometers with bulb lengths of 5 and 20 cm. If the shorter and longer Faden thermometers register 25 and 45°C respectively, then, using equation (4.3), the average emergent liquid column temperature is 41°C.

Once the average temperature for the elc has been assessed the correction to the indicated temperature can be determined using

$$C = kN(T_{\text{elc}} - T_1)$$
(4.4)

where: k = the apparent cubic thermal expansion coefficient of the liquid in the particular glass from which the stem is made (°C⁻¹)
  T_{elc} = the average elc temperature (°C)
  T_1 = indicated temperature on the test thermometer (°C)
  N = the number of degrees Celsius equivalent to the length of the emergent liquid column and is the difference between the thermometer indication and the actual or extrapolated value corresponding to the specified immersion level.

Partial immersion thermometers maintain less of the thermometric liquid at the temperature of the application and therefore tend to have an uncertainty associated with the reading double that of total immersion thermometers. Complete immersion thermometers are designed to be used with the entire thermometer immersed in the medium of interest. As a result the thermometers tend to be sensitive to pressure and the consequent errors are not easy to assess.

4.2.4 Pressure effects

The glass wall of a thermometer bulb is intentionally thin, of the order of 0.35–0.45 mm thick to provide reasonable speed of response. It is, as a result, relatively flexible and sensitive to changes in pressure such as atmospheric pressure or internal pressures. A change in atmospheric pressure of 3 kPa, for example, can change a thermometer reading by 0.005°C and a typical external pressure coefficient is 0.15°C per bar. The effect is therefore small and need only be considered when low uncertainty is necessary or when high-pressure pressure applications are involved. The correction for a particular thermometer can be determined by supporting it in a closed transparent tube maintained at a constant temperature and constant pressure. The external pressure coefficient can be evaluated by measuring a series of corrections over a range of pressures.

4.2.5 Stiction

Stiction is the tendency of a liquid to stick to the glass in the capillary. This can be a problem in some high-resolution fine-bore thin-walled bulb thermometers. To the observer the column of mercury appears to jump up and down in sudden steps over a period of time without coming to an equilibrium position even if the temperature of the application appears to be stable. This effect is a consequence of the properties of glass and surface tension. With rising temperature the meniscus for mercury becomes more convex and this increases the pressure in the bulb. The thin glass of the bulb, which is not rigid, expands in response. The expansion may occur in sudden jumps and the mercury in the column will respond by moving rapidly to a new level. The pressure in the bulb is reduced and the process is repeated. The magnitude of the oscillation depends on the thermometer design but can be between 0.005°C and 0.01°C.

4.2.6 Malfunctions and construction faults

A variety of faults can occur in the manufacture and use of liquid-in-glass thermometers. Faults such as uneven or missing scale divisions, oxidation of the mercury, foreign objects in the bulb or capillary, significant strain in the stem or bulb and a distorted or tapering capillary cannot be rectified without significant manufacturing capability and therefore usually result in the return or controlled discard of the thermometer. Oxidation of mercury in the bulb and capillary is caused by contamination with air or water. This can be identified by telltale rings on the inside of the bore. These tend to interfere with the smooth movement of the column and therefore the thermometer reading. Foreign objects, such as chips of glass, in the bulb and bore can result from manufacturing faults or significant strain on the thermometer in use. A chip of glass can distort the meniscus, impede the movement of the thermometric liquid or distort the bulb volume.

Some less significant faults can, however, often be rectified. These include bubbles of gas in the bulb, a separated column of thermometric fluid and wear on the scale causing the scale lines to be faint.

Liquid columns can become separated, for example, if the thermometer is jarred when it is horizontal. The separated portion or portions may then move up and down relative to the main column motion. Rejoining of the column of liquid is essential in order to maintain measurement validity. Some techniques for achieving this, in order of preference, include:

- Lightly tapping the thermometer whilst holding it in the vertical position. Tapping the bulb onto a paper pad is, for example, sometimes successful.
- Using centrifugal force. This can be achieved by holding the bulb between thumb and index finger, with the stem along your arm and raising your arm above your head and bringing it down rapidly to alongside your leg.
- Cooling the bulb so that all the thermometric liquid descends into the bulb followed by tapping to rejoin the mercury. Useful cooling mediums include salt, ice and water (to −18°C), solid CO_2 (−78°C) and liquid nitrogen (−196°C). No harm results if the mercury freezes during this process as it contracts on freezing. However, the bulb should be warmed gently from the top so that the mercury at the top of the bulb melts first to avoid trapping an expanding fluid.
- If the thermometer includes a contraction or an expansion chamber, careful heating of the ends so that both the main column and the separated column of liquid are in the same expansion chamber followed by careful tapping of the thermometer until the columns rejoin can be successful. Care must be taken not to exceed the safe temperature limit for the thermometer.

Following successful rejoining of the thermometric liquid a reference check at, say, the ice point should be undertaken.

A bubble within the bulb will distort the measurement as it will cause the liquid column to be artificially high. The bubble should be persuaded to move to the top of the bulb, ideally immediately underneath the capillary, by tapping the thermometer whilst it is vertical. One of the techniques described for rejoining a separated column can then be used to move the bubble up to the top of the capillary.

Faint scale lines can be improved by rubbing the scale with a paste-like mixture of sodium silicate, water and a black oxide such as MnO_2 followed by wiping to remove any excess.

4.3 Calibration

In order to make accurate measurements using a liquid-in-glass thermometer it should be calibrated and checked against a reference temperature in use. Calibration involves immersing the thermometer in a medium at a known temperature and recording the reading given by the thermometer in a report associated with that thermometer. The known temperature can be defined by a physical process such as the triple point of water, which is known to occur at a unique temperature, or by comparison against another temperature-measuring device which has itself been calibrated. Ideally a calibration process should be traceable back to the International Temperature Scale of 1990 (ITS-90).

Most practical measurement devices are not actually calibrated by national standards laboratories. Instead a liquid-in-glass thermometer will be compared with the results from a device such as a PRT (platinum resistance thermometer) that has itself been calibrated against a SPRT (standard PRT). This route provides the traceability to ensure that a temperature indicated by one device has relevance to the temperature indicated by a different device but whilst ensuring that calibration procedures and costs are not prohibitive. The calibration procedure for a liquid-in-glass thermometer usually involves inspection of the thermometer, conditioning of the thermometer to avoid temporary depression errors, a reference point check and comparing the results of the thermometer against a calibrated temperature-measuring device across a range of temperatures or at a distinct set of temperatures. All aspects of the calibration should be recorded in a report.

Because of the effects of secular change, liquid-in-glass thermometers should be checked regularly, to ensure that the reference temperature indicated is as expected. The reference temperature selected by the manufacturer is any temperature within the range of the thermometer that can be reproduced to better than the calibration uncertainty. The most convenient reference temperature is the ice point but alternatives include the steam point. The ice point is the melting point of pure ice at standard atmospheric pressure (101325 Pa). The temperature under these conditions is 273.15 K. It is

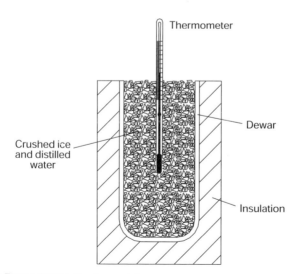

Figure 4.8 Reproducing the ice-point

relatively straightforward to generate the ice point temperature as illustrated in Figure 4.8 and with care it is reproducible to within 1 mK. A well-insulated wide-mouth container such as a dewar (or a vacuum flask) should be used which is deep enough to permit the immersion of the thermometer to be calibrated. For general-purpose use, Wise (1976) recommends a container 36 cm deep and 8 cm in diameter. The container should be filled with a mixture of crushed or shaved ice and distilled water. The ice used must have been made using distilled water and the size of the ice particles should be approximately 2–5 mm in diameter. Care should be taken to avoid contamination of the mixture with other substances as this tends to lower the temperature. Stainless steel or glass stirrers and containers are recommended. An ice point apparatus set up in this way can maintain a temperature of 0.00°C within ±0.001°C for several hours. This time can be extended by siphoning off excess water, which, because it is more dense than ice, tends to form a pool at the bottom of the dewar, and adding fresh crushed ice that has been made from distilled water. Periodic stirring of the mixture is necessary in order to ensure a uniform temperature. There is an alternative calibration point, which is now preferred, and this is the commercially available water triple point cell (Figure 4.9) with its lower uncertainty.

In order to perform a calibration a variable temperature enclosure is used such as a cryostat, stirred liquid bath or furnace. The choice depends on the temperature range; a cryostat may be used for temperatures to 0°C, a stirred liquid bath for temperatures between −50°C and 600°C and a furnace for

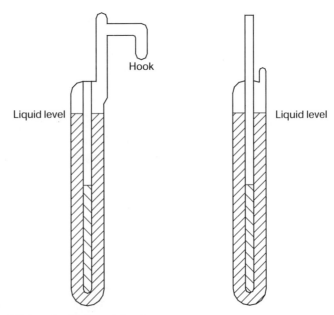

Figure 4.9 A water triple point cell

temperatures from 600°C to the temperature limit of furnaces, of the order of 1800°C. Because of the temperature range of most liquid-in-glass thermometers, for these devices stirred liquid baths are most commonly used. The thermometer or thermometers to be calibrated should be placed in the bath to the prescribed depth of immersion according to their type. For example, total immersion thermometers should be immersed so that just a few divisions of the scale around the test point are visible. Ideally the bulb of the thermometer should be at the same depth as the reference thermometer. Liquid-in-glass thermometers cannot be read accurately with the naked eye. Instead a viewing system such as a stem-mounted 10× microscope or a rack and pinion-mounted telescope can be used. With experience and possibly using a graticule a thermometer can be read to better than 1/10th of a division. Liquid in glass thermometers are best calibrated with the temperature in the bath rising slowly to minimize the effect of stiction. If the bath is controlled by a programmable logic controller, then the ramp function should be set at a low rate. Depending on the response rate of the reference and the test thermometers a rate equivalent to the expected uncertainty per minute is usually suitable (Horrigan, 1998). Once the temperature in the bath is rising at a constant rate the readings on both the test thermometer and reference temperature device can be taken and recorded as a pair. Readings are taken across the range of interest for the thermometer application. Following a calibration, the thermometer should be carefully conditioned to avoid rapid strain of the glass. For partial immersion thermometers the calibration procedure requires the use of Faden thermometers to determine the emergent liquid column temperature during the calibration and correction of the reading using equation (4.4).

4.4 Special types

A number of specialist thermometers have been developed for specific applications such as recording the maximum or minimum temperature reached over a period of time or monitoring a very small temperature difference. The text by Knowles Middleton (1966) provides a comprehensive review of developments in liquid thermometry and reviews some of the many specialist thermometer types. Here just two of the specialist thermometers, the Beckmann and the clinical thermometer, are described because of the principles involved and widespread use.

The Beckmann thermometer, for instance, is designed to enable a small temperature difference to be measured but with the versatility of an adjustable zero so it can be used across a wide range of temperatures. A typical Beckmann thermometer is illustrated in Figure 4.10. A fine needle capillary is used to move small drops of mercury from the main section of the thermometer into a reservoir to enable the zero indication to be set at an

Reservoir

Figure 4.10 The Beckmann thermometer

arbitrary temperature, within the range of the device. This allows a relatively short stem to be used to monitor a restricted temperature difference with reasonable resolution.

Measurement of body temperature is essential in routine medical practice. The range of body temperature in healthy individuals is $36.8 \pm 0.4°C$ but is influenced by many factors. It is dependent on the site of measurement with oral temperature measured in the sublingual pocket close to the lingual artery generally the lowest and rectal temperature the highest. Tympanic membrane temperature is assumed to best reflect the core body temperature. Body temperature should therefore be measured in a standardized fashion, preferably at rest in the morning on waking. The use of mercury-in-glass thermometers to measure rectal temperature still remains the reference in France, for example (Mari *et al.*, 1997). However, it is inconvenient and associated with a risk of rectal injury and cross-infection. Mercury-in-glass thermometers are associated with high levels of mercury pollution and the tendency is to replace them by electronic thermometers to measure oral temperature and infrared thermometers for measuring the tympanic membrane temperature. The typical design for a clinical thermometer is illustrated in Figure 4.11. They are designed to respond quickly and so have a small bulb volume and in order to provide reasonably low uncertainty and resolution a very fine bore. It is, however, difficult to see the mercury column in a fine diameter capillary tube so a lens is frequently incorporated into the thermometer stem, which magnifies the bore when the thermometer is viewed from an appropriate angle. A constriction is used in the capillary bore to break the mercury column as it cools following removal from a body. After reading

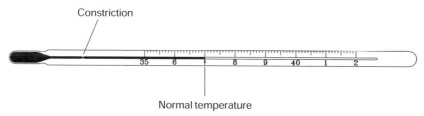

Constriction

Normal temperature

Figure 4.11 Typical clinical thermometer design

the measurement the mercury must then be forced back into the bulb by shaking the thermometer. The constriction design is critical as it must provide a clean break in the mercury column but still allow the mercury to be forced back into the bulb when shaken. A number of national standards are available for the design of clinical thermometers including AS 2190, BS 691 and BS 6985.

References

Books and papers

Horrigan, C. Resistance and liquid glass thermometry. In Bentley, R.E. (Editor), *Handbook of Temperature Measurement*, Vol. 2, Springer, 1998.

Knowles Middleton, W.E. *A History of the Thermometer and its Use in Metrology.* Johns Hopkins University Press, 1966.

Liberatore, L.C. and Whitcomb, H.J. Density changes in thermometer glasses. *J. Am. Ceram. Soc.*, **35**, 67, 1952.

Mari, I., Pouchot, J. and Vinceneux, P. Measurement of body temperature in clinical practice. *Revue de Medecine Interne*, **18**, 30–36, 1997.

Nicholas, J.V. Liquid-in-glass thermometers. Section 32.8 in Webster, J.G. (Editor), *The Measurement Instrumentation and Sensors Handbook.* CRC Press, 1999.

Nicholas, J.V. and White, D.R. *Traceable Temperatures.* Wiley, 1994.

Van Dijk, S.J., Hall, J.A. and Leaver, V.M. The influence of rate of cooling on the zeros of mercury-in-glass thermometers. *Journal of Scientific Instruments*, **35**, 334–338, 1958.

Wise, J.A. Liquid-in-glass thermometry. *National Bureau of Standards Monograph 150*, 1976.

Standards

AS 2190. Clinical maximum thermometers. Australian Standard. 1978.

BS ISO 4795:1996. Laboratory glassware. Glass for thermometer bulbs.

BS 593:1989. Specification for laboratory thermometers.

BS 692:1990. Specification for meteorological thermometers.

BS 791 1990. Specification for solid-stem calorimeter thermometers.

BS 1041: Section 2.1: 1985. British standard temperature measurement. Part 2 Expansion thermometers.

Section 2.1 Guide to selection and use of liquid-in-glass thermometers.

BS 1365:1990. Specification for short-range short-stem thermometers.

BS 1704: 1985, (ISO 1770–1981). British standard specification for solid-stem general purpose thermometers.

BS 1900: 1976. Secondary reference thermometers.

BS 5074:1974. Specification for short and long solid-stem thermometers for precision use.

BS 2000–0.1:1996. Methods of test for petroleum and its products. IP Standard thermometers.

BS 691:1987. Specification for solid-stem clinical maximum thermometers (mercury-in-glass).

BS 6985:1989. Specification for enclosed-scale clinical maximum thermometers (mercury-in-glass).

ISO 386–1977. Liquid-in-glass thermometer – Principles of design construction and use.

ISO 651–1975. Solid stem calorimeter thermometers.

ISO 653 1980. Long solid stem thermometers for precision use.

ISO 654 1980. Short solid stem thermometers for precision use.

ISO 1770 1981. Solid stem general purpose thermometer.

ASTM Vol. 14.03 E1–95. Specification for ASTM thermometers.

ASTM Vol. 14.03 E77–92. Test method for inspection and verification of liquid-in-glass thermometers.

JIS Z 8705 1992. Method of temperature measurement by liquid-in-glass thermometers.

Web sites

At the time of going to press the world wide web contained useful information relating to this chapter at the following sites.

http://fluid.nist.gov/836.05/greenbal/liquid.html

http://www.bdcanada.com/Fever/Glasstherm

http://www.brannan.co.uk/

http://www.britglass.co.uk/

http://www.csiworld.com/moeller/moliquid.htm

http://www.cstl.nist.gov/div836/836.05/greenbal/liquid.html

http://www.hartscientific.com/products/ligtherms.htm

http://www.lab-glass.com/html/nf/Thermometers.html

http://www.main.nc.us/amthermometer/

http://www.npl.co.uk/npl/cbtm/thermal/publications/ds_pm0100.html

http://www.npl.co.uk/npl/cbtm/thermal/service/ms_tm_lig.html

http://www.omega.com/toc_asp/section.asp?book=temperature&section=e

http://www.plowden-thompson.com/

http://www.schott.de/english/rohrglas/anwendungen.htm?e=N

http://www.scienceproducts.corning.com/

http://www.thermometry.inms.nrc.ca/thermometry/Calibrations.htm

http://www.wekslerglass.com/

Nomenclature

d	=	number of degrees equivalent length (m)
elc	=	emergent liquid column
k	=	the apparent cubic thermal expansion coefficient of the liquid in the particular glass from which the stem is made ($°C^{-1}$)
$L_{\text{Faden } n}$	=	length of the bulb of Faden thermometer n (m)
L_{elc}	=	length of the emergent liquid column (m)
N	=	the number of degrees Celsius equivalent to the length of the emergent liquid column
T	=	temperature (°C)
T_{elc}	=	average emergent liquid column temperature (°C)
T_{Faden}	=	temperature reading of Faden thermometer (°C)
V	=	volume of liquid at temperature T (m^3)
V_0	=	volume of liquid at 0°C (m^3)
α	=	first coefficient of thermal expansion ($°C^{-1}$)
β	=	second coefficient of thermal expansion ($°C^{-2}$)

5

Thermocouples

Thermocouples are a common choice for temperature measurement because of their self-energization, low cost, robust nature and wide temperature range. There are many types of thermocouples and different choices often need to be made for different applications. This chapter introduces the physical phenomena involved, the various types of thermocouple, their installation and practical use.

5.1 Introduction

In its simplest practical form a thermocouple can consist of two dissimilar wires connected together at one end with a voltage measurement device connected across the free ends (Figure 5.1). A net emf will be indicated by the voltmeter, which is a function of the temperature difference between the join and the voltmeter connections. Such devices are commercially available from as little as about $15 (year 2001 prices) giving a direct display of temperature (Figure 5.2). Thermocouples are widely used in industry and research with applications from −272°C to 2000°C. The merits of thermocouples are their relatively low cost, small size, rugged nature, versatility, reasonable stability, reproducibility, reasonable uncertainty and fast speed of response. Although PRTs are more accurate and stable and thermistors are more sensitive, thermocouples are generally a more economical solution than PRTs and their temperature range is greater than thermistors. The main disadvantage of

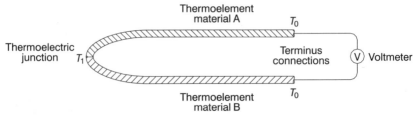

Figure 5.1 Simple thermocouple circuit. T_1 = the temperature at the thermoelectric junction. T_0 = temperature at the terminus connections

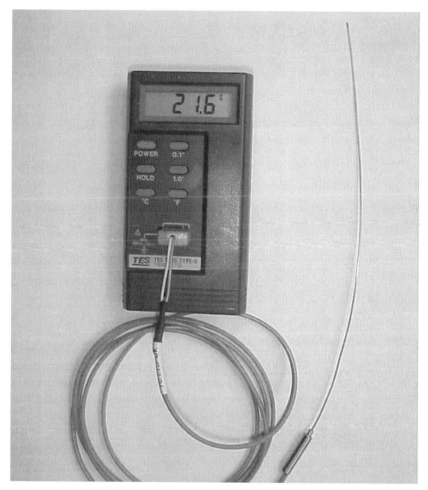

Figure 5.2 A commercial device providing a direct display of the temperature reading from a thermocouple

thermocouples is their relatively weak signal, ~4.1 mV at 100°C for a type K thermocouple. This makes their reading sensitive to corruption from electrical noise. In addition, their output is non-linear and requires amplification and their calibrations can vary with contamination of the thermocouple materials, cold-working and temperature gradients.

The fundamental physical phenomenon exploited in thermocouples is that heat flowing in a conductor produces a movement of electrons and thus an electromotive force (emf). This was originally demonstrated by Johann Seebeck who discovered that a small current flowed through the circuit shown in Figure 5.3 when the temperature of the two junctions was different (Seebeck, 1823). The emf produced is proportional to the temperature

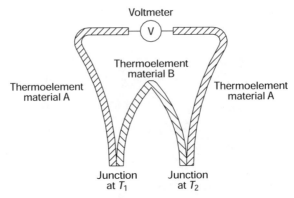

Figure 5.3 A current will flow through the above circuit that is proportional to the temperature difference between the two junctions

difference and is called the Seebeck emf or thermoelectric potential. As well as being a function of the net temperature difference, the magnitude of the emf produced is also a function of the materials used.

Thermocouples can be easy to use, but it is also possible to make errors in installation and interpreting a reading. Careful attention to the details of a thermocouple installation is therefore necessary. Some of the errors that occur in practice are due to a lack of appreciation of the behavior of thermocouples, and in order to overcome this it is helpful to develop a basic understanding of how a thermocouple generates a signal and this is described in Section 5.2. In describing thermocouples the following terminology can be used, although it should be noted that the terms described are not rigidly adhered to in common practice and are often subsumed into the general phrases thermocouple wire, junctions and joins:

- A *thermocouple* is any pair of dissimilar electrically conducting materials coupled at an interface.
- A *thermoelement*, or leg, is an electrically conducting material used to form a thermocouple.
- An electrical interface between dissimilar electric conductors is a *thermoelectric junction*.
- A free end of a thermoelement is a *terminus*.
- Couplings between identical thermoelements are *joins*.

5.2 Thermocouple analysis

Thermocouple circuits can consist of the simple forms illustrated in Figures 5.1 and 5.3 or much more complex forms resulting from considerations of data logging and long distances between the point of interest and emf measurement. In this case the use of thermocouple wire for the whole length may be prohibitively expensive and alternative materials used to convey the emf. In order to ensure correct assessment of the signal produced the circuit must be analysed. Analysis of a given thermocouple circuit can be undertaken using fundamental physical relationships. This can be highly complex and to date is not necessarily fully understood (see, for example, Pollock, 1991; Barnard, 1972; Blatt *et al.*, 1976; Bourassa *et al.*, 1978; Mott and Jones, 1958). Alternatively, thermocouple behaviour can be modelled in most applications using a number of laws or an algebraic technique. Thermocouple laws have been developed by a number of authors (e.g. Roeser, 1940). Here three laws from Doebelin (1990) will be described, as these can be particularly useful in practical temperature measurement (see Sections 5.2.1 to 5.2.3).

5.2.1 The law of interior temperatures

The thermal emf of a thermocouple with the junction at T_1 and terminus connections at T_0 is unaffected by temperature elsewhere in the circuit provided the properties of the two thermoelements used are homogenous. This requirement means that the physical properties of the wires must be constant with length. If the wire is stretched or strained in a region or the chemical makeup of the wire varies along its length, this will affect the thermoelectric output and could invalidate the law. The law of interior temperatures is illustrated in Figure 5.4. Provided the wire is uniform and homogeneous on both sides of the hot spot then no net emf is generated by the hot spot. The thermocouple will respond only to the temperature difference between the thermoelectric junction and the terminus connections. This result is particularly useful as it means that the emf from a thermocouple is not dependent on intermediate temperatures along a thermoelement.

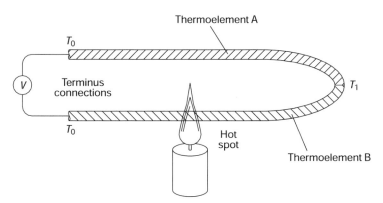

Figure 5.4 Illustration of the law of interior temperatures. The thermocouple is unaffected by hot spots along the thermoelement and the reading is only a function of T_1 and T_0

5.2.2 The law of inserted materials

If a third homogenous material C is inserted into either thermoelement A or B, then, as long as the two new thermoelectric junctions are at the same temperature, the net emf of the circuit is unchanged irrespective of the temperature in material C away from the thermoelectric junctions. This law is illustrated in Figure 5.5 where a third material is inserted into thermoelement A and then heated locally. Provided the thermoelectric junctions between C and A are both at the same temperature, the net emf of the thermocouple is unaffected by the presence of the inserted material and any local hot spot as the emf excursion between 2 and 3 is cancelled by that between 3 and 4.

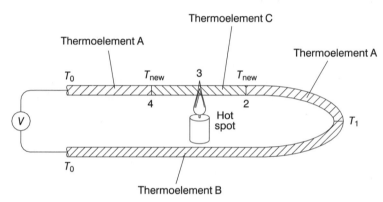

Figure 5.5 Illustration of the law of inserted materials. The thermocouple is unaffected by the presence of the inserted material and any local hot spot

5.2.3 The law of intermediate materials

If material C is inserted between A and B, the temperature of C at any point away from the junctions A–C, B–C is not significant. This law is illustrated in Figure 5.6. Here an intermediate thermoelectric material is inserted between the two thermoelements. As there is no thermal gradient across the new thermoelectric junctions then the presence of the inserted material does not contribute to the net emf produced by the thermocouple. This law is of great practical significance as it allows us to model the implications of manufacturing techniques used to form thermoelectric junctions. Provided there is no thermal gradient across the thermoelectric junction it does not matter if the thermoelements are joined by a third material such as solder or if local thermoelectric properties are changed at the junction by, for instance, welding.

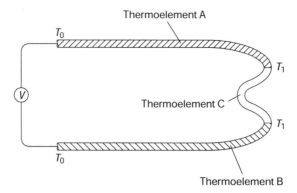

Figure 5.6 Illustration of the law of intermediate materials. As there is no thermal gradient across the new thermoelectric junctions then the presence ot the inserted material does not contribute to the net emf produced by the thermocouplo

As stated, a given thermocouple circuit can be analysed in a number of ways. The laws, however, allow a quick commonsense approach to be taken.

5.2.4 Modelling fundamental thermoelectric phenomena

Alternatively an assessment of the fundamental thermoelectric phenomena can be made. These are the Seebeck, Peltier and Thompson effects but in practical thermocouple circuits the contribution of the Peltier and Thompson effects is insignificant. The Seebeck effect is the generation of emf in a conductor whenever there is heat transfer and is a consequence of electron movements when heat transfer occurs. Thus an emf will be generated in a material whenever there is a temperature difference in that material. The magnitude of the emf will be a function of the temperature difference and the type of material. The Seebeck coefficient is a measure of how the electrons are coupled to the metal lattice and grain structure. It is sensitive to changes in the chemical and physical structure of the solid and will alter if the material is contaminated, oxidized, strained or heat treated. The Seebeck coefficient is defined by equation (5.1) and is a transport property of all electrically conducting materials:

$$S(T) = \lim_{\Delta T \to 0} \frac{\Delta E}{\Delta T} = \frac{dE}{dT} \tag{5.1}$$

where: S = Seebeck coefficient (μV/K)

ΔT = temperature difference across a segment of a conductor (K)

ΔE = absolute Seebeck emf (μV).

The Seebeck coefficient cannot be measured directly. Instead it must be determined by measuring the Thomson coefficient and using a thermodynamic relationship. The Seebeck coefficient varies with temperature so it must be mathematically defined by the gradient dE/dT at a specific temperature.

Rearranging equation (5.1) allows us to model the net thermoelectric emf generated by a practical thermocouple circuit:

$$dE = S(T)dT \tag{5.2}$$

If the circuit comprises two materials, A and B, then

$$E = \int_{T_1}^{T_2} S_A dT - \int_{T_1}^{T_2} S_B dT = \int_{T_1}^{T_2} (S_A - S_B)dT \tag{5.3}$$

where: E = thermoelectric emf (μV)
S_A = the Seebeck coefficient for material A (μV/K)
S_B = the Seebeck coefficient for material B (μV/K).

In equation (5.3) the difference in the Seebeck coefficients in the two thermoelements appears. It is this difference that is of practical interest in thermocouple thermometry and it is called the relative Seebeck coefficient. This coefficient is normally determined with respect to a reference material such as platinum:

$$S_{APt} = S_A - S_{Pt} \tag{5.4}$$

$$S_{BPt} = S_B - S_{Pt} \tag{5.5}$$

$$S_{AB} = S_A - S_B = S_{APt} - S_{BPt} \tag{5.6}$$

where: S_{APt} = Seebeck coefficient for material A relative to platinum (μV/K),
S_{BPt} = Seebeck coefficient for material B relative to platinum (μV/K).

Values for the Seebeck coefficient relative to platinum are listed in Table 5.1 for a variety of materials.

Substituting for S_A and S_B in equation (5.3) gives

$$E = \int_{T_1}^{T_2} S_{AB} dT \tag{5.7}$$

and if the relative Seebeck coefficient can be taken as constant over the temperature range then

$$E = S_{AB} (T_2 - T_1) \tag{5.8}$$

Equation (5.8) is particularly useful as it allows analysis of a wide range of circuits as illustrated in the following examples. This can be achieved using

Table 5.1 Values of the Seebeck coefficient for various materials. After Bentley (1998)

Material	20°C	1000°C
Chromel	22.2	9.4
Fe	13.3	-7
Nicrosil	11.8	8.8
Au	2.0	4
Cu	1.9	7
Ag	1.7	
W	1.3	20.3
Pt	4.7	21.4
Nisil	-14.8	-29.8
Alumel	-18.2	-29.6
Ni	-19.5	-35.4
Constantan	-38.3	-65.6

loop analysis. In loop analysis the contribution of emf due to the thermal gradient across each element is summed together to produce the total emf that would be indicated by a voltmeter.

The influence of connection leads shown in Figure 5.7 on the output can be analysed using equation (5.8). If identical conductors are used, then by using loop analysis

$$E = S_C (T_1 - T_0) + S_A (T_2 - T_1) + S_B (T_1 - T_2) + S_C (T_0 - T_1) \quad (5.9)$$

This simplifies to

$$E = S_{AB}(T_2 - T_1) \quad (5.10)$$

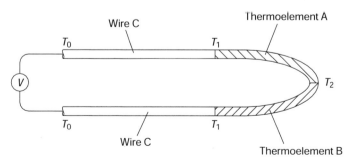

Figure 5.7 Use of connection leads for a thermocouple

The output emf is therefore dependent only on the temperature difference between the thermoelement junction and the terminus junctions. In other words, the temperature of the connections at the multimeter or data logger does not contribute to the emf produced by the thermocouple circuit. In effect, the reference temperature T_1 has been moved from the data logger to the terminus connections.

Use of the circuit illustrated in Figure 5.7 requires the temperature T_1 to be known. This can be achieved by using an alternative method of temperature measurement at this junction such as a thermistor or a PRT. Alternatively, the terminus connections can be submerged in a distilled ice/water bath as illustrated in Figure 5.8. The temperature at T_1 will then be known and at 0°C.

Another option to ensure that the temperature of the reference junction is known is illustrated in Figure 5.9. Here one of the thermoelements is

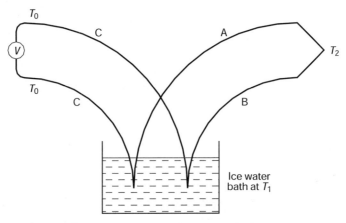

Figure 5.8 Connection leads submerged in an ice bath

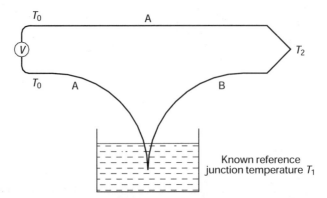

Figure 5.9 Lower leg placed at a known reference temperature

connected directly to the multimeter and the same type of thermoelement is used between the ice bath and the multimeter. Using loop analysis gives:

$$E = S_A(T_2 - T_0) + S_B(T_1 - T_2) + S_A(T_0 - T_1) \tag{5.11}$$

which simplifies to

$$E = S_A(T_2 - T_1) + S_B(T_1 - T_2) \tag{5.12}$$

or

$$E = S_{AB}(T_2 - T_1) \tag{5.13}$$

So if the thermoelement junction in the lower leg is placed in ice water then the output is referenced to 0°C or temperature T_1 if it is at some different value.

In practical use it is uncommon to need values for the Seebeck coefficients. Instead the emf temperature characteristics of certain selected thermocouples have been identified under controlled conditions and standardized and this data can be used. If, however, you are involved in the design of a new type of thermocouple or some unusual application, data for Seebeck coefficients of a wide range of materials is tabulated in Pollock (1991) and Mott and Jones (1958).

5.3 Thermocouple types

There are hundreds of types of thermocouple that have been developed. In principle almost any dissimilar metals and even semiconductors can be used to form thermocouples. The wide scope of alloys indicates the range possible. Kinzie (1973), for example, lists over 300 combinations of materials for thermocouples. A series of international and national standards are, however, available for thermocouples listing just eight combinations of materials that are widely used. Stringent guidelines are provided for these thermocouples to ensure that devices that are manufactured to be compatible with the specific standard concerned from one company perform in a similar fashion to those manufactured by another company.

The eight standardized thermocouples fall into three general categories:

- rare-metal thermocouples (types B, R and S)
- nickel-based thermocouples (types K and N)
- constantan negative thermocouples (types E, J and T).

The rare-metal group, types B, R and S, are based on platinum and its alloys with rhodium. These are the most stable of the standard thermocouples and can be used at high temperatures (up to 1750°C) but are generally more expensive and sensitive to contamination. The nickel-based thermocouples,

Table 5.2 Standardized thermocouples. Data from Kinzie (1973), Bedwell (1996), Guildner and Burns (1979), Finch (1962), Quinn (1990). Table after Childs *et al.* (2000)

Temperature range (°C)	Output (μV/°C)	Cost	Stability over the temperature range specified	Cable specification	Common name	Brief description
−262 → 850	15@−200°C 60@350°C	Low	Low	T	Copper/a copper–nickel alloy (constantan)	Type T, copper constantan, thermocouples are useful for the −250°C to 350°C range in oxidizing or inert atmospheres. Above 400°C the copper arm rapidly oxidizes. Care needs to be applied to avoid problems arising from the high thermal conductivity of the copper arm. As one lead of this thermocouple is copper, there is no need for special compensation cable. Note that constantan is a general term for a copper–nickel alloy with between 35% and 50% copper. The thermoelectric characteristics of each alloy will vary according to the alloying proportions. Although constantan is used in type T, J and E thermocouples the actual material for each is slightly different.
−196 → 700	26@−190°C 63@800°C	Low	Low	J	Iron/a copper–nickel alloy (constantan)	Type J thermocouples are commonly called iron constantan thermocouples and are popular due to their high Seebeck coefficient and low price. These can be safely used in reducing atmospheres from 0°C up to 550°C beyond which degradation is rapid. The maximum temperature for continuous operation is 800°C.
−268 → 800	68@100°C 81@500°C 77@900°C	Low	low–mid	E	Nickel–chromium alloy (chromel)/a copper–nickel alloy (constantan)	Type E, chromel constantan, thermocouples give high output for the range −250°C to 900°C. They are ideally suited to temperature measurement around ambient because of the large Seebeck coefficient, low thermal conductivity and corrosion resistance.

Type	Temperature range (°C)	emf (mV)			Composition	Comments
K	−250 → 1100	40 from 250–1000°C 35@1300°C	Low	Low	Nickel–chromium alloy (chromel)/ nickel–aluminium alloy (alumel)	The type K thermocouple is commonly called chromel alumel. It is the most commonly used thermocouple and is designed for use in oxidizing atmospheres. Maximum continuous use is limited to 1100°C although above 800°C oxidation causes drift and decalibration. Note that the type K thermocouple is unstable with hysteresis between 300°C and 600°C which can result in errors of several degrees.
N	0→1250	37@1000°C	Low	Mid–high	Nickel–chromium silicon (nicrosil)/ nickel–silicon– magnesium alloy (nisil)	Type N thermocouples have been developed to address the instability of type K (Burley, 1971). These trade linear response for stability and an algorithm is required for conversion between the generated emf and temperature. The voltage temperature curve for type N thermocouples is slightly lower than that for type K thermocouples.
B	100→1750	5@1000°C	High	High	Platinum–30% rhodium/platinum– 6% rhodium	Type B thermocouples can be used continuously to 1600°C and intermittently to 1800°C. However, due to a local minimum in its thermoelectric emf this thermocouple exhibits a double value ambiguity between 0°C and 42°C. The emf below 100°C is very low and little concern need be given to cold junction compensation.
S	0→1500	6 from 0–100°C	High	High	Platinum–10% rhodium/platinum	Type S thermocouples can be used in oxidizing or inert atmospheres continuously at temperatures up to 1400°C and for brief periods up to 1650°C.
R	0→1600	10@1000°C	High	High	Platinum–13% rhodium/platinum	Type R thermocouples give a similar performance to type S, but give slightly higher output and improved stability.

types N and K, are commonly used for applications not requiring the elevated temperature range of the rare-metal thermocouples. The constantan negative leg thermocouples, types E, J and T, have high emf outputs as constantan has a strong negative Seebeck coefficient.

The criteria for thermocouple selection include cost, maximum and minimum operating temperatures, chemical stability, material compatibility, atmospheric protection, mechanical limitations, duration of exposure, sensor lifetime, sensitivity and output emf. Descriptions of the various commonly available standardized thermocouples are presented in Table 5.2.

Standards organizations throughout the world have specified allowable tolerances for the deviation of the thermocouple output from standardized tables. Many of the standards are based on the results and research at national laboratories and there is similarity between the magnitude of tolerances in the standards from one country to another. Tables 5.3 and 5.4 give the manufacturing tolerances for ASTM and British Standard thermocouples. Many countries use the term Class 1 for special tolerance and Class 2 for standard tolerance.

It should be noted that a standard grade, Class 2, thermocouple will be just as stable as one produced to a special tolerance. The tolerance refers to the maximum deviation of the thermocouple output from the standardized tables when the thermocouple is new. The stability of the wire and any drift is independent of its class.

The data for thermoelectric emf versus temperature for the standardized thermocouples according to IEC 584.1: 1995 and BS EN 60584.1: 1996 is reproduced in Tables 5.5–5.12 and plotted in Figure 5.10. Here just part of the data available is presented in 10°C steps. Complete data sets, with listings in

Table 5.3 Manufacturing tolerances for ASTM thermocouples (ASTM E 230–98)

Type	Temperature range (°C)	Standard tolerance (°C)	Special tolerance (°C)
B	870 to 1700	±0.5%	±0.25%
E	0 to 870	±1.7 or ±0.5%	±1.0 or ±0.4%
J	0 to 760	±2.2 or ±0.75%	±1.1 or ±0.4%
K	0 to 1260	±2.2 or ±0.75%	±1.1 or ±0.4%
N	0 to 1260	±2.2 or ±0.75%	±1.1 or ±0.4%
R	0 to 1480	±1.5 or ±0.25%	±0.6 or ±0.1%
S	0 to 1480	±1.5 or ±0.25%	±0.6 or ±0.1%
T	0 to 370	±1.0 or ±0.75%	±0.5 or ±0.4%
E	−200 to 0	±1.7 or ±1.0%	
K	−200 to 0	±2.2 or ±2.0%	
T	−200 to 0	±1.0 or ±1.5%	

Table 5.4 Manufacturing tolerances for thermocouples (BS EN 60584.2)

Type	Tolerance Class 1 (°C)	Tolerance Class 2 (°C)
B		±0.25% (600 to 1700)
E	±1.5 (−40 to 375) ±0.4% (375 to 800)	±2.5 (−40 to 333) ±0.75% (333 to 900)
J	±1.5 (−40 to 375) ±0.4% (375 to 750)	±2.5 (−40 to 333) ±0.75% (333 to 750)
K	±1.5 (−40 to 375) ±0.4% (375 to 1000)	±2.5 (−40 to 333) ±0.75% (333 to 1200)
N	±1.5 (−40 to 375) ±0.4% (375 to 1000)	±2.5 (−40 to 333) ±0.75% (333 to 1200)
R	±1.0 (0 to 1100) ±(1+0.003(t-1100)) (1100 to 1600)	±1.5 (0 to 600) ±0.25% (600 to 1600)
S	±1.0 (0 to 1100) ±(1+0.003(t-1100)) (1100 to 1600)	±1.5 (0 to 600) ±0.25% (600 to 1600)
T	±0.5 (−40 to 125) ±0.4% (125 to 350)	±1.0 (−40 to 133) ±0.75% (133 to 350)

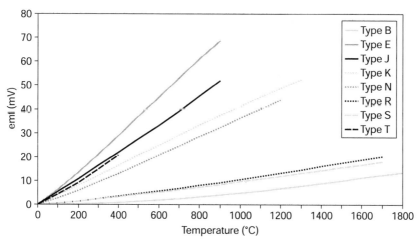

Figure 5.10 Emf versus temperature characteristics for the standard thermocouples

Table 5.5 Standard reference data to IEC 584.1:1995 and BS EN 60584.1 Part 7: 1996 for type B platinum–30% rhodium/platinum–6% rhodium thermocouples giving thermocouple emf in μV for various tip temperatures assuming a cold junction at 0°C

Type B										
°C	0	10	20	30	40	50	60	70	80	90	100
0	0	-2	-3	-2	0	2	6	11	17	25	33
100	33	43	53	65	78	92	107	123	141	159	178
200	178	199	220	243	267	291	317	344	372	401	431
300	431	462	494	527	561	596	632	669	707	746	787
400	787	828	870	913	957	1002	1048	1095	1143	1192	1242
500	1242	1293	1344	1397	1451	1505	1561	1617	1675	1733	1792
600	1792	1852	1913	1975	2037	2101	2165	2230	2296	2363	2431
700	2431	2499	2569	2639	2710	2782	2854	2928	3002	3078	3154
800	3154	3230	3308	3386	3466	3546	3626	3708	3790	3873	3957
900	3957	4041	4127	4213	4299	4387	4475	4564	4653	4743	4834
1000	4834	4926	5018	5111	5205	5299	5394	5489	5585	5682	5780
1100	5780	5878	5976	6075	6175	6276	6377	6478	6580	6683	6786
1200	6786	6890	6995	7100	7205	7311	7417	7524	7632	7740	7848
1300	7848	7957	8066	8176	8286	8397	8508	8620	8731	8844	8956
1400	8956	9069	9182	9296	9410	9524	9639	9753	9868	9984	10099
1500	10099	10215	10331	10447	10563	10679	10796	10913	11029	11146	11263
1600	11263	11380	11497	11614	11731	11848	11965	12082	12199	12316	12433
1700	12433	12549	12666	12782	12898	13014	13130	13246	13361	13476	13591
1800	13591	13706	13820								

Table 5.6 Standard reference data to IEC 584.1:1995 and BS EN 60584.1 Part 6: 1996 for type E nickel–chromium/copper–nickel thermocouples giving thermocouple emf in μV for various tip temperatures assuming a cold junction at 0°C

Type E										
°C	0	10	20	30	40	50	60	70	80	90	100
-200	-8825	-9063	-9274	-9455	-9604	-9718	-9797	-9835			
-100	-5237	-5681	-6107	-6516	-6907	-7279	-7632	-7963	-8273	-8561	-8825
0	0	-582	-1152	-1709	-2255	-2787	-3306	-3811	-4302	-4777	-5237
0	0	591	1192	1801	2420	3048	3685	4330	4985	5648	6319
100	6319	6998	7685	8379	9081	9789	10503	11224	11951	12684	13421
200	13421	14164	14912	15664	16420	17181	17945	18713	19484	20259	21036
300	21036	21817	22600	23386	24174	24964	25757	26552	27348	28146	28946
400	28946	29747	30550	31354	32159	32965	33772	34579	35387	36196	37005
500	37005	37815	38624	39434	40243	41053	41862	42671	43479	44286	45093
600	45093	45900	46705	47509	48313	49116	49917	50718	51517	52315	53112
700	53112	53908	54703	55497	56289	57080	57870	58659	59446	60232	61017
800	61017	61801	62583	63364	64144	64922	65698	66473	67246	68017	68787
900	68787	69554	70319	71082	71844	72603	73360	74115	74869	75621	76373

Table 5.7 Standard reference data to IEC 584.1:1995 and BS EN 60584.1 Part 3: 1996 for type J iron/copper–nickel thermocouples giving thermocouple emf in μV for various tip temperatures assuming a cold junction at 0°C

						Type J					
°C	0	10	20	30	40	50	60	70	80	90	100
−200	−7890	−8095									
−100	−4633	−5037	−5426	−5801	−6159	−6500	−6821	−7123	−7403	−7659	−7890
0	0	−501	−995	−1482	−1961	−2431	−2893	−3344	−3786	−4215	−4633
0	0	507	1019	1537	2059	2585	3116	3650	4187	4726	5269
100	5269	5814	6360	6909	7459	8010	8562	9115	9669	10224	10779
200	10779	11334	11889	12445	13000	13555	14110	14665	15219	15773	16327
300	16327	16881	17434	17986	18538	19090	19642	20194	20745	21297	21848
400	21848	22400	22952	23504	24057	24610	25164	25720	26276	26834	27393
500	27393	27953	28516	29080	29647	30210	30788	31362	31909	32519	33102
600	33102	33689	34279	34873	35470	36071	36675	37284	37896	38512	39132
700	39132	39755	40382	41012	41645	42281	42919	43559	44203	44848	45494
800	45494	46141	46786	47431	48074	48715	49353	49989	50622	51251	51877
900	51877	52500	53119	53735	54347	54956	55561	56164	56763	57360	57953
1000	57953	58545	59134	59721	60307	60890	61473	62054	62634	63214	63792
1100	63792	64370	64948	65525	66102	66679	67255	67831	68406	68980	69553

Table 5.8 Standard reference data to IEC 584.1:1995 and BS EN 60584.1 Part 4: 1996 for type K nickel–chromium/nickel–aluminium thermocouples giving thermocouple emf in μV for various tip temperatures assuming a cold junction at 0°C

						Type K					
°C	0	10	20	30	40	50	60	70	80	90	100
−200	−5891	−6035	−6158	−6262	−6344	−6404	−6441	−6458			
−100	−3554	−3852	−4138	−4411	−4669	−4913	−5141	−5354	−5550	−5730	−5891
0	0	−392	−778	−1156	−1527	−1889	−2243	−2587	−2920	−3243	−3554
0	0	397	798	1203	1612	2023	2436	2851	3267	3682	4096
100	4096	4509	4920	5328	5735	6138	6540	6941	7340	7739	8138
200	8138	8539	8940	9343	9747	10153	10561	10971	11382	11795	12209
300	12209	12624	13040	13457	13874	14293	14713	15133	15554	15975	16397
400	16397	16820	17243	17667	18091	18516	18941	19366	19792	20218	20644
500	20644	21071	21497	21924	22350	22776	23203	23629	24055	24480	24905
600	24905	25330	25755	26179	26602	27025	27447	27869	28289	28710	29129
700	29129	29548	29965	30382	30798	31213	31628	32041	32453	32865	33275
800	33275	33685	34093	34501	34908	35313	35718	36121	36524	36925	37326
900	37326	37725	38124	38522	38918	39314	39708	40101	40494	40885	41276
1000	41276	41665	42053	42440	42826	43211	43595	43978	44359	44740	45119
1100	45119	45497	45873	46249	46623	46995	47367	47737	48105	48473	48838
1200	48838	49202	49565	49926	50286	50644	51000	51355	51708	52060	52410
1300	52410	52759	53106	53451	53795	54138	54479	54819			

Table 5.9 Standard reference data to IEC 584.1:1995 and BS EN 60584.1 Part 8: 1996 for type N nickel–chromium–silicon/nickel–silicon–magnesium thermocouples giving thermocouple emf in μV for various tip temperatures assuming a cold junction at 0°C

						Type N					
°C	0	10	20	30	40	50	60	70	80	90	100
−200	−3990	−4083	−4162	−4226	−4277	−4313	−4336	−4345			
−100	−2407	−2612	−2808	−2994	−3171	−3336	−3491	−3634	−3766	−3884	−3990
0	0	−260	−518	−772	−1023	−1269	−1509	−1744	−1972	−2193	−2407
0	0	261	525	793	1065	1340	1619	1902	2189	2480	2774
100	2774	3072	3374	3680	3989	4302	4618	4937	5259	5585	5913
200	5913	6245	6579	6916	7255	7597	7941	8288	8637	8988	9341
300	9341	9696	10054	10413	10774	11136	11501	11867	12234	12603	12974
400	12974	13346	13719	14094	14469	14846	15225	15604	15984	16366	16748
500	16748	17131	17515	17900	18286	18672	19059	19447	19835	20224	20613
600	20613	21003	21393	21784	22175	22566	22958	23350	23742	24134	24527
700	24527	24919	25312	25705	26098	26491	26883	27276	27669	28062	28455
800	28455	28847	29239	29632	30024	30416	30807	31199	31590	31981	32371
900	32371	32761	33151	33541	33930	34319	34707	35095	35482	35869	36256
1000	36256	36641	37027	37411	37795	38179	38562	38944	39326	39706	40087
1100	40087	40466	40845	41223	41600	41976	42352	42727	43101	43474	43846
1200	43846	44218	44588	44958	45326	45694	46060	46425	46789	47152	47513

Table 5.10 Standard reference data to IEC 584.1:1995 and BS EN 60584.1 Part 2: 1996 for type R platinum–13% rhodium/platinum thermocouples giving thermocouple emf in μV for various tip temperatures assuming a cold junction at 0°C

						Type R					
°C	0	10	20	30	40	50	60	70	80	90	100
0	0	54	111	171	232	296	363	431	501	573	647
100	647	723	800	879	959	1041	1124	1208	1294	1381	1469
200	1469	1558	1648	1739	1831	1923	2017	2112	2207	2304	2401
300	2401	2498	2597	2696	2796	2896	2997	3099	3201	3304	3408
400	3408	3512	3616	3721	3827	3933	4040	4147	4255	4363	4471
500	4471	4580	4690	4800	4910	5021	5133	5245	5357	5470	5583
600	5583	5697	5812	5926	6041	6157	6273	6390	6507	6625	6743
700	6743	6861	6980	7100	7220	7340	7461	7583	7705	7827	7950
800	7950	8073	8197	8321	8446	8571	8697	8823	8950	9077	9205
900	9205	9333	9461	9590	9720	9850	9980	10111	10242	10374	10506
1000	10506	10638	10771	10905	11039	11173	11307	11442	11578	11714	11850
1100	11850	11986	12123	12260	12397	12535	12673	12812	12950	13089	13228
1200	13228	13367	13507	13646	13786	13926	14066	14207	14347	14488	14629
1300	14629	14770	14911	15052	15193	15334	15475	15616	15758	15899	16040
1400	16040	16181	16323	16464	16605	16746	16887	17028	17169	17310	17451
1500	17451	17591	17732	17872	18012	18152	18292	18431	18571	18710	18849
1600	18849	18988	19126	19264	19402	19540	19677	19814	19951	20087	20222
1700	20222	20356	20488	20620	20749	20877	21003				

Table 5.11 Standard reference data to IEC 584.1:1995 and BS EN 60584.1 Part 1: 1996 for type S platinum–10% rhodium/platinum thermocouples giving thermocouple emf in μV for various tip temperatures assuming a cold junction at 0°C

						Type S					
°C	0	10	20	30	40	50	60	70	80	90	100
0	0	55	113	173	235	299	365	433	502	573	646
100	646	720	795	872	950	1029	1110	1191	1273	1357	1441
200	1441	1526	1612	1698	1786	1874	1962	2052	2141	2232	2323
300	2323	2415	2507	2599	2692	2786	2880	2974	3069	3164	3259
400	3259	3355	3451	3548	3645	3742	3840	3938	4036	4134	4233
500	4233	4332	4432	4532	4632	4732	4833	4934	5035	5137	5239
600	5239	5341	5443	5546	5649	5753	5857	5961	6065	6170	6275
700	6275	6381	6486	6593	6699	6806	6913	7020	7128	7236	7345
800	7345	7454	7563	7673	7783	7893	8003	8114	8226	8337	8449
900	8449	8562	8674	8787	8900	9014	9128	9242	9357	9472	9587
1000	9587	9703	9819	9935	10051	10168	10285	10403	10520	10638	10757
1100	10757	10875	10994	11113	11232	11351	11471	11590	11710	11830	11951
1200	11951	12071	12191	12312	12433	12554	12675	12796	12917	13038	13159
1300	13159	13280	13402	13523	13644	13766	13887	14009	14130	14251	14373
1400	14373	14494	14615	14736	14857	14978	15099	15220	15341	15461	15582
1500	15582	15702	15822	15942	16062	16182	16301	16420	16539	16658	16777
1600	16777	16895	17013	17131	17249	17366	17483	17600	17717	17832	17947
1700	17947	18061	18174	18285	18395	18503	18609				

Table 5.12 Standard reference data to IEC 584.1:1995 and BS EN 60584.1 Part 5: 1996 for type T copper/copper–nickel thermocouples giving thermocouple emf in μV for various tip temperatures assuming a cold junction at 0°C

						Type T					
°C	0	10	20	30	40	50	60	70	80	90	100
−200	−5603	−5753	−5888	−6007	−6105	−6180	−6232	−6258			
−100	−3379	−3657	−3923	−4177	−4419	−4648	−4865	−5070	−5261	−5439	−5603
0	0	−383	−757	−1121	−1475	−1819	−2153	−2476	−2788	−3089	−3379
0	0	391	790	1196	1612	2036	2468	2909	3358	3814	4279
100	4279	4750	5228	5714	6206	6704	7209	7720	8237	8759	9288
200	9288	9822	10362	10907	11458	12013	12574	13139	13709	14283	14862
300	14862	15445	16032	16624	17219	17819	18422	19030	19641	20255	20872
400	20872										

1°C steps, are available in the standards and have also been reproduced in many manufacturers' catalogues.

The reference tables (5.5–5.12) give data that applies to a thermocouple with a reference junction at 0°C. The tables allow the emf measured under these conditions to be converted to the temperature experienced at the tip of the thermocouple.

Example 5.1

The voltage across a thermocouple circuit of the form illustrated in Figure 5.8 using a type T thermocouple is 12.43 mV. Determine the temperature.

Solution

12.43 mV = 12430 μV.

Examination of Table 5.12 shows that the nearest emfs listed are 12013 μV at 250°C and 12574 μV at 260°C. These values dictate the bounds on the temperature, which can be determined more closely by linear interpolation:

$$T = 250 + (260 - 250) \left(\frac{12430 - 12013}{12574 - 12013} \right) = 257.4°C$$

Example 5.2

The voltage across a thermocouple circuit of the form illustrated in Figure 5.9 with the reference junction at 0°C using a type S thermocouple is 5.87 mV. Determine the temperature.

Solution

5.87 mV = 5870 μV.

Examination of Table 5.11 shows that the nearest emfs listed are 5857 μV at 660°C and 5961 μV at 670°C. Using linear interpolation gives the temperature as

$$T = 660 + (670 - 660) \left(\frac{5870 - 5857}{5961 - 5857} \right) = 661.3°C$$

If the reference junction is at some temperature other than 0°C then the reference tables cannot be used directly and a further calculation must be undertaken. The temperature of the reference junction must be measured by an independent means or its value estimated. This value must then be used to determine the emf that would occur for the thermocouple if the tip was at the reference temperature and the reference junction was at 0°C. This value of emf must then be added to the measured value and the tables used to determine the temperature of the tip.

The procedure is as follows:

1 Measure the emf across terminations.
2 Measure the reference temperature at the terminations by some independent means such as using a thermistor.
3 Convert the reference temperature into an equivalent emf using the tables.
4 Add the measured emf to the equivalent emf to give a total emf.
5 Convert the total emf to a temperature using the tables. This is the temperature of the thermoelectric junction (also commonly referred to as the tip).

Example 5.3

A circuit with a type T thermocouple of the form illustrated in Figure 5.7 indicates a voltage of 8.54 mV. The temperature of the reference junction was measured to be 21.1°C using a thermistor. Determine the temperature of the thermoelectric junction.

Solution

$$8.54 \text{ mV} = 8540 \, \mu\text{V}$$

The reference junction temperature is 21.1°C. Examination of Table 5.12 gives the nearest temperatures to this value as 20°C and 30°C with corresponding emfs of 790 μV and 1196 μV respectively.

The emf equivalent to the temperature of the reference junction can be found by

$$E|_{\text{equivalent at } 21.1°C} = 790 + (1196 - 790) \left(\frac{21.1 - 20}{30 - 20} \right) = 838 \, \mu\text{V}$$

The total emf is given by

$$E_{\text{total}} = E_{\text{terminations}} + E_{\text{equivalent}} = 8540 + 838 = 9378 \, \mu\text{V}$$

From Table 5.12 the nearest emfs listed are 9288 μV at 200°C and 9822 μV at 210°C. Using linear interpolation gives the temperature as

$$T = 200 + (210 - 200) \left(\frac{9378 - 9288}{9822 - 9288} \right) = 201.7°C$$

As an alternative to using the look-up tables provided, the polynomials from which this data were derived could be used. These as listed in BS EN 60584.1 are given in Table 5.13. The polynomials are particularly useful for

Table 5.13 Power series expansions and polynomials for the standard thermocouples

K	T	J	N
Temperature range −270°C to 0°C	*Temperature range* −270°C to 0°C	*Temperature range* −210°C to 760°C	*Temperature range* −270°C to 0°C
$E = \sum_{i=1}^{n} a_i(t_{90})^i \, \mu V$	$E = \sum_{i=1}^{n} a_i(t_{90})^i \, \mu V$	$E = \sum_{i=1}^{n} a_i(t_{90})^i \, \mu V$	$E = \sum_{i=1}^{n} a_i(t_{90})^i \, \mu V$
where	**where**	**where**	**where**

K	T	J	N
$a_1 = 3.945\ 012\ 802\ 5 \times 10^1$	$a_1 = 3.874\ 810\ 636\ 4 \times 10^1$	$a_1 = 5.038\ 118\ 781\ 5 \times 10^1$	$a_1 = 2.615\ 910\ 596\ 2 \times 10^1$
$a_2 = 2.362\ 237\ 359\ 8 \times 10^{-2}$	$a_2 = 4.419\ 443\ 434\ 7 \times 10^{-2}$	$a_2 = 3.047\ 583\ 693\ 0 \times 10^{-2}$	$a_2 = 1.095\ 748\ 422\ 8 \times 10^{-}$
$a_3 = -3.285\ 890\ 678\ 4 \times 10^{-4}$	$a_3 = 1.184\ 432\ 310\ 5 \times 10^{-4}$	$a_3 = -8.568\ 106\ 572\ 0 \times 10^{-5}$	$a_3 = -9.384\ 111\ 155\ 4 \times 10$
$a_4 = -4.990\ 482\ 877\ 7 \times 10^{-6}$	$a_4 = 2.003\ 297\ 355\ 4 \times 10^{-5}$	$a_4 = 1.322\ 819\ 529\ 5 \times 10^{-7}$	$a_4 = -4.641\ 203\ 975\ 9 \times 10$
$a_5 = -6.750\ 905\ 917\ 3 \times 10^{-8}$	$a_5 = 9.013\ 801\ 955\ 9 \times 10^{-7}$	$a_5 = -1.705\ 295\ 833\ 7 \times 10^{-10}$	$a_5 = -2.630\ 335\ 771\ 6 \times 10$
$a_6 = -5.741\ 032\ 742\ 8 \times 10^{-10}$	$a_6 = 2.265\ 115\ 659\ 3 \times 10^{-8}$	$a_6 = 2.094\ 809\ 069\ 7 \times 10^{-13}$	$a_6 = -2.265\ 343\ 800\ 3 \times 10$
$a_7 = -3.108\ 887\ 289\ 4 \times 10^{-12}$	$a_7 = 3.607\ 115\ 420\ 5 \times 10^{-10}$	$a_7 = -1.253\ 839\ 533\ 6 \times 10^{-16}$	$a_7 = -7.608\ 930\ 079\ 1 \times 10$
$a_8 = -1.045\ 160\ 936\ 5 \times 10^{-14}$	$a_8 = 3.849\ 393\ 988\ 3 \times 10^{-12}$	$a_8 = 1.563\ 172\ 569\ 7 \times 10^{-20}$	$a_8 = -9.341\ 966\ 783\ 5 \times 10$
$a_9 = -1.988\ 926\ 687\ 8 \times 10^{-17}$	$a_9 = 2.821\ 352\ 192\ 5 \times 10^{-14}$		
$a_{10} = -1.632\ 269\ 748\ 6 \times 10^{-20}$	$a_{10} = 1.425\ 159\ 447\ 9 \times 10^{-16}$		
	$a_{11} = 4.876\ 866\ 228\ 6 \times 10^{-19}$		
	$a_{12} = 1.079\ 553\ 927\ 0 \times 10^{-21}$		
	$a_{13} = 1.394\ 502\ 706\ 2 \times 10^{-24}$		
	$a_{14} = 7.979\ 515\ 392\ 7 \times 10^{-28}$		

K	T	J	N
0°C to 1372°C	**0°C to 400°C**	**760°C to 1200°C**	**0°C to 1300°C**
$E = b_0 + \sum_{i=1}^{n} b_i (t_{90})^i +$ $c_0 \exp[c_1 (t_{90} - 126.9686)^2] \mu V$	$E = \sum_{i=1}^{n} a_i(t_{90})^i \, \mu V$	$E = \sum_{i=0}^{n} a_i(t_{90})^i \, \mu V$	$E = \sum_{i=1}^{n} a_i(t_{90})^i \, \mu V$
where	**where**	**where**	**where**

K	T	J	N
$b_0 = -1.760\ 041\ 368\ 6 \times 10^1$	$a_1 = 3.874\ 810\ 636\ 4 \times 10^1$	$a_0 = 2.964\ 562\ 568\ 1 \times 10^5$	$a_1 = 2.592\ 939\ 460\ 1 \times 10$
$b_1 = 3.892\ 120\ 497\ 5 \times 10^1$	$a_2 = 3.329\ 222\ 788\ 0 \times 10^{-2}$	$a_1 = -1.497\ 612\ 778\ 6 \times 10^3$	$a_2 = 1.571\ 014\ 188\ 0 \times 10$
$b_2 = 1.855\ 877\ 003\ 2 \times 10^{-2}$	$a_3 = 2.061\ 824\ 340\ 4 \times 10^{-4}$	$a_2 = 3.178\ 710\ 392\ 4$	$a_3 = 4.382\ 562\ 723\ 7 \times 10$
$b_3 = -9.945\ 759\ 287\ 4 \times 10^{-5}$	$a_4 = -2.188\ 225\ 684\ 6 \times 10^{-6}$	$a_3 = -3.184\ 768\ 670\ 1 \times 10^{-3}$	$a_4 = -2.526\ 116\ 979\ 4 \times 1$
$b_4 = 3.184\ 094\ 571\ 9 \times 10^{-7}$	$a_5 = 1.099\ 688\ 092\ 8 \times 10^{-8}$	$a_4 = 1.572\ 081\ 900\ 4 \times 10^{-6}$	$a_5 = 6.431\ 181\ 933\ 9 \times 10$
$b_5 = -5.607\ 284\ 488\ 9 \times 10^{-10}$	$a_6 = -3.081\ 575\ 877\ 2 \times 10^{-11}$	$a_5 = -3.069\ 136\ 905\ 6 \times 10^{-10}$	$a_6 = -1.006\ 347\ 151\ 9 \times 1$
$b_6 = 5.607\ 505\ 905\ 9 \times 10^{-13}$	$a_7 = 4.547\ 913\ 529\ 0 \times 10^{-14}$		$a_7 = 9.974\ 533\ 899\ 2 \times 10$
$b_7 = -3.202\ 072\ 000\ 3 \times 10^{-16}$	$a_8 = -2.751\ 290\ 167\ 3 \times 10^{-17}$		$a_8 = -6.086\ 342\ 560\ 7 \times 1$
$b_8 = 9.715\ 114\ 715\ 2 \times 10^{-20}$			$a_9 = 2.084\ 922\ 933\ 9 \times 10$
$b_9 = -1.210\ 472\ 127\ 5 \times 10^{-23}$			$a_{10} = -3.068\ 219\ 615\ 1 \times 1$

$c_0 = 1.185\ 976 \times 10^2$

$c_1 = -1.183\ 432 \times 10^{-4}$

3S EN 60584.1)

E	R	S	B
Temperature range $-270°C$ to $0°C$	Temperature range $-50°C$ to $1064.18°C$	Temperature range $-50°C$ to $1064.18°C$	Temperature range $0°C$ to $630.615°C$
$E = \sum_{i=1}^{n} a_i(t_{90})^i \, \mu V$	$E = \sum_{i=1}^{n} a_i(t_{90})^i \, \mu V$	$E = \sum_{i=1}^{n} a_i(t_{90})^i \, \mu V$	$E = \sum_{i=1}^{n} a_i(t_{90})^i \, \mu V$

where (E, $-270°C$ to $0°C$):

$= 5.866\ 550\ 870\ 8 \times 10^1$
$_2 = 4.541\ 097\ 712\ 4 \times 10^{-2}$
$_3 = -7.799\ 804\ 868\ 6 \times 10^{-4}$
$_4 = -2.580\ 016\ 084\ 3 \times 10^{-5}$
$_5 = -5.945\ 258\ 305\ 7 \times 10^{-7}$
$= -9.321\ 405\ 866\ 7 \times 10^{-9}$
$= -1.028\ 760\ 553\ 4 \times 10^{-10}$
$= -8.037\ 012\ 362\ 1 \times 10^{-13}$
$= -4.397\ 949\ 739\ 1 \times 10^{-15}$
$= -1.641\ 477\ 635\ 5 \times 10^{-17}$
$= -3.967\ 361\ 951\ 6 \times 10^{-20}$
$_2 = -5.582\ 732\ 872\ 1 \times 10^{-23}$
$_3 = -3.465\ 784\ 201\ 3 \times 10^{-26}$

where (R, $-50°C$ to $1064.18°C$):

$a_1 = 5.289\ 617\ 297\ 65$
$a_2 = 1.391\ 665\ 897\ 82 \times 10^{-2}$
$a_3 = -2.388\ 556\ 930\ 17 \times 10^{-5}$
$a_4 = 3.569\ 160\ 010\ 63 \times 10^{-8}$
$a_5 = -4.623\ 476\ 662\ 98 \times 10^{-11}$
$a_6 = 5.007\ 774\ 410\ 04 \times 10^{-14}$
$a_7 = -3.731\ 058\ 861\ 91 \times 10^{-17}$
$a_8 = 1.577\ 164\ 823\ 67 \times 10^{-20}$
$a_9 = -2.810\ 386\ 252\ 51 \times 10^{-24}$

where (S, $-50°C$ to $1064.18°C$):

$a_1 = 5.403\ 133\ 086\ 31$
$a_2 = 1.259\ 342\ 897\ 40 \times 10^{-2}$
$a_3 = -2.324\ 779\ 686\ 89 \times 10^{-5}$
$a_4 = 3.220\ 288\ 230\ 36 \times 10^{-8}$
$a_5 = -3.314\ 651\ 963\ 89 \times 10^{-11}$
$a_6 = 2.557\ 442\ 517\ 86 \times 10^{-14}$
$a_7 = -1.250\ 688\ 713\ 93 \times 10^{-17}$
$a_8 = 2.714\ 431\ 761\ 45 \times 10^{-21}$

where (B, $0°C$ to $630.615°C$):

$a_1 = -2.465\ 081\ 834\ 6 \times 10^{-1}$
$a_2 = 5.904\ 042\ 117\ 1 \times 10^{-3}$
$a_3 = -1.325\ 793\ 163\ 6 \times 10^{-6}$
$a_4 = 1.566\ 829\ 190\ 1 \times 10^{-9}$
$a_5 = -1.694\ 452\ 924\ 0 \times 10^{-12}$
$a_6 = 6.299\ 004\ 709\ 4 \times 10^{-16}$

$0°C$ to $1000°C$	$1064.18°C$ to $1664.5°C$	$1064.18°C$ to $1664.5°C$	$630.615°C$ to $1820°C$
$E = \sum_{i=1}^{n} a_i(t_{90})^i \, \mu V$	$E = \sum_{i=0}^{n} a_i(t_{90})^i \, \mu V$	$E = \sum_{i=0}^{n} a_i(t_{90})^i \, \mu V$	$E = \sum_{i=0}^{n} a_i(t_{90})^i \, \mu V$

where (E, $0°C$ to $1000°C$):

$= 5.866\ 550\ 871\ 0 \times 10^1$
$= 4.503\ 227\ 558\ 2 \times 10^{-2}$
$= 2.890\ 840\ 721\ 2 \times 10^{-5}$
$= -3.305\ 689\ 665\ 2 \times 10^{-7}$
$= 6.502\ 440\ 327\ 0 \times 10^{-10}$
$= -1.919\ 749\ 550\ 4 \times 10^{-13}$
$= -1.253\ 660\ 049\ 7 \times 10^{-15}$
$= 2.148\ 921\ 756\ 9 \times 10^{-18}$
$= -1.438\ 804\ 178\ 2 \times 10^{-21}$
$= 3.596\ 089\ 948\ 1 \times 10^{-25}$

where (R, $1064.18°C$ to $1664.5°C$):

$a_0 = 2.951\ 579\ 253\ 16 \times 10^3$
$a_1 = -2.520\ 612\ 513\ 32$
$a_2 = 1.595\ 645\ 018\ 65 \times 10^{-2}$
$a_3 = -7.640\ 859\ 475\ 76 \times 10^{-6}$
$a_4 = 2.053\ 052\ 910\ 24 \times 10^{-9}$
$a_5 = -2.933\ 596\ 681\ 73 \times 10^{-13}$

where (S, $1064.18°C$ to $1664.5°C$):

$a_0 = 1.329\ 004\ 440\ 85 \times 10^3$
$a_1 = 3.345\ 093\ 113\ 44$
$a_2 = 6.548\ 051\ 928\ 18 \times 10^{-3}$
$a_3 = -1.648\ 562\ 592\ 09 \times 10^{-6}$
$a_4 = 1.299\ 896\ 051\ 74 \times 10^{-11}$

where (B, $630.615°C$ to $1820°C$):

$a_0 = -3.893\ 816\ 862\ 1 \times 10^3$
$a_1 = 2.857\ 174\ 747\ 0 \times 10^1$
$a_2 = -8.488\ 510\ 478\ 5 \times 10^{-2}$
$a_3 = 1.578\ 528\ 016\ 4 \times 10^{-4}$
$a_4 = -1.683\ 534\ 486\ 4 \times 10^{-7}$
$a_5 = 1.110\ 979\ 401\ 3 \times 10^{-10}$
$a_6 = -4.451\ 543\ 103\ 3 \times 10^{-14}$
$a_7 = 9.897\ 564\ 082\ 1 \times 10^{-18}$
$a_8 = -9.379\ 133\ 028\ 9 \times 10^{-22}$

1664.5°C to 1768.1°C

$E = \sum_{i=0}^{n} a_i(t_{90})^i \, \mu V$

where (R):

$a_0 = 1.522\ 321\ 182\ 09 \times 10^5$
$a_1 = -2.688\ 198\ 885\ 45 \times 10^2$
$a_2 = 1.712\ 802\ 804\ 71 \times 10^{-1}$
$a_3 = -3.458\ 957\ 064\ 53 \times 10^{-5}$
$a_4 = -9.346\ 339\ 710\ 46 \times 10^{-12}$

1664.5°C to 1768.1°C

$E = \sum_{i=0}^{n} a_i(t_{90})^i \, \mu V$

where (S):

$a_0 = 1.466\ 282\ 326\ 36 \times 10^5$
$a_1 = -2.584\ 305\ 167\ 52 \times 10^2$
$a_2 = 1.636\ 935\ 746\ 41 \times 10^{-1}$
$a_3 = -3.304\ 390\ 469\ 87 \times 10^{-5}$
$a_4 = -9.432\ 236\ 906\ 12 \times 10^{-12}$

microprocessor data logger applications where either the temperature must be calculated from the measured voltage or it is necessary to determine to what voltage a particular temperature corresponds.

5.4 Thermocouple assemblies and installation

Temperature measurement can rarely be undertaken using bare thermocouple wire. Often the thermocouple wires must be electrically isolated from the application and protected from the environment. When measuring the temperature of a moving fluid it may also be necessary to locate the thermocouple within a specialized assembly to produce a measurement that can be related to the temperature of the flow (e.g. see Figure 2.11). A thermocouple assembly therefore involves consideration of joining the wires at the tip to form the thermoelectric junction, electrical isolation of the wires, protection of the wires, installation of the assembly into the application and connection of the thermocouple to the voltage-measuring device.

5.4.1 Bead formation

As indicated by the law of intermediate materials, the precise form of connection of the two thermoelements does not significantly affect the output provided the temperature of the bead formed is relatively uniform. Thermocouple beads can therefore be formed by twisting the end few millimetres of the wires together at the tip which provides some strength, crimping or welding as illustrated in Figure 5.11. The choice of connection depends upon the requirements of the application such as the need for strength or high speed of response. For applications requiring high strength of the bead or resistance to vibration then a welded bead may be most appropriate. The

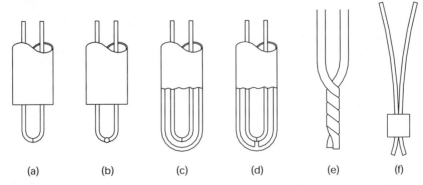

(a)   (b)   (c)   (d)   (e)   (f)

Figure 5.11 Bead formation options: (a) butt welded, (b) welded bead, (c) butt welded and sheathed, (d) grounded and sheathed, (e) twisted wires, (f) crimped

vast majority of commercial thermocouples are welded during manufacture and this need not be a concern for the bulk of users. If, however, thermocouples are being assembled from wire then the bead can be formed by using a discharge welding machine.

5.4.2 Insulation and protection

Many applications require the thermocouple wires to be electrically or chemically isolated from the environment or medium of interest. Examples of insulation materials include PVC (polyvinyl chloride) for temperatures between −30°C and 105°C, Teflon for −273°C to +250°C, glass fibre for −50°C to 400°C and polyimide for −269°C to 400°C. Higher temperatures can be achieved using ceramic sheaths. The requirements of electrical or chemical isolation and good thermal contact are often in conflict, giving rise to thermal disturbance errors. The range of typical options for insulation is illustrated in Figure 5.12.

In addition to the need to electrically isolate the thermocouple it is also necessary in some applications to protect the thermocouple from exposure to the local environment that could otherwise impair the function of the measurement device. An example is the immersion of a thermocouple into a corrosive fluid. Levels of protection can be achieved by use of a protection tube or sheath around the insulation. An example of a thermocouple assembly with a protecting tube and head are illustrated in Figure 5.13 along with some

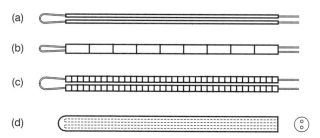

Figure 5.12 Typical forms of insulation. (a) PVC (polyvinyl chloride), polyimide or asbestos sleeving. (b) Double-bore insulators. (c) Fish-spine insulators. (d) Hard-fired ceramic double-bore insulator

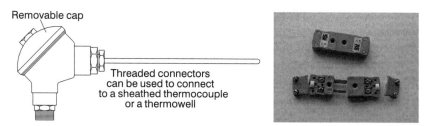

Removable cap

Threaded connectors can be used to connect to a sheathed thermocouple or a thermowell

Figure 5.13 Some protection tube and connector designs

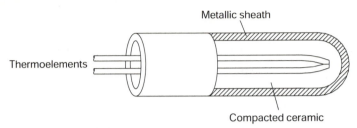

Figure 5.14 A MIMS (mineral insulated metal sheathed) thermocouple for high temperature capability and protection from the local environment

connectors. These tubes and connectors are commercially available. Choice of the insulator and sheath materials depends on the application. The ASTM manual on the use of thermocouples in temperature measurement (1993) lists a wide range of protecting tube materials for different applications. The MIMS (mineral insulated metal sheathed) thermocouple attempts to combine high temperature capability and protection from the environment in a single assembly. In these devices a mineral such as magnesium oxide is compacted around the thermocouple wires to electrically isolate and support them and this assembly is encapsulated within a metal sheath, (Figure 5.14). MIMS thermocouples can be used for the range –200°C to 1250°C.

5.4.3 Installation

Thermocouples are useful for determining the temperature for a wide range of applications from surfaces and solid bodies to stationary and flowing fluids. As a thermocouple consists simply of the thermoelectric wires it is usually necessary to insulate the wires and sometimes encapsulate the wires in a protective sheath. In addition, in order to avoid errors caused by heat transfer, it may be necessary to install the thermocouple within a specialized assembly to ensure appropriate measurement of the parameter of interest.

The temperature of a solid surface can be measured using a thermocouple by a variety of means depending on the type of surface and uncertainty desired. Figure 5.15 illustrates a range of possibilities. Each type of mounting has its merits and disadvantages and the choice is often limited by the uncertainty desired, practicality and costs. Stick-on thermocouples, (a), are probably the cheapest form of installation but can be easily dislodged and the tape and insulation will cause the temperature of the surface to be disturbed. If the temperature of the fluid is higher than the surface and the thermal conductivity of the tape is lower than that of the undisturbed surface then the measured temperature will be elevated in comparison to the temperature of the surface in the undisturbed condition. The use of thin metal shims to hold the thermocouple onto the surface, (b), is useful if the surface experiences high g forces as on a rotating component in a gas turbine engine. Machining a recess

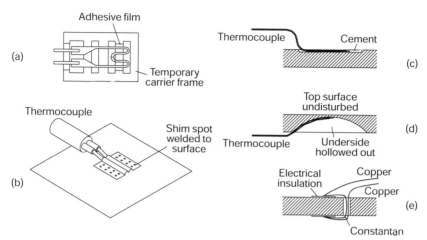

Figure 5.15 Installation of surface thermocouples. (a) Stick-on thermocouple. (b) Shim-welded thermocouple. (c) Flush-mounted thermocouple. (d) Cavity-mounted thermocouple. (e) Surface-mounted thermocouple (after Smout, 1996)

in the surface so that the thermocouple is flush with the surface, (c), avoids disturbances to the boundary layer flow but is expensive in terms of extra machining time. In addition, cement must be used to hold the wire within the channel or, alternatively, the surface can be peened over so that the wire is encapsulated within the channel. The material under the surface can be removed as illustrated in (d) and the thermocouple installed so that it is just under the surface but not exposed to the fluid. This can be useful in monitoring the temperature of a chemically aggressive fluid. The surface-mounted thermocouple, (e), with some form of flying lead, whilst cheap, is susceptible to errors induced by conduction along the cable.

Solid temperatures and the temperature of a liquid flowing in a pipe are sometimes measured with the thermocouple immersed in a thermowell. A thermowell is a protecting tube to prevent or minimize damage from harmful atmospheres, corrosive fluids or mechanical damage and common commercial forms are illustrated in Figure 5.16. As a general rule of thumb a thermowell should be immersed in the liquid to a depth of ten times the diameter. Material property requirements for thermowells are documented in ASME PTC 19.3 (1974).

A thermocouple placed in a gas environment will experience heat transfer by conduction along its wires and support, convective heat transfer with the surrounding gas both at the tip and along the length of the wires and support, and radiative heat exchange with the surroundings. The contribution due to conduction is usually relatively small in comparison to convection. Radiative heat transfer tends to be a function of the fourth power of the absolute temperature and can therefore be significant if the temperature of the gas is

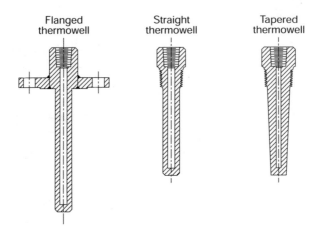

Figure 5.16 Thermowells (ASTM, 1993)

distinctly different from the temperature of the enclosure. Because of these exchanges of heat the measurement of gas temperature needs to be carefully thought through and undertaken. The principles of measuring the temperature of a fluid in motion are considered in Chapter 2, Section 2.2.3. A gas temperature measurement, for example, may require the use of a stagnation probe and radiation shield and interpretation of the results using a dynamic correction factor.

5.4.4 Extension leads

In some applications the measurement location and readout instrumentation are separated by a considerable length. An example is the measurement of temperatures of various different components in an engine on test with, say, a hundred thermocouples all connected to a data-acquisition system some 30 m away. In most applications the thermal gradient and resultant emf is most significant in the first few metres of wire. The remaining wire length serves the purpose of transmitting the emf to the data-acquisition system. The thermoelectric properties of this length of wire are less critical than the part in the region of large thermal gradient. The drive for economy can make the use of a lower specification and hence cheaper wire for this region of wire attractive and these lengths of wire are known as extension leads or extension wires. There are two types of extension wire: those with similar physical composition to the thermocouple wire itself but manufactured to a less stringent specification and those manufactured using a different material altogether. Some people use the term 'extension leads' for wires manufactured using the same material as the thermocouple and the term 'compensation lead' for wires manufactured using different materials. However, this terminology

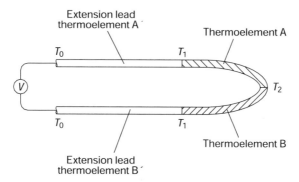

Figure 5.17 The use of extension leads in a thermocouple circuit

is not universal and the terms extension and compensation lead, wire and cable are used interchangeably. The use of extension leads in a thermocouple circuit is illustrated in Figure 5.17.

For the configuration shown in Figure 5.17 loop analysis can again be used to analyse the output. If the Seebeck coefficients of the extension leads are $S_{A'}$ and $S_{B'}$ respectively, then,

$$E = S_{A'}(T_1 - T_0) + S_A(T_2 - T_1) + S_B(T_1 - T_2) + S_{B'}(T_0 - T_1) \quad (5.14)$$

This simplifies to

$$E = S_{A'B'}(T_1 - T_0) + S_{AB}(T_2 - T_1) \quad (5.15)$$

If the wire pair $A'B'$ is selected so that it has approximately the same relative Seebeck coefficient as AB then

$$S_{A'B'} \approx S_{AB} \quad (5.16)$$

and

$$E \sim S_{AB}(T_2 - T_0) \quad (5.17)$$

Therefore the reference junction is effectively located at the connections between the voltmeter or data logger and the extension leads.

Example 5.4

A type K thermocouple with extension leads indicates a voltage of 4.233 mV. The temperature of the terminus connections (the reference junction) was measured to be 15.2°C using a thermistor. Determine the temperature of the thermoelectric junction.

Solution

The reference junction temperature is 15.2°C. Examination of Table 5.8 gives the nearest temperatures to this value as 10°C and 20°C with corresponding emfs of 397 μV and 798 μV respectively.

The emf equivalent to the temperature of the reference junction can be found by

$$E|_{15.2°C} = 397 + (798 - 397) \left(\frac{15.2 - 10}{20 - 10} \right) = 605.5 \, \mu V$$

The total emf is given by

$$E_{total} = E_{thermocouple} + E_{equivalent} = 4233 + 606 = 4831 \, \mu V$$

From Table 5.8 the nearest emfs listed are 4509 μV at 110°C and 4920 μV at 120°C. Using linear interpolation gives the temperature as

$$T = 110 + (120 - 110) \left(\frac{4831 - 4509}{4920 - 4509} \right) = 118°C$$

Extension wires using the same material as the thermocouple are widely used for E, J, K and T thermocouples. Compensation cables using a different wire from that of the thermocouple tend to be used for R, S and B thermocouples. The use of extension leads instead of compensation cable for type E, J, K and T thermocouples is recommended as any cost advantage gained by the use of cheaper cable is usually outweighed by the improvements gained in uncertainty (Kerlin, 1999). This is because the materials used in compensation cable (normally copper and constantan) generate emf and hence cause errors in the reading; these errors can be as high as 5°C if the cable is at 100°C and 29°C if it is at 200°C. These errors show the importance of using the correct type of extension for the application and if compensation cables are used to ensure that the thermal gradient across them is small.

Polarity of extension and compensation leads is important. Normally the contribution of extension leads to the overall emf is small. As a result, incorrect connection of the polarity of the leads can initially go unnoticed. It will, however, cause an error in the measurement. For example, if the polarity of the extension leads for a type K thermocouple with the bead temperature 20°C above the cold junction temperature are incorrectly connected the thermocouple assembly could read 40°C too low. Confirmation that the extension leads have been correctly attached can be achieved by two checks with reference to Figure 5.18.

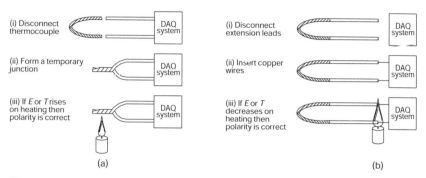

Figure 5.18 Checking polarity of extension leads. (a) Disconnect thermoelements and twist ends of extension leads together. If on heating *T* or *E* rises then polarity is correct. (b) Insert copper leads between extension leads and the logger. If on heating *T* or *E* decreases then polarity is correct

1 Disconnect the extension leads from the thermocouple but keep the connections to the data logger. Form a temporary junction with the loose extension lead ends by twisting them together. Heat this junction and if the instrument indicates an increase in emf or temperature shown then the polarity is correct.

2 If the above method is impracticable then a test at the cold junction can be undertaken. Disconnect the extension leads from the data logger and insert two fine-gauge copper wires so that the temperature of the new cold junction (at the join between the copper leads and the extension leads) can be easily controlled by heating. This can be achieved using heat from fingers or a hot-air gun. Heat the copper extension wire junctions simultaneously and if the polarity is correct the logger will indicate a fall in temperature. Conversely, if the junctions are simultaneously cooled by, say, touching onto an ice cube the data logger will indicate a rise in temperature if the polarity is correct.

5.5 EMF measurement

Most thermocouple measurement systems do not involve the user in the process of converting the emf generated by a thermocouple into an indication of temperature. If a thermocouple and associated panel meter are purchased together as illustrated in Figure 5.2 then all the user needs to do is connect the thermocouple and switch on the device. The process becomes a little more involved when using multiple thermocouples and connecting them to a data logger. Here care needs to be taken to avoid extraneous signals and to ensure that the cold junction temperature has been correctly accounted for within the system. Some thermocouple users however may choose to measure the emf directly. There are generally three steps involved in converting the emf generated by a thermocouple into a measure of voltage:

1 measure the emf developed by the thermocouple
2 establish the cold junction temperature and convert this into an equivalent emf
3 convert the combination of these emfs into a value of temperature.

The emf developed by a thermocouple is independent of its length and diameter; it is only dependent on the temperature difference across the thermocouple. In order to avoid circuit losses the emf generated should be measured by a device with a high-impedance input stage. The most convenient instrument is usually a digital voltmeter or data logger card with a thermocouple cartridge. Digital voltmeters are commonly available with a resolution of $1\,\mu V$ or better and a stability of better than 0.01%, and this usually results in an uncertainty less than that of the calibration. The reason a digital voltmeter with a resolution of $1\,\mu V$ is adequate can be seen from an examination of the sensitivity this provides. A type K thermocouple, for example, gives an output of $4096\,\mu V$ at 100°C. Therefore, $1\,\mu V \equiv 100/4096 = 0.024°C$ and thus the best possible uncertainty that can be achieved using a voltmeter with a resolution of $1\,\mu V$ is ±0.012°C. This figure is perfectly suitable for the vast majority of applications.

Example 5.5

A handheld thermometer for use with Class 1 type K thermocouples is required with an uncertainty of better than ±2°C at a temperature of 300°C. One device considered uses ice-point reference junction compensation with a stated uncertainty of 0.5°C and a resolution of 0.1°C. Determine whether the thermometer will provide the desired level of uncertainty.

Solution

The resolution is 0.1°C. The corresponding error associated with this can be taken as one half of the stated resolution. The bilateral tolerance would therefore be ±(0.1/2) = ±0.05°C. The corresponding standard deviation, taking with a 95% coverage (±2σ) would be

$$\sigma_{resolution} = 0.05/2 = 0.025°C$$

The standard deviation to account for the uncertainty of the reference junction compensation, taking a 95% percentage coverage (±2σ) is given by

$$\sigma_{compensation} = 0.5/2 = 0.25°C$$

The uncertainty associated with the indicator can be combined as

$$\sigma_{indicator} = \sqrt{\sigma^2_{stated\ uncertainty} + \sigma^2_{resolution}} = \sqrt{0.25^2 + 0.025^2} = 0.251°C$$

As the thermocouple is a Class 1 device the uncertainty associated with the tolerance of the thermocouple, ±1.5°C for a Class 1 type K thermocouple up to 375°C, must also be accounted for. The standard deviation to account for the uncertainty of the thermomocouple, taking a 95% percentage coverage (±2σ) is given by

$$\sigma_{thermocouple} = 1.5/2 = 0.75°C$$

The overall standard deviation for the thermometer is a function of both the thermocouple and the indicator uncertainty and is given by

$$\sigma_{thermometer} = \sqrt{\sigma^2_{thermocouple} + \sigma^2_{indicator}} = \sqrt{0.75^2 + 0.251^2} = 0.791°C$$

The bilateral tolerance using a 95% coverage factor would be ±2σ = ±(2×0.791) = ±1.58°C. This is within the level of desired uncertainty, so the thermometer combination is acceptable.

5.6 Grounding and noise

The signal from a thermocouple is relatively small: 0.798 mV at 20°C for a type K thermocouple. Because the signal is so small thermocouples are sensitive to errors caused by noise. Noise can be defined as any unwanted alteration of the signal and can appear due to drift, internally generated random signals and signals picked up from external sources.

Drift can occur due to changes in the gain of dc amplifiers caused by variations in the power supply and temperature of the amplifier. It can also occur due to ageing of components within an amplifier. The effects of drift are minimized by use of stabilized power supplies and maintaining any circuitry at a constant temperature.

Internally generated random noise occurs due to two main sources: thermal and shot noise. Thermal noise arises from electron movements due to the thermal vibration of the atoms within a conductor and is classed as white noise, which means that it is made up of all frequencies. The rms voltage of thermal noise is proportional to the square root of the absolute temperature of the conductor. Shot noise arises from random fluctuations in the passage of charge carriers through semiconductors and is also a white noise. Usually the signal-to-noise ratio is high enough to make the effect of thermal and shot noise insignificant.

Noise can be picked up from external sources by capacitance from electrostatic fields and induction from electromagnetic fields. This can be reduced by screening or by positioning instrumentation cables away from sources of noise such as 50 or 60 Hz power cables, electric motors, generators, solenoids, switches and switchgear. Screening involves enclosing cables and amplifiers and any other instrumentation circuitry in metal that is earthed so that voltages picked up by the screen are short-circuited to earth. In order to screen against capacitative noise the material used for the screen needs to be a good conductor, whilst in order to screen against inductive noise the screen must be made of iron or some other magnetic material. To effectively guard against magnetic fields the screen must be thick enough to dissipate the magnetic field. In practice a thickness of 2 or 3 mm may be necessary to screen against 50 or 60 Hz magnetic fields. Figure 5.19 illustrates the use of a screen for a cable consisting of two thermoelements twisted together. Twisting of the instrumentation cables serves the purpose of reducing the inductive pick-up as the wires pick up nearly identical fields, which effectively cancel out within each loop because each time the wire is twisted the flux-induced current is inverted. For bare wire thermocouples in insulated sleeves the traditional method for twisting the wires is to clamp one end of the pair and insert the other ends into the chuck of a drill, turn the drill and thereby twist the wires. The loose ends can then be welded to form the thermoelectric junction at one end.

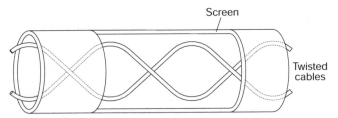

Figure 5.19 Screening of thermocouples

If connections to earth are made at more than one point in a circuit it is possible to generate unwittingly large sources of noise due to earth loops. Earth loops can become the source of noise by two mechanisms. If current is flowing through the material to which the earth is connected then some of the current will flow through the screening and could transmit noise by capacitance to the instrumentation. The earth loop can act as a winding within a magnetic field with currents induced in it by the alternating magnetic field of an external noise source. The noise is transmitted to the instrumentation cables by capacitive pick-up. It is therefore important to carefully site connections to earth to avoid or minimize these effects (Benedict and Russo, 1972).

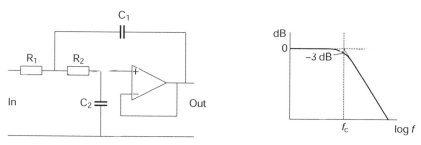

Figure 5.20 Simple active low-pass filter for excluding high-frequency noise

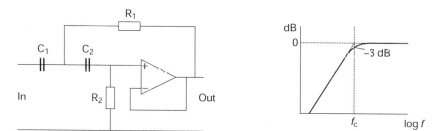

Figure 5.21 Simple active high-pass filter for excluding low-frequency noise

The noise on a thermocouple signal can be excluded by use of filters. A low-pass filter, (Figure 5.20) can be used to exclude high frequencies, while a high-pass filter, (Figure 5.21), can be used to exclude low-frequency noise. A combination of these can be used to exclude low- and high-frequency noise.

As a general principle, an amplification system for a thermocouple should be designed so that the arms of the circuit are balanced and any noise induced in one arm is equal to that in the other. If the circuit is then connected to a differential amplifier, induced emfs are equal in each branch and do not affect the output of the differential amplifier which indicates the emf output of the thermocouple. If the branches are not balanced, then elaborate circuitry is necessary to redress the balance; see, for example, Morrison (1998).

5.7 Calibration

Calibration of thermocouples is sometimes necessary in order to confirm or establish the emf-to-temperature relationship for the device concerned. Not all thermocouples are calibrated nor do they necessarily need to be calibrated. Use of standardized thermocouples allows the user the option of relying on the tolerances stated by the manufacturer and use of standard tables to convert the emf generated into a temperature.

Calibration can be taken to mean the establishment of the relationship between the transducer output and the transducer temperature within a tolerance or band of uncertainty. Calibrations can be considered to fall into one of four categories:

1 acceptance tests
2 batch calibration
3 calibration of a single thermocouple for a fixed application
4 calibration of reference standards thermocouples.

Figure 5.22 A large-volume liquid calibration bath enabling simultaneous calibration of several devices with a temperature range of 5–200°C using water to 90°C and silicon oil to 200°C. Photograph courtesy of Isothermal Technology Ltd

There are a number of ways of calibrating thermocouples. The emf of the thermocouple can be determined at a relatively small number of fixed points and interpolation performed using agreed formulae or by considering the difference from a standard table. An alternative method is to compare the emf of the thermocouple being calibrated with that of the same type of standard thermocouple for a large number of temperatures and then fit a curve or algebraic relationship to the emf versus temperature data using a least squares method.

For non-standard thermocouples a calibration can be undertaken against another thermocouple or another type of sensor that has itself already been

Figure 5.23 An oil or salt calibration bath for calibration of devices between 150°C and 600°C. Photograph courtesy of Isothermal Technology Ltd. Note that salt baths are generally considered hazardous due to nitrate emissions and splash burns

calibrated. Ideally a calibration should be traceable to the ITS-90. This can be achieved by a hierarchy of calibration activity between the end user and the national standards laboratory or accredited service. Calibration should be undertaken to a specified uncertainty and within a range of temperatures relevant to the application. There is no point calibrating a device to an uncertainty of ±0.001°C if the application only requires an uncertainty of ±1°C. Similarly, calibrating a thermocouple to its maximum possible temperature is pointless if the application only requires measurement at a fraction of this value.

For calibration by comparison methods, a liquid bath can be used up to 600°C (Figures 5.22 and 5.23) and a furnace, for higher temperatures up to a limit of about 1800°C (Figure 5.24). Fully automatic systems are commercially available that control a heater across a programmable temperature range and allow measurements to be taken from a calibrated temperature sensor and the thermocouple to be tested and provides the facility for the data to be logged in a data file (Figure 5.25).

A specialized method of calibrating thermocouples makes use of the melting point of certain metals such as gold, palladium and platinum. The temperature at which these metals melt in their pure form is known with low uncertainty. A piece of one of these metals is used to form the junction

Figure 5.24 Thermocouple calibration furnace with a temperature range between 100°C and 1300°C. Photograph courtesy of Isothermal Technology Ltd

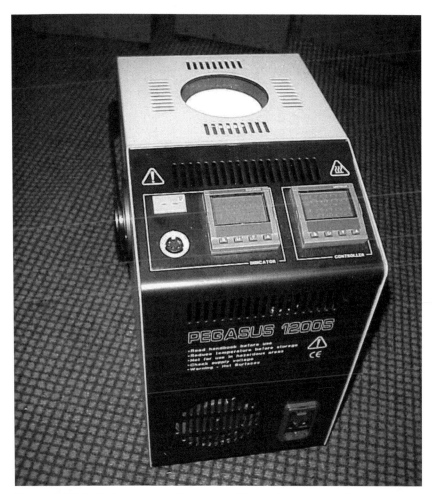

Figure 5.25 Digitally controlled furnace calibrator for operation in the range 150°C to 1200°C. Photograph courtesy of Isothermal Technology Ltd

between the two thermocouple wires. When this piece of material melts a discontinuity in the emf occurs and this indicates that the melting point temperature has been reached and hence provides a calibration point. This method is sometimes used for type R, S and B thermocouples.

5.8 Troubleshooting/thermocouple diagnostics

If a measurement is worth taking, and it is always sensible to question whether it makes any difference if it is taken, then it is necessary to ensure that the measurement process is undertaken to the quality required. Thermocouples can malfunction and produce unreliable results. It may however not be obvious without a bit of investigation that a thermocouple is not functioning correctly but a number of checks can be made to ensure confidence in the measurement or to identify that the thermocouple needs to be replaced.

Common occurrences in the use of thermocouples include:

- change in the loop resistance
- change in the insulation resistance
- sudden steps in open current voltage
- fluctuations.

The loop resistance of a thermocouple depends on the length, type and diameter of the thermoelement wires, the length, type and diameter of the extension wires, temperatures along the circuit and the contact resistance at any connections. If on installation, and at regular intervals in use, a measurement is made of the loop resistance then a change in this value can be used to indicate wire thinning due to chemical attack, loose or corroded connections, increased contact resistance due to broken but touching wires or electrical shunting due to loss of insulation at some location along the wire as illustrated in Figure 5.26.

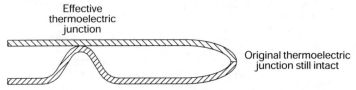

Figure 5.26 Electrical shunting of a thermocouple due to loss of insulation

A resistance measurement of the thermocouple should be undertaken by averaging the results of two measurements, one with the wires of the resistance-measuring device connected one way round and then one where they are reversed. This serves to nullify the effect of any thermoelectric emf on the resistance measurement.

The resistance of insulation can decrease if the insulation is contaminated by, for example, moisture. This causes shunting of the electrical signal and degradation of the thermocouple performance and resultant difficulties in assessing what the signal is actually indicating. A well-insulated thermocouple should have an insulation resistance of at least $100\,\mathrm{M}\Omega$ at room temperature. This can be checked by measuring the resistance using a megohmeter between the thermoelement wire and some point that is electrically common to the structure in which the thermocouple is installed.

Sudden steps in a sensor output provide a diagnostic warning to a user. This can indicate electrical shorting, possibly due to contamination of the insulation by moisture or inhomogenity in a thermoelement. If during a heat transfer process a section of inhomogeneous wire experiences a sudden thermal gradient, this can cause a shift in the thermoelectric output, even if the thermoelectric junction is at a constant temperature. Monitoring of the emf can reveal these problems.

Fluctuations in the thermoelectric emf at a frequency dissimilar to the expected rate of change of temperature can give an indication that a thermocouple wire is broken and only providing a signal by intermittent contact of the broken junction. Such contact will be closely related to any vibrational frequencies experienced in the application. Rapid changes in output can be detected by use of an oscilloscope, high-speed data logger or spectrum analyser.

5.9 Selection

As stated, there is a large choice of thermocouple materials, possible forms of insulation and various methods of installation. The choice of thermocouple types can be significantly reduced if you wish to restrict the choice to standardized thermocouples that are widely available from a variety of thermocouple manufacturers. Factors that should be considered when selecting a thermocouple for a given application include:

- performance
- desired level of uncertainty
- operating temperature and temperature range
- life
- cost
- speed of response
- stability
- size
- compatibility
- environment
- installation
- emf measurement.

The principal decisions that must be made in specifying a thermocouple include: the thermocouple type, wire exposure (sheathed or bare), isolated or grounded, wire diameter for bare thermocouples, sheath diameter for sheathed thermocouples, thermocouple assembly diameter, tip formation, sensor length, type of connectors and method of cold junction compensation. Once the type of thermocouple has been identified the specification can be determined. Figure 5.27 provides a guide in the selection of a thermocouple and the various criteria listed above are described in more detail in Sections 5.9.1 to 5.9.8.

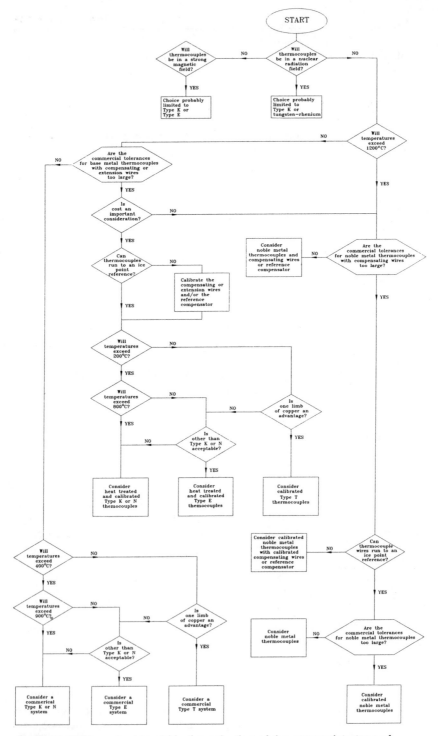

Figure 5.27 Flow chart to aid in the selection of the appropriate type of thermocouple to use for a given application. After ESDU 82035 (1982)

5.9.1 Performance

The general criterion of performance includes consideration of signal output, sensitivity, maximum and minimum operating temperature capability and temperature range capability, uncertainty, speed of response, ability to operate in the environment of the specific application (compatibility) and chemical and physical stability. An indication of the signal levels for the standard thermocouples is given in Figure 5.10 and Tables 5.5–5.12. The maximum and minimum temperature ranges for each of the standard thermocouples are listed in Table 5.1. The sensitivity of the standard thermocouples can be identified over the specific range of temperature of interest by calculating the ratio of emf/temperature using the data in Tables 5.5–5.12. Considerations of uncertainty, speed of response, stability and compatibility are considered in Sections 5.9.4, 5.9.5, 5.9.6 and 5.9.7 respectively.

5.9.2 Costs

The relative costs of different thermocouples is best identified by a survey of manufacturers' prices. However, the cost of a measurement using a thermocouple does not just consist of the thermocouple wire. The cost of extension leads, special requirements for installation, the emf measurement device or data-acquisition system, need for periodic calibration and cost of replacement must also be assessed. Generally, high-temperature thermocouples such as W and W–Rh cost five to ten times as much as base thermocouple wire. Sheath material and insulation contribute significantly to this cost. The cost of R, S and B type thermocouples fluctuates with the price of platinum. It is worth considering and costing an alternative to a thermocouple for a given application to identify that the most appropriate sensor has been identified. Alternatives typically include thermistors and PRTs. The relative cost depends, of course, on the sensor package but as an indication the relative cost to a thermocouple for a $\frac{1}{4}$-inch (6.25 mm) outside diameter sheathed device is two or three times for a PRT and half for a thermistor, although the thermistor is likely to have a temperature range limited to −80°C to 150°C.

The type of connection wire used for a transducer can dominate the total cost. PRTs can be connected to a data-acquisition system using copper wires. Thermocouples, however, require compensation cable, which can be five times as expensive as copper cable. An alternative to the use of compensation cable is to connect the thermocouple wires to copper within a controlled environment container, known as an isothermal box, close to the application. The aim of the isothermal box is to maintain the temperature of the connections at a uniform temperature and acts as the cold junction for the thermocouple. The temperature of the isothermal box can be measured by an independent device, such as a thermistor. This arrangement, however, adds complexity to the measurement system and its cost−benefit may therefore not be so straightforward.

5.9.3 Life

The specification for service life can range from a few seconds, compatible with measuring the temperature in a combustion application, to as long as possible. The service life of a thermocouple can be defined as the time before it experiences catastrophic failure, for instance it disintegrates, or its degradation causes unacceptable uncertainty. In order to maximize service life for bare thermocouples the materials chosen must be compatible with their environment. A common failure is wire tip breakage caused by chemical attack or thermal stress. For these reasons the largest convenient diameter wires should be chosen but this must be balanced by consideration of the speed of response; the larger the diameter, the slower the speed of response.

5.9.4 Uncertainty

The uncertainty tolerances for thermocouples supplied by manufacturers to international standards are listed in Tables 5.2 and 5.3. Generally the measurement uncertainty for new thermocouples is between 1°C and 13°C. These figures will deteriorate as the thermocouples age. The lowest uncertainties are obtained with the noble metal thermocouples. Careful annealing can be performed to ensure homogenity along the length of the wire and individual calibration of each thermocouple undertaken to minimize errors in new thermocouples.

5.9.5 Response

How quickly a thermocouple responds to a change of temperature in the surrounding medium depends on the mode of heat transfer, the nature of the temperature change, the size of the thermocouple assembly and the materials used. Generally the speed of response will be faster as the size of the thermocouple is reduced. The methods described in Chapter 2, Section 2.2.4, can be used to analyse many thermocouple installations modelling the thermocouple assembly as a cylinder or bead.

5.9.6 Stability

Modification of the thermoelectric characteristics for a thermocouple can occur due to changes in the composition of the materials used and in the physical structure. The composition and purity of thermoelements is carefully controlled by manufacturers to ensure that the voltage–temperature characteristic conforms to the tolerance specified. This applies to the device when it is supplied but the use of a thermocouple in an application can cause this to change. For example, platinum can combine with a wide range of metals and contaminants such as oils at high temperatures or even oxidize. Base metal thermoelements can be contaminated by sulphur and oil. Nickel chromium thermoelements are susceptible to contamination in conditions where the supply of oxygen is limited and the chromium content is preferentially oxidized causing a reduction in output. Copper and iron will oxidize significantly at temperatures above 400°C and 800°C respectively.

Some thermocouples exhibit a hysteresis affect where the output on cycling around a temperature varies depending on whether the thermocouple temperature is rising or falling. Type K thermocouples suffer from hysteresis particularly in the temperature range above 400°C. The type N thermocouple was developed to minimize these effects (see Burley, 1971; Burley et al., 1978) and some organizations have set up strategies to replace existing type K thermocouples by type N.

The physical structure or shape of a thermocouple can be altered by plastic strain induced by cold working (Pollock and Finch, 1962). Cold working can inadvertently be introduced by coiling, twisting, stretching and general manipulation of the thermoelements. Thermocouples should be operated in a fully annealed condition and they are normally supplied in this form. The effects of cold work on base thermoelements is generally small and recovery from cold work can be achieved by heating materials such as Ni–Cr, Cu–Ni and Ni–Al at 450°C for a few hours. Ni–Cr and Ni–Al can be fully annealed by heating at about 950°C for 15 minutes and cooling slowly. Pt thermoelements can be annealed by heating for a few minutes at around 1300°C and Ir thermoelements must be heated to 1750°C for approximately five hours (ESDU, 1982).

The effects of nuclear radiation can alter the composition of thermo-elements. Nickel-based thermoelements are the least affected (Browning and Miller, 1962; Carpenter et al., 1972) and type K thermocouples are regularly used in nuclear reactors.

The theory of thermoelectricity indicates that the emf produced in a thermocouple is dependent on pressure because of contractions in the lattice structure of the material. The effect, however, appears to be small. Getting and Kennedy (1970) reported a 160 μV correction at 1000°C for a Pt 105 Rh/Pt thermocouple experiencing 30 kbar.

5.9.7 Size, environment, compatibility and installation

Once a type of thermocouple has been selected its dimensions must be selected. For a bare thermocouple this involves specification of the diameter of the thermoelements and selection of appropriate insulation. Alternatively, the overall diameter for sheathed and insulated thermocouples must be considered. Generally the robustness of a thermocouple will increase as its diameter increases but the speed of response will decrease and there must therefore be some compromise between these two criteria. Thermoelements must be chemically compatible with the insulation selected to avoid any chemical reaction over the temperature range for the application and the thermocouple assembly must not react with the application. Guidance on the selection of appropriate insulation materials is provided in Section 5.4.2. Installation may involve taping or welding the thermocouple to the surface or mounting the thermocouple in a special assembly. The considerations for installation are described in Section 5.4.3.

5.9.8 EMF measurement

A number of options are possible for the measurement of the thermoelectric emf ranging from use of a voltmeter to data logging systems. For a single thermocouple a handheld commercial panel meter type display may well be appropriate, but for multiple measurements a dedicated data logger card connected to a PC provides good voltage measurement uncertainty and data recording capability. Such systems are available from £70 ($100) for an eight-channel system. The price increases significantly with sophistication and a 100-channel system with the capacity to log data at up to 500 kHz can cost £14 300 ($20 000) (year 2000 prices).

References

Books and papers

Barnard, R.D. *Thermoelectricity in Metals and Alloys*. Taylor and Francis, 1972.

Bedwell D.R. *Introduction to Thermocouple Measurements*. In Temperature Measurements, von Karman Institute for Fluid Mechanics, Lecture Series 1996–07, 1996.

Benedict R.P. and Russo R.J. A note on grounded thermocouple circuits. *Journal of Basic Engineering*, 337–380, 1972.

Bentley, R.E. *Handbook of Temperature Measurement*. Vol. 3, *Theory and Practice of Thermoelectric Thermometry*. Springer, 1998.

Blatt, F.J., Schroeder, P.A., Foiles, C.L. and Greig, D. *Thermoelectric Power in Metals*. Plenum, 1976.

Bourassa, R.R., Wang, S.Y. and Lengeler, B. Energy dependence of the Fermi surface and thermoelectric power of the noble metals. *Phys. Rev. B*, **18**, 1533–1536, 1978.

Browning, W.E. and Miller, C.E. Calculated radiation induced changes in thermocouple composition. In Herzfield C.M. (Editor), *Temperature. Its Measurement and Control in Science and Industry*, Vol. 3, Part 2 pp. 271–276. Reinhold, 1962.

Burley N.A. Nicrosil and Nisil. Highly stable nickel base alloys for thermocouples. Fifth Symposium on Temperature, Washington DC, Paper T-4, pp. 1677–1695, 1971.

Burley, N.A., Powell, R.L., Burns, G.W. and Scroger, M.G. *The Nicrosil versus Nisil Thermocouple: Properties and Thermoelectric Reference Data*. National Bureau of Standards (US) Monograph 161, 1978.

Carpenter, F.D., Sandefur, N.L., Grenda, R.J. and Steibel, J.S. EMF stability of chromel/alumel and tungsten-3% rhenium/tungsten-25% rhenium sheathed thermocouples in neutron environment. In Plumb, H.H. (Editor), *Temperature. Its*

Measurement and Control in Science and Industry, Vol. 4, Part 1, pp. 1927–1934. ISA, 1972.

Childs, P.R.N., Greenwood, J.R. and Long, C.A. Review of Temperature measurement. *Review of Scientific Instruments*, **71**, 2959–2978, 2000.

Doebelin, E. O. *Measurement Systems: Application and design*, 4th edition. McGraw-Hill, 1990.

ESDU, Engineering Sciences Data Unit. ESDU 84035. *Temperature Measurement: Thermocouples*, 1982.

Finch D.I. General principles of thermoelectric thermometry. In Herzfield C.M. (Editor), *Temperature. Its Measurement and Control in Science and Industry*, Vol. 3, Part 2, pp. 3–32. Reinhold, 1962.

Getting, I.C. and Kennedy, G.C. The effect of pressure on the emf of chromel–alumel and platinum–platinum 10% rhodium thermocouples. *J. Appl. Phys.*, **41**, 4552–4562, 1970.

Guildner L.A. and Burns G.W. Accurate thermocouple thermometry. *High Temperatures and Pressures*, **11**, 173–192, 1979.

Kerlin, T.W. and Shepard, R.L. *Industrial Temperature Measurement*. Instrument Society of America, 1982.

Kerlin, T.W. *Practical Thermocouple Thermometry*. ISA, 1999.

Kinzie, P.A. *Thermocouple Temperature Measurement*. Wiley-Interscience, 1973.

Kittel, C. *Introduction to Solid State Physics*, 6th edition. Wiley, 1986.

Morrison, R. *Grounding and Shielding Techniques*, 4th Edition, Wiley, 1998.

Mott, N.F. and Jones, H. *Theory and Properties of Metals and Alloys*. Dover, 1958.

Pollock, D.D. *Thermocouples: Theory and properties*. CRC Press, 1991.

Pollock, D.D. and Finch, D.I. The effect of cold working upon thermoelements. In Herzfield C.M. (Editor), *Temperature. Its Measurement and Control in Science and Industry*, Vol. 3, Part 2, pp. 237–241, Reinhold, 1962.

Quinn, T.J. *Temperature*, 2nd Edition. Academic Press, 1990.

Roeser, W.F. Thermoelectric thermometry. *Journal of Applied Physics*, **11**, 388–407, 1940.

Seebeck, T.J. Evidence of the thermal current of the combination Bi–Cu by its action on magnetic needle. Royal Academy of Science, Berlin, p. 265, 1823.

Smout, P.D. Probe installation and design criteria. In *Temperature Measurements*, von Karman Institute for Fluid Dynamics, Lecture Series 1996–07, 1996.

Standards

ASME PTC 19.3. Performance test codes supplement on instruments and apparatus Part 3 Temperature measurement. ASME 1974.

ASTM E230–98. Standard specification and temperature-electromotive force (emf) tables for standardised thermocouples. American Society for Testing and Materials, 1999.

ASTM, American Society for Testing and Materials. Manual on the use of thermocouples in temperature measurement. 4th edition. ASTM PCN 28–012093–04, 1993.

BS EN 60584: 1996. Thermocouples. Part 1: Reference tables. Part 2: Tolerances.

IEC 584–1: Thermocouples – Part 1: Reference tables. Second edition 1995–09, International Electrotechnical Commission.

Web sites

At the time of going to press the world wide web contained useful information relating to this chapter at the following sites.

http://clay.justnet.com/toolbox/tcouple.htm
http://instserv.com/orphn/rmocoupl.htm
http://wilkes1.wilkes.edu/~varora/PROJECT/table1.htm
http://www.axiomatic.com/thermo_b.html
http://www.biccthermoheat.co.uk/temp/index.shtml
http://www.bubthermo.com/
http://www.ceramics.com/vesuvius/
http://www.cgindustrial.com/hart9112.htm
http://www.cstl.nist.gov/div836/836.05/greenbal/thermo.html
http://www.engr.orst.edu/~aristopo/temper.html
http://www.exhaustgas.com/product.htm
http://www.hartscientific.com/products/5629.htm
http://www.hoskinsmfgco.com/main/pr_thermo.htm
http://www.inotek.com/Catalog/pyromation2th.html
http://www.iotechsensors.com/
http://www.isotech.co.uk/
http://www.its.org/
http://www.jimnel.com/thermo.html
http://www.mcgoff-bethune.com/sensors/sens-bar.htm
http://www.microlink.co.uk/isothermal.html
http://www.nist.gov/srd/webguide/nist60/60down.htm#EMF VERSUS
    TEMPERATURE TABLES
http://www.omega.com/techref/themointro.html
http://www.pyromation.com/products/tc.html
http://www.sanjac.ttnet.net/
http://www.temperatures.com/tcread.html
http://www.wahlco.com/
http://www.wici.com/technote/tmprmch1.htm#thermometers
http://www.windmillsoft.com/acatalog/593.html

Nomenclature

E = thermoelectric emf (μV)
S = Seebeck coefficient (μV/K)
S_A = the Seebeck coefficient for material A (μV/K)
S_B = the Seebeck coefficient for material B (μV/K)
S_{Apt} = Seebeck coefficient for material A relative to platinum (μV/K)
S_{BPt} = Seebeck coefficient for material B relative to platinum (μV/K)
T = temperature (°C or K)
ΔT = temperature difference (°C)

6
Resistance temperature detectors

A variety of devices is available based on the variation of resistance with temperature, including platinum resistance thermometers, thermistors and semiconductor-based sensors. They are used in a wide range of applications. This chapter describes the fundamental principles of their operation and their practical use.

6.1 Introduction

Resistance temperature detectors (RTDs) exploit the variation in electrical resistance of conductors and semiconductors with temperature. Any conductor could in theory be used for a resistance temperature detector. Practical considerations, however, such as cost, electrical resistivity (a large value leads to a more sensitive instrument), ability to resist oxidation and manufacturing constraints limit the choice. Copper, nickel and platinum are the most widely used for metal-based transducers. Copper is sometimes used for the range $-100°C$ to $100°C$ and is relatively cheap. Nickel and its alloys are also relatively low in cost, have high resistivities and values of temperature coefficient of resistance. However, the variation in electrical resistance with temperature is non-linear and sensitive to strain. The resistivity of platinum ($\approx 10\,\mu\Omega/cm$) is six times that of copper, platinum is relatively unreactive, has a high melting point ($1769°C$) and, because it has a well-established temperature coefficient of resistance, it is a common choice for the precise measurement of temperatures between $-260°C$ and $1000°C$. For cryogenic applications, certain carbon radio resistors make excellent inexpensive temperature sensors. Although less common, the resistance temperature characteristics of germanium, rhodium–iron alloys and ruthenium oxide make them particularly suitable for cryogenic applications. If uncertainty is less critical then a cheaper form of RTD than PRTs is the thermistor. Thermistors consist of a semiconductor whose resistance is sensitive to temperature. Modern thermistors are usually mixtures of oxides (such as oxides of nickel, manganese, iron, copper, cobalt and titanium), other metals and doped ceramics. Transistor- and diode-based temperature sensors are also possible and are easy to incorporate into an integrated circuit or as a stand-alone package.

RTDs are a popular choice for temperature measurement. Applications range from temperature measurement in the food industry, medical practice and industrial usage monitoring pipelines and storage tanks. The provision of a relatively simple signal makes them an excellent choice for process control or data monitoring. PRTs are described in Section 6.2 and copper and nickel-based RTDs in Section 6.3. RTDs for cryogenic applications are discussed in Section 6.4. The principles, types and use of thermistors are described in Section 6.5. Semiconductor temperature sensors such as transistor- and diode-based thermometers are considered in Section 6.6. The principal selection criteria for an RTD are explored in Section 6.7.

6.2 Platinum resistance thermometers

A platinum resistance thermometer measurement system will comprise a resistance element made from platinum supported in some form of mounting, possibly a sheath to protect the sensor, electrical connection leads and a method of measuring the resistance and converting this quantity to a meaningful measure of temperature. Platinum resistance thermometers can be capable of very low uncertainty measurements and they are used to define part of the International Temperature Scale of 1990 (ITS-90) between the triple point of hydrogen, 13.8023 K, and the freezing point of silver, 1234.93 K. An uncertainty of $\pm 2 \times 10^{-3}$ K at the lower end of the scale and $\pm 7 \times 10^{-3}$ K at the upper end of the scale is possible (Mangum and Furukawa, 1990). A PRT used to realise the ITS-90 is referred to as standard platinum resistance thermometer (SPRT). These devices are not particularly robust as the platinum sensing element must be mounted so that it is strain free and they are on the whole unsuitable for industrial applications except where low uncertainty is necessary and the resulting cost of the measurement justifiable. For industrial applications the sensing element needs to be firmly mounted and such devices are available from a large number of commercial suppliers and are manufactured in a range of geometries for different applications. Such devices are referred to as industrial platinum resistance thermometers (IPRTs). The achievable uncertainty for a commercially available IPRT is generally of the order of $\pm 0.01°C$ to $\pm 0.2°C$ over the range $0°C$ to $300°C$ (Hashemian and Petersen, 1992).

6.2.1 Temperature resistance characteristics

Electrical conduction in a material occurs due to the presence of an electrical field modifying the normal random motion of electrons and causing a net flow in a particular direction. The net flow is impeded by scattering due to imperfections and distortions in the crystal lattice and due to thermally induced effects such as lattice vibration and phonon scattering. This impeding of the electrons can be quantified by the material's electrical resistance or resistivity. Resistivity is defined as the resistance of a cubic volume of material between two faces and is a measure of the impedance to motion of electrons in an electrical field. For metals the resistivity can be modelled by Matthiessen's rule:

$$\rho = \rho_t + \rho_i + \rho_d \tag{6.1}$$

where: ρ = overall resistivity = RA/L ($\Omega \cdot$m)
R = resistance (Ω)
A = cross-sectional area (m^2)
L = length (m)
ρ_t = component of resistivity due to thermal effects ($\Omega \cdot$m)
ρ_i = component of resistivity due to the presence of impurities ($\Omega \cdot$m)
ρ_d = component of resistivity due to lattice deformation arising from mechanical stress ($\Omega \cdot$m).

In order to produce a device whose output varies only with temperature the effects of strain and impurities must be minimized in design and production. For metals the relationship between voltage and current follows Ohm's law:

$$I = \frac{V}{R} \tag{6.2}$$

where: I = the current (A)
V = the voltage (V)
R = the resistance (Ω).

For a resistor as the temperature increases, the lattice vibrations and hence the resistance to the flow of electrons increase in proportion to the absolute temperature. The relationship between resistance and temperature can be approximated by the relation

$$R_T = R_0 (1 + \alpha T) \tag{6.3}$$

where: R_0 = resistance at 0°C (Ω)
R_T = resistance at temperature T (Ω)
α = the temperature coefficient of resistance ($\Omega\Omega^{-1}$ °C^{-1} or °C^{-1})
T = temperature (°C).

The temperature coefficient of resistance, α, is calculated from

$$\alpha = \frac{R_{100} - R_0}{100°C \times R_0} \tag{6.4}$$

where R_{100} = resistance at 100°C (Ω).

The materials used most commonly in resistance thermometry are platinum, copper, nickel and semiconductors. The temperature coefficient of resistance for platinum, copper and nickel are approximately equal to $1/273.15 = 3.66 \times 10^{-3}\,K^{-1}$, $0.00427\,K^{-1}$ and $0.00672\,K^{-1}$ respectively. The deviation from the actual coefficient of resistance for platinum is small and is illustrated in Figure 6.1, which also shows data for nickel and copper. The presence of impurities in the metal lattice tends to increase the resistance by a constant amount as indicated by equation (6.1) and the effect on the resistance can be modelled by

$$R_T' = R_T + \Delta R \tag{6.5}$$

where ΔR = the resistance due to impurities (Ω) and

$$R_T' = R_0' \,(1 + \alpha'T) \tag{6.6}$$

$$\alpha' = \alpha \,\frac{R_0}{R_0 + \Delta R} \tag{6.7}$$

where: R_T' = resistance in presence of impurities (Ω)
R_0' = ice point resistance in presence of impurities (Ω).

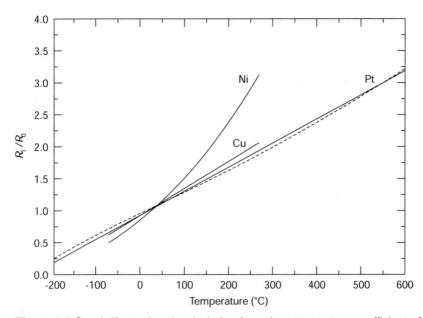

Figure 6.1 Graph illustrating the deviation from the temperature coefficient of resistance for platinum, copper and nickel. (Reproduced from Nicholas and White, 1994)

Examination of equation (6.7) shows that impurities increase the ice-point resistance, R_0, and decrease the temperature coefficient of resistance, α. In other words, the purer the substance, the higher the temperature coefficient of resistance.

Much of the original work on resistance thermometry was undertaken by Siemens (1871) and Callendar (1887, 1891, 1899). Callendar found that the resistance of platinum could fairly accurately be described by a quadratic in the form

$$R_T = R_0 (1 + AT + BT^2) \tag{6.8}$$

where A and B are constants. This equation has traditionally been written in the form given in equation (6.9) which simplifies the calculations necessary to determine the calibration constants α and δ. Equation (6.9) is known as Callendar's equation:

$$R_T = R_0 \left[1 + \alpha T + \alpha \delta \left(\frac{T}{100} \right) \left(1 - \frac{T}{100} \right) \right] \tag{6.9}$$

where δ is a constant. For temperatures below 0°C, additional terms are needed and the temperature resistance characteristic is, with relatively low uncertainty, given by:

$$R_T = R_0(1 + AT + BT^2 + C(T - 100)T^3) \tag{6.10}$$

where C = a constant and is zero above 0°C.

Equation (6.10) is called the Callendar–Van Dusen equation and was the basis of the International Practical Temperature Scales of 1927, 1948 and 1968. This equation has now been superseded by reference functions in the ITS-90 (Preston Thomas, 1990) but continues to be used to define the resistance temperature relationship for some industrial resistance thermometers. Typical values for the coefficients in equation (6.10) for a standard PRT and an industrial PRT are listed in Table 6.1.

Standards (e.g. BS 1041: Part 3, ASTM E1137 and EN 60751) have been produced for PRTs. Table 6.2 lists selected values of resistance and

Table 6.1 Typical values for the constants in the Callendar-Van Dusen equation for an SPRT and an IPRT (Nicholas and White, 1994)

Constant	SPRT	IPRT
A	$3.985 \times 10^{-3}/°C$	$3.908 \times 10^{-3}/°C$
B	$-5.85 \times 10^{-7}/°C^2$	$-5.80 \times 10^{-7}/°C^2$
C	$-4.27 \times 10^{-12}/°C^4$	$-4.27 \times 10^{-12}/°C^4$
α	$3.927 \times 10^{-3}/°C$	$3.85 \times 10^{-3}/°C$

Table 6.2 Temperature resistance relationship EN 60751: 1995

°C ITS-90	0	1	2	3	4	5	6	7	8	9	10	°C ITS-90
−200	18.52											−200
−190	22.83	22.40	21.97	21.54	21.11	20.68	20.25	19.82	19.38	18.95	18.52	−190
−180	27.10	26.67	26.24	25.82	25.39	24.97	24.54	24.11	23.68	23.25	22.83	−180
−170	31.34	30.91	30.49	30.07	29.64	29.22	28.80	28.37	27.95	27.52	27.10	−170
−160	35.54	35.12	34.70	34.28	33.86	33.44	33.02	32.60	32.18	31.76	31.34	−160
−150	39.72	39.31	38.89	38.47	38.05	37.64	37.22	36.80	36.38	35.96	35.54	−150
−140	43.88	43.46	43.05	42.63	42.22	41.80	41.39	40.97	40.56	40.14	39.72	−140
−130	48.00	47.59	47.18	46.77	46.36	45.94	45.53	45.12	44.70	44.29	43.88	−130
−120	52.11	51.70	51.29	50.88	50.47	50.06	49.65	49.24	48.83	48.42	48.00	−120
−110	56.19	55.79	55.38	54.97	54.56	54.15	53.75	53.34	52.93	52.52	52.11	−110
−100	60.26	59.85	59.44	59.04	58.63	58.23	57.82	57.41	57.01	56.60	56.19	−100
−90	64.30	63.90	63.49	63.09	62.68	62.28	61.88	61.47	61.07	60.66	60.26	−90
−80	68.33	67.92	67.52	67.12	66.72	66.31	65.91	65.51	65.11	64.70	64.30	−80
−70	72.33	71.93	71.53	71.13	70.73	70.33	69.93	69.53	69.13	68.73	68.33	−70
−60	76.33	75.93	75.53	75.13	74.73	74.33	73.93	73.53	73.13	72.73	72.33	−60
−50	80.31	79.91	79.51	79.11	78.72	78.32	77.92	77.52	77.12	76.73	76.33	−50
−40	84.27	83.87	83.48	83.08	82.69	82.29	81.89	81.50	81.10	80.70	80.31	−40
−30	88.22	87.83	87.43	87.04	86.64	86.25	85.85	85.46	85.06	84.67	84.27	−30
−20	92.16	91.77	91.37	90.98	90.59	90.19	89.80	89.40	89.01	88.62	88.22	−20
−10	96.09	95.69	95.30	94.91	94.52	94.12	93.73	93.34	92.95	92.55	92.16	−10
0	100.00	99.61	99.22	98.83	98.44	98.04	97.65	97.26	96.87	96.48	96.09	0
0	100.00	100.39	100.78	101.17	101.56	101.95	102.34	102.73	103.12	103.51	103.90	0
10	103.90	104.29	104.68	105.07	105.46	105.85	106.24	106.63	107.02	107.40	107.79	10
20	107.79	108.18	108.57	108.96	109.35	109.73	110.12	110.51	110.90	111.29	111.67	20

30	115.54	115.15	114.77	114.38	114.00	113.61	113.22	112.83	112.45	112.06	111.67	30
40	119.40	119.01	118.63	118.24	117.86	117.47	117.08	116.70	116.31	115.93	115.54	40
50	123.24	122.86	122.47	122.09	121.71	121.32	120.94	120.55	120.17	119.78	119.40	50
60	127.08	126.69	126.31	125.93	125.54	125.16	124.78	124.39	124.01	123.63	123.24	60
70	130.90	130.52	130.13	129.75	129.37	128.99	128.61	128.22	127.84	127.46	127.08	70
80	134.71	134.33	133.95	133.57	133.18	132.80	132.42	132.04	131.66	131.28	130.90	80
90	138.51	138.13	137.75	137.37	136.99	136.61	136.23	135.85	135.47	135.09	134.71	90
100	142.29	141.91	141.54	141.16	140.78	140.40	140.02	139.64	139.26	138.88	138.51	100
110	146.07	145.69	145.31	144.94	144.56	144.18	143.80	143.43	143.05	142.67	142.29	110
120	149.83	149.46	149.08	148.70	148.33	147.95	147.57	147.20	146.82	146.44	146.07	120
130	153.58	153.21	152.83	152.46	152.08	151.71	151.33	150.96	150.58	150.21	149.83	130
140	157.33	156.95	156.58	156.20	155.83	155.46	155.08	154.71	154.33	153.96	153.58	140
150	161.05	160.68	160.31	159.94	159.56	159.19	158.82	158.45	158.07	157.70	157.33	150
160	164.77	164.40	164.03	163.66	163.29	162.91	162.54	162.17	161.80	161.43	161.05	160
170	168.48	168.11	167.74	167.37	167.00	166.63	166.26	165.89	165.51	165.14	164.77	170
180	172.17	171.80	171.43	171.07	170.70	170.33	169.96	169.59	169.22	168.85	168.48	180
190	175.86	175.49	175.12	174.75	174.38	174.02	173.65	173.28	172.91	172.54	172.17	190
200	179.53	179.16	178.79	178.43	178.06	177.69	177.33	176.96	176.59	176.22	175.86	200
210	183.19	182.82	182.46	182.09	181.72	181.36	180.99	180.63	180.26	179.89	179.53	210
220	186.84	186.47	186.11	185.74	185.38	185.01	184.65	184.28	183.92	183.55	183.19	220
230	190.47	190.11	189.75	189.38	189.02	188.66	188.29	187.93	187.56	187.20	186.84	230
240	194.10	193.74	193.37	193.01	192.65	192.29	191.92	191.56	191.20	190.84	190.47	240
250	197.71	197.35	196.99	196.63	196.27	195.91	195.55	195.18	194.82	194.46	194.10	250
260	201.31	200.95	200.59	200.23	199.87	199.51	199.15	198.79	198.43	198.07	197.71	260
270	204.90	204.55	204.19	203.83	203.47	203.11	202.75	202.39	202.03	201.67	201.31	270
280	208.48	208.13	207.77	207.41	207.05	206.70	206.34	205.98	205.62	205.26	204.90	280
290	212.05	211.70	211.34	210.98	210.63	210.27	209.91	209.56	209.20	208.84	208.48	290
300	215.61	215.25	214.90	214.54	214.19	213.83	213.48	213.12	212.76	212.41	212.05	300
310	219.15	218.80	218.44	218.09	217.74	217.38	217.03	216.67	216.32	215.96	215.61	310

°C ITS-90	0	1	2	3	4	5	6	7	8	9	10	°C ITS-90
320	219.15	219.51	219.86	220.21	220.57	220.92	221.27	221.63	221.98	222.33	222.68	320
330	222.68	223.04	223.39	223.74	224.09	224.45	224.80	225.15	225.50	225.85	226.21	330
340	226.21	226.56	226.91	227.26	227.61	227.96	228.31	228.66	229.02	229.37	229.72	340
350	229.72	230.07	230.42	230.77	231.12	231.47	231.82	232.17	232.52	232.87	233.21	350
360	233.21	233.56	233.91	234.26	234.61	234.96	235.31	235.66	236.00	236.35	236.70	360
370	236.70	237.05	237.40	237.74	238.09	238.44	238.79	239.13	239.48	239.83	240.18	370
380	240.18	240.52	240.87	241.22	241.56	241.91	242.26	242.60	242.95	243.29	243.64	380
390	243.64	243.99	244.33	244.68	245.02	245.37	245.71	246.06	246.40	246.75	247.09	390
400	247.09	247.44	247.78	248.13	248.47	248.81	249.16	249.50	249.85	250.19	250.53	400
410	250.53	250.88	251.22	251.56	251.91	252.25	252.59	252.93	253.28	253.62	253.96	410
420	253.96	254.30	254.65	254.99	255.33	255.67	256.01	256.35	256.70	257.04	257.38	420
430	257.38	257.72	258.06	258.40	258.74	259.08	259.42	259.76	260.10	260.44	260.78	430
440	260.78	261.12	261.46	261.80	262.14	262.48	262.82	263.16	263.50	263.84	264.18	440
450	264.18	264.52	264.86	265.20	265.53	265.87	266.21	266.55	266.89	267.22	267.56	450
460	267.56	267.90	268.24	268.57	268.91	269.25	269.59	269.92	270.26	270.60	270.93	460
470	270.93	271.27	271.61	271.94	272.28	272.61	272.95	273.29	273.62	273.96	274.29	470
480	274.29	274.63	274.96	275.30	275.63	275.97	276.30	276.64	276.97	277.31	277.64	480
490	277.64	277.98	278.31	278.64	278.98	279.31	279.64	279.98	280.31	280.64	280.98	490
500	280.98	281.31	281.64	281.98	282.31	282.64	282.97	283.31	283.64	283.97	284.30	500
510	284.30	284.63	284.97	285.30	285.63	285.96	286.29	286.62	286.95	287.29	287.62	510
520	287.62	287.95	288.28	288.61	288.94	289.27	289.60	289.93	290.26	290.59	290.92	520
530	290.92	291.25	291.58	291.91	292.24	292.56	292.89	293.22	293.55	293.88	294.21	530
540	294.21	294.54	294.86	295.19	295.52	295.85	296.18	296.50	296.83	297.16	297.49	540
550	297.49	297.81	298.14	298.47	298.80	299.12	299.45	299.78	300.10	300.43	300.75	550
560	300.75	301.08	301.41	301.73	302.06	302.38	302.71	303.03	303.36	303.69	304.01	560
570	304.01	304.34	304.66	304.98	305.31	305.63	305.96	306.28	306.61	306.93	307.25	570

580	310.49	310.16	309.84	309.52	309.20	308.87	308.55	308.23	307.90	307.58	307.25	580
590	313.71	313.39	313.06	312.74	312.42	312.10	311.78	311.45	311.13	310.81	310.49	590
600	316.92	316.60	316.28	315.96	315.64	315.31	314.99	314.67	314.35	314.03	313.71	600
610	320.12	319.80	319.48	319.16	318.84	318.52	318.20	317.88	317.56	317.24	316.92	610
620	323.30	322.98	322.67	322.35	322.03	321.71	321.39	321.07	320.75	320.43	320.12	620
630	326.48	326.16	325.84	325.53	325.21	324.89	324.57	324.26	323.94	323.62	323.30	630
640	329.64	329.32	329.01	328.69	328.38	328.06	327.74	327.43	327.11	326.79	326.48	640
650	332.79	332.48	332.16	331.85	331.53	331.22	330.90	330.59	330.27	329.96	329.64	650
660	335.93	335.62	335.31	334.99	334.68	334.36	334.05	333.74	333.42	333.11	332.79	660
670	339.06	338.75	338.44	338.12	337.81	337.50	337.18	336.87	336.56	336.25	335.93	670
680	342.18	341.87	341.56	341.24	340.93	340.62	340.31	340.00	339.69	339.37	339.06	680
690	345.28	344.97	344.66	344.35	344.04	343.73	343.42	343.11	342.80	342.49	342.18	690
700	348.38	348.07	347.76	347.45	347.14	346.83	346.52	346.21	345.90	345.59	345.28	700
710	351.46	351.15	350.84	350.54	350.23	349.92	349.61	349.30	348.99	348.69	348.38	710
720	354.53	354.22	353.92	353.61	353.30	353.00	352.69	352.38	352.08	351.77	351.46	720
730	357.59	357.28	356.98	356.67	356.37	356.06	355.76	355.45	355.14	354.84	354.53	730
740	360.64	360.33	360.03	359.72	359.42	359.12	358.81	358.51	358.20	357.90	357.59	740
750	363.67	363.37	363.07	362.76	362.46	362.16	361.85	361.55	361.25	360.94	360.64	750
760	366.70	366.40	366.10	365.79	365.49	365.19	364.89	364.58	364.28	363.98	363.67	760
770	369.71	369.41	369.11	368.81	368.51	368.21	367.91	367.60	367.30	367.00	366.70	770
780	372.71	372.41	372.11	371.81	371.51	371.21	370.91	370.61	370.31	370.01	369.71	780
790	375.70	375.41	375.11	374.81	374.51	374.21	373.91	373.61	373.31	373.01	372.71	790
800	378.68	378.39	378.09	377.79	377.49	377.19	376.90	376.60	376.30	376.00	375.70	800
810	381.65	381.35	381.06	380.76	380.46	380.17	379.87	379.57	379.28	378.98	378.68	810
820	384.60	384.31	384.01	383.72	383.42	383.13	382.83	382.54	382.24	381.95	381.65	820
830	387.55	387.25	386.96	386.67	386.37	386.08	385.78	385.49	385.19	384.90	384.60	830
840	390.48	390.19	380.90	389.60	389.31	389.02	388.72	388.43	388.14	387.84	387.55	840
850											390.48	850

corresponding temperatures from EN 60751 for a transducer manufactured with a nominal resistance of 100 Ω at 0°C. Two classes of uncertainty are defined for IPRTs. For the temperature T measured in °C, these are:

- Class A devices where the uncertainty is within ±(0.15 + 0.002 |T|) and
- Class B devices, with an uncertainty of ±(0.3 + 0.005 |T|).

These tolerances are illustrated graphically in Figure 6.2.

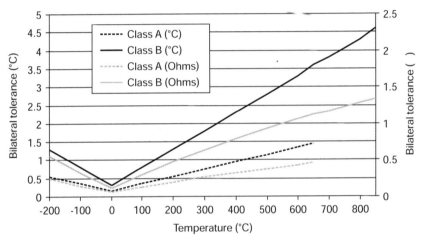

Figure 6.2 Tolerance as a function of temperature for 100 Ω PRTs. (After EN 60751)

To accurately represent the resistance temperature relationship of a given SPRT by a polynomial, the necessary resulting degree would exceed the number of fixed points available. As a result the behaviour of an SPRT is defined in the ITS-90 using a reference function that specifies the intrinsic functional dependence of R on T and a deviation function that expresses the difference between the calibration for an individual PRT and the reference function. These functions are defined in terms of the ratio of the resistance of the PRT at a temperature T, R_T, and the resistance of the PRT at the water triple point, R_{TP},

$$W = \frac{R_T}{R_{TP}} \tag{6.11}$$

The ITS-90 uses predefined functions $W_r(T_{90})$, given by equations (6.12), (6.14), (6.16) and (6.17), from which deviations are calculated in order to describe the temperature resistance relationship (Table 6.3).

Table 6.3 Coefficients for use in the ITS-90 predefined reference functions, equations (6.12)–(6.15) (Preston–Thomas, 1990)

A_0	−2.13534729	B_0	0.183324722	C_0	2.78157254	D_0	439.932854
A_1	3.18324720	B_1	0.240975303	C_1	1.64650916	D_1	472.418020
A_2	−1.80143597	B_2	0.209108771	C_2	−0.13714390	D_2	37.684494
A_3	0.71727204	B_3	0.190439972	C_3	−0.00649767	D_3	7.472018
A_4	0.50344027	B_4	0.142648498	C_4	−0.00234444	D_4	2.920828
A_5	−0.61899395	B_5	0.077993465	C_5	0.00511868	D_5	0.005184
A_6	−0.05332322	B_6	0.012475611	C_6	0.00187982	D_6	−0.963864
A_7	0.28021362	B_7	−0.032267127	C_7	−0.00204472	D_7	−0.188732
A_8	0.10715224	B_8	−0.075291522	C_8	−0.00046122	D_8	0.191203
A_9	−0.29302865	B_9	−0.056470670	C_9	0.00045724	D_9	0.049025
A_{10}	0.04459872	B_{10}	0.076201285				
A_{11}	0.11868632	B_{11}	0.123893204				
A_{12}	−0.05248134	B_{12}	−0.029201193				
		B_{10}	−0.091173542				
		B_{14}	0.001317696				
		B_{15}	0.026025526				

For the range 13.8033 K to 273.16 K,

$$\ln\left[W_r(T_{90})\right] = A_0 + \sum_{i=1}^{12} A_i \left\{ [\ln(T_{90}/273.16\,K) + 1.5]/1.5 \right\}^i \qquad (6.12)$$

The corresponding inverse function is

$$T_{90}/273.16\,K = B_0 + \sum B_i \left\{ [W_r(T_{90})^{1/6} - 0.65]/0.35 \right\}^i \qquad (6.13)$$

For the range from 0°C to 961.78°C,

$$W_r(T_{90}) = C_0 + \sum_{i=1}^{9} C_i \left[(T_{90}/K - 754.15)/481\right]^i \qquad (6.14)$$

and the corresponding inverse function is

$$T_{90}/K - 273.15 = D_0 + \sum_{i=1}^{9} D_i \left\{ [W_r(T_{90}) - 2.64]/1.64 \right\}^i \qquad (6.15)$$

The coefficients A_0, A_i, B_0, B_i, C_0, C_i, D_0, D_i are listed in Table 6.3.

The deviation function for the range from 0°C to the freezing point of argon is

$$W(T_{90}) - W_r(T_{90}) = a\left[W(T_{90}) - 1\right] + b\left[W(T_{90}) - 1\right]^2 + c\left[W(T_{90}) - 1\right]^3$$
$$+ d\left[W(T_{90}) - 1\right]^4 - [W(660.323°C)]^2 \qquad (6.16)$$

For 0°C to the freezing point of aluminium, 660.323°C

$$W(T_{90}) - W_r(T_{90}) = a_8 \left[W(T_{90}) - 1\right] + b_8[W(T_{90}) - 1]^2 \qquad (6.17)$$

where the coefficients a_8 and b_8 are obtained from resistance ratio deviation measurements at the freezing points of zinc and tin.

6.2.2 PRT construction

The aim in PRT design is to ensure that the sensing element responds to the temperature of the application whilst being unaffected by other environmental factors such as corrosive fluids, vibration, pressure and humidity. Generally the most serious concern in PRT design is strain of the platinum element due to mechanical shock or thermal expansion. In its simplest form a resistance thermometer could comprise a coil of wire mounted on an electrical insulating support. Such a design, however, would be sensitive to shock and vibrations. Knocks could cause unsupported sections of the wire to flex and strain causing work hardening of the wire, which introduces defects to the crystal structure and increases the resistance. An alternative solution is to mount the coil on a solid former. This design is, however, sensitive to differential expansion between the former and the coil which also serves to work harden the platinum with subsequent uncertainties in the resistivity. Practical PRT designs are a compromise between mechanical robustness and minimization of strain.

The three principal types of SPRTs are capsule, long stem and high temperature. Although the PRT is specified for realizing the ITS-90, no single instrument can cover the whole range. Instead each of the SPRT designs listed are appropriate for particular ranges of temperature.

Capsule type SPRTs should be used for low temperatures to 13.8 K and a typical design is illustrated in Figure 6.3. Capsule SPRTs consist of the sensor only and the lead wires generally must be attached by the user. Lead wires should be kept short and are usually welded to thin copper or constantan extension leads. The principle of the design is to minimize thermal conduction errors along any connections. At cryogenic temperatures such conduction errors can be significant. The sensor typically has a nominal resistance of 25 Ω at 0°C and thermal contact is achieved by means of a helium gas fill with a fraction of oxygen to aid heat transfer at a pressure of 30 kPa at room temperature and a platinum sheath. In order to produce a resistance of 25 Ω about 60 cm of 0.07 mm diameter coiled wire is necessary. The capsule, which is usually less than 50 mm long, must be totally immersed in the medium of interest. This is normally achieved by inserting the gauge in a thermowell filled with a suitable high thermal conductivity grease or alternative substance. For low uncertainty applications these devices should not be used above 30°C but can be used to 250°C if reduced lifetime and increased uncertainty are acceptable.

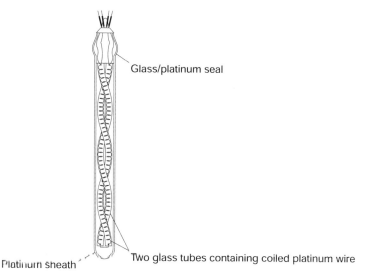

Glass/platinum seal

Platinum sheath

Two glass tubes containing coiled platinum wire

Figure 6.3 A typical capsule PRT

For use at higher temperatures SPRTs need longer stems so that the seal is close to room temperature in order to avoid significant thermal gradients and the accepted interpolation instrument for the temperature range $-189.352°C$ to $419.527°C$ is the long-stem SPRT, (Figure 6.4). These devices consist of platinum wire wound onto a former, which is contained within a glass or quartz sheath. In traditional designs the former or mandrel used to support the platinum wire is a cross, made from thin mica sheet. The four edges of the mica cross are cut with fine teeth to locate the platinum wire. The grooves can be designed to form a double helix so that one half of the wire is wound onto the former leaving alternate grooves empty and the wire is then wound back located in the vacant grooves. The result is known as a bifilar winding which has the advantage that the electrical fields in adjacent wires approximately cancel to minimize inductive effects. The total resistance is commonly

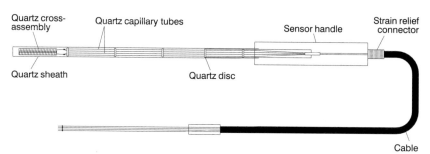

Quartz cross-assembly

Quartz capillary tubes

Sensor handle

Strain relief connector

Quartz sheath

Quartz disc

Cable

Figure 6.4 A standard platinum resistance thermometer

designed to be 25.5 Ω at 0°C and this value provides a resistance change of approximately 0.1 Ω per degree Celsius. The use of mica limits the upper temperature at which the device can be used as it is bound with water and will deteriorate and flake at temperatures, which drive off the water (Sostmann and Metz, 1991). Phlogopitic mica as used in quartz-tube designs is limited to use below 600°C and ruby mica using pyrex tubes is limited to below 500°C.

Glass or quartz sheaths are used to limit conduction along the stem. Thermal radiation, however, can be transmitted along the stem and room lighting, for example, can raise the indicated temperature for the triple point of water by 0.2 mK (Nicholas and White, 1994). The quartz tubes of most thermometers are clear at the hot end to enable visualization of the element and the rest of the element is roughened by sand blasting or coated with a black material such as Aquadag to minimize the transmission of thermal radiation.

High-temperature SPRTs (sometimes referred to as HTSPRTs) must be designed to resist chemical changes. This has implications for the platinum wire, the former and the sheath. Platinum although relatively stable does oxidize (Berry, 1982a,b) and undergoes crystal growth (Berry, 1972). Oxidization ultimately limits the upper useful temperature for PRTs and also the long-term stability and uncertainty for a particular thermometer. Mica cannot be used at sustained temperatures above about 600°C and high-purity quartz is usually used as an alternative. Unfortunately the insulation properties of substances reduce with temperature and the quartz mandrel serves as a shunt resistance across the platinum windings introducing an uncertain error source. Examples of high temperature capability PRTs are presented by Li Xumo et al. (1982) and Sawada and Mochizuki (1972).

SPRTs must be manufactured using high-purity platinum and great care taken to ensure the assembly is strain free (Mangum and Furukawa, 1990). The sensing element for a SPRT typically consists of a coil of fine gauge (around 0.075 mm diameter) platinum wire wound onto a structure made from either mica or pure quartz glass (Riddle et al., 1973; Meyers, 1932). SPRTs compatible with ITS-90 are constructed from platinum with a temperature coefficient of resistance of 3.986×10^{-3} K^{-1} at 273.16 K (Mangum and Furukawa, 1990). In order to ensure that the calibration equations for all SPRTs are similar, the ratio of the resistance of the sensor at the melting point of gallium to the resistance of the sensor at the water triple point is defined by:

$$W(29.7646°C) \geq 1.11807 \qquad (6.18)$$

or

$$W(-38.8344°C) \leq 0.844235 \qquad (6.19)$$

In addition if the SPRT is to be used up to the silver point then,

$$W(961.78°C) \geq 4.2844 \qquad (6.20)$$

A variation of SPRTs are secondary SPRTs. These are designed for laboratory environments and can withstand some handling although the sensing element is still mounted so that it is strain free and mechanical shocks will cause damage. They are manufactured using less expensive materials such as reference-grade high-purity platinum wire. They are intended for use over a limited temperature range, normally −200°C to 500°C, and have an uncertainty of ±0.03°C.

The strain-free designs of an SPRT would not survive the shock and vibration encountered in the industrial environment. The industrial platinum resistance thermometer (IPRT) typically comprises a platinum wire encapsulated within a ceramic housing or a thick film sensor coated onto a ceramic surface (Figure 6.5). The actual sensing element of an IPRT can be further protected from the environment by a metal (e.g. Inconel™) sheath. In the design of an IPRT the overriding consideration is robustness and the constraint for mounting the platinum wire so that strain is minimized is relaxed. The degree of constraint on the wire depends on the design and intended applications. Most designs have just two wires sealed within the sensor body as this reduces the risk of short circuits and makes the device more robust. The ceramic encapsulation used should be designed so that its thermal coefficient of expansion matches, as nearly as possible, that of the platinum and any former assembly over the range of use of the PRT. The ceramic material used for encapsulation can include compounds and impurities that react with the platinum and change the transducer's temperature coefficient. Both of these effects cause some variability in the stability and performance of IPRTs.

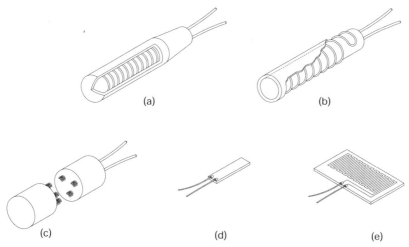

Figure 6.5 Typical configurations for IPRTs. (a) Cylindrical PRT with wire set in grooves. (b) Bifilar wound coil. (c) Multi-bore alumina tubing support. (d) Flat pack. (e) Thin-film PRT. (Figures adapted from Johnston, 1975)

An alternative to the use of a platinum coils and wound wires is the use of thin- or thick-film platinum tracks deposited on an insulating substrate such as alumina. The size of a thick-film sensor depends on the actual device but the surface area can be of the order of 25 mm^2 with an active area of 5 mm^2. They are mass produced and therefore comparatively cheap with each sensor individually trimmed to produce the desired resistance. The thick-film structure means that these devices are robust and capable of operating in a vibrating application. They are also non-inductive and therefore insensitive to stray electrical fields. Thick-film PRTs are normally designated to Class B of IEC-751. An innovative form of temperature measurement is the application of PRTs to polyimide sheet. A massed array of PRTs can be formed on a polyimide sheet by vacuum depositing platinum followed by depositing gold leads using a magnetron. The resulting sheet is flexible and can be glued to the geometry of interest (Jones, 1995).

For the lowest uncertainty applications quartz spacers and supports are used. For partially supported PRTs four-bore alumina insulators are sometimes used. Alumina powder can be used to restrict the movement of the leads in the sheath. Magnesia is commonly used in MIMS cable. Magnesia, however, has the unfortunate drawback of absorbing moisture which reduces the insulation resistance and can cause moisture-induced hysteresis effects where the indicated temperature is elevated below its real value on heating as the water boils off and is high on cooling as the magnesia is yet to reabsorb moisture.

For high-temperature applications the only suitable lead wire is platinum but for most applications, the high cost of platinum wire cannot be justified. For temperatures up to 250°C glass-insulated copper or silver wire is used. For temperatures above 250°C nickel alloy or platinum-coated nickel wires are often used. The markings c, C, t and T are commonly used for four lead thermometers. The cables marked C and T are the potential leads.

6.2.3 Installation

The installation procedures and requirements will depend on the specific type of sensor involved. For SPRTs the primary concern is that the resistance wire remains uncontaminated and free from strain. Contamination can occur if the protective sheath or the end seal with the connection leads is damaged providing a leakage path. In order to avoid strain of the platinum element, SPRTs must be mounted so that they do not experience mechanical shocks and vibrations. Generally even low levels of acceleration can cause the sensing element to stretch or become detached from its mountings. SPRTs are usually supplied ready for use; the user may only need to attach the connection leads in some cases. In order to protect the sensor against contaminants that would cause damage to the stem, a thermowell or other suitable pocket in the application can be used. SPRTs have a specified temperature range of operation and procedures should be adopted to ensure that these limits are not exceeded.

IPRTs are generally much more robust than SPRTs. Nevertheless, appropriate precautions should be taken to avoid damage. The geometry of IPRT may be cylindrical, with a diameter from one to tens of millimetres, or rectangular. Sensors are available comprising the sensing element contained within a ceramic or glass former. These can be installed directly in the application by bonding using an adhesive or cement. Some sensors are supplied with a metal coating applied to one surface to enable them to be soldered to the application thereby providing better thermal contact. Such sensors may not be hermetically sealed and it is possible for the platinum element to be contaminated. If the environment is corrosive then the sensor and its leads must be protected. This can be achieved by using a protective covering which comprises a tube and a connection head similar to the protection tubes for thermocouples illustrated in Figure 5.13. A wide variety of protection tubes and heads are available. Stainless steel is a frequent choice for the protection tubes and heads although many other alloys are also used.

IPRTs are also available with an integral protective sheath similar to a MIMS thermocouple. Here the platinum sensor is embedded in ceramic powder and contained at the tip of a steel cable. The steel cable provides excellent protection for many environments and may be flexible or rigid. For severe environments such as in chemical plant and refineries where there is a clear risk of contamination and degradation of the sensor, the PRT must be fitted in a pocket or thermowell. The thermowell serves to protect the sensor from contact with the fluid and also to protect against mechanical forces and abrasion. The disadvantage of using a thermowell is that the additional mass of the thermowell reduces the transient response and increases thermal disturbance.

6.2.4 Resistance measurement and data acquisition

In order to obtain a measure of temperature using an RTD, the resistance must be measured. Simplistically this can be achieved by passing a current through the sensing resistor and measuring the potential across it. If the current is known, then the ratio of potential to current provides a measurement of the resistance from Ohm's law and the temperature can be determined from the resistance temperature characteristic. However, even if an uncertainty of only 1°C is desired the resistance must be measured to better than 0.4% and as a result must be undertaken with care (Nicholas and White, 1994). There are two basic methods for measuring resistance commonly used in platinum resistance thermometry: potentiometric and bridge methods.

Potentiometric methods utilize the relationship between the voltage dropped across a known resistor and that across the resistance element at the unknown temperature. The principle is illustrated in Figure 6.6. A current is driven through both the standard resistor R_s and the unknown resistance, R_T.

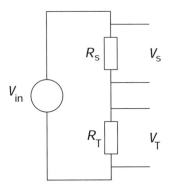

Figure 6.6 Measurement of resistance by a potentiometric method

Since the current is the same through both resistors, then using Ohm's law the unknown resistance can be determined from

$$R_T = \frac{V_T}{V_s} R_s \tag{6.21}$$

This requires measurements of V_T and V_s, which can be achieved using a high-impedance voltmeter.

The Wheatstone bridge can also be used as the basis for resistance measurement. The principle of this method is the comparison of the output voltage from two voltage dividers one of which contains the resistance thermometer as illustrated in Figure 6.7. The output voltage for the bridge is given by

$$V_{output} = V_1 - V_2 = \frac{R_2 R_T - R_3 R_1}{(R_2 + R_3)(R_1 + R_T)} V_{in} \tag{6.22}$$

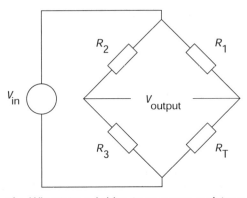

Figure 6.7 Use of a Wheatstone bridge to measure resistance

A Wheatstone bridge can be operated in two ways in order to determine the unknown voltage:

1 balanced mode
2 unbalanced mode.

In the balanced mode, one of the resistances R_1, R_2 or R_3 is adjusted until the output voltage from the two arms of the bridge, V_1 and V_2, are equal. For this condition V_1-V_2 is zero and the bridge is said to be balanced. The advantage of this method is that a voltmeter connected to read V_1-V_2 has to detect a null or zero reading which eases the requirements for the voltmeter. The unknown resistance R_T can then be determined from

$$R_T = \frac{R_3}{R_2} R_1 \tag{6.23}$$

The disadvantages of the balanced bridge method is that it is not easy to eliminate the lead resistance of connections to the resistances and also the variable resistor must have a low uncertainty and as a result tends to be an expensive component in the system.

An alternative is to operate the Wheatstone bridge in an unbalanced mode. In this mode the variable resistor is adjusted once until a balance is achieved at a single temperature, normally the ice point, $0°C$. The output voltage at other temperatures then becomes the measure of the temperature. From equation (6.22),

$$V_{\text{output}} = \frac{R_T - R_0}{(R_1 + R_0)(R_1 + R_T)} R_1 V_{\text{in}} \tag{6.24}$$

If R_1 is large relative to R_T then the output can be approximated by

$$V_{\text{output}} \approx \frac{V_{\text{in}}}{R_1} R_0 \alpha T \tag{6.25}$$

This approximation is a common approach in devices that need an uncertainty of only $\pm 1°C$.

The disadvantage of both balanced and unbalanced Wheatstone bridge methods is the difficulty of eliminating the effects of lead wire resistances. Often the sensing platinum element may be several metres away from the voltage-measuring instrument and the resistance of the lead wires can become indistinguishable from that of the platinum resistance element. For a two-lead arrangement as illustrated in Figure 6.8 with each lead having a resistance of R_L then the error in the temperature measurement is approximately:

$$\Delta T_{\text{error}} = \frac{2R_L}{\alpha R_0} \tag{6.26}$$

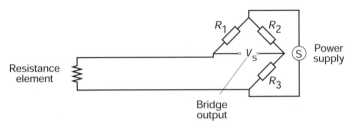

Figure 6.8 Two-lead wire arrangement

From this equation the error for a 100 Ω nominal resistance with a lead wire resistance of 1 Ω is 2.5°C. One option to compensate for this error is to treat the platinum resistance element as a 101 Ω resistor and a reduced temperature coefficient but this does not allow for variations in the lead wire of platinum resistance with time. As a result the uncertainty of most two lead IPRTs is ±0.3°C at best.

There are a number of ways of reducing the problem of lead wire resistance. The ideal solution is to measure the resistance by a four-lead method such as the potentiometric method illustrated in Figure 6.6 (Nicholas and White, 1994). Using this principle because there is almost no current flowing in the leads to the voltmeter there is no voltage drop in these leads and therefore no or minimal error. Unfortunately it is not easy to incorporate four-lead arrangements into Wheatstone bridge circuits. Two practical approaches are used in bridge circuits utilizing a three-leg or four-leg configuration as illustrated in Figures 6.9 and 6.10.

In the three-leg RTD circuit, the lead resistance of the middle leg is common to both halves of the bridge. Its resistance cancels out when the bridge is balanced and reduces lead wire error considerably. If, however, there is any imbalance or $R_3 \neq R_T$ or if the lead resistances are not equal then small errors will be introduced.

The four-leg configuration gives the best performance, eliminating lead wire resistance uncertainties. In this configuration a set of dummy leads is inserted into the other arm of the bridge to make it balance correctly. This

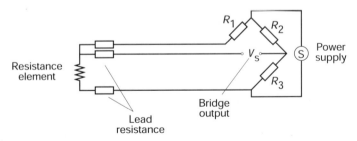

Figure 6.9 Three-leg RTD circuit

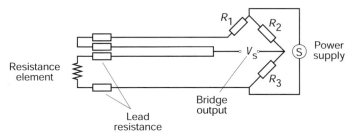

Figure 6.10 Four-leg RTD circuit

design is also sensitive to variations in the resistance of one lead that does not occur in the others and to both lead resistances when the bridge is unbalanced. Depending on the type of circuit used and application, the output voltage for a platinum resistance thermometer (PRT) will typically be 250 mV to 750 mV at 0°C and 800°C respectively.

It is possible to measure the voltage using DC or AC methods. In DC systems there are three main sources of errors associated with the voltage measurement: thermoelectric effects, offset effects and electrolytic effects. Thermoelectric electromotive forces are generated whenever there is a thermal gradient across a conductor comprising dissimilar metals connected together. Different materials will inevitably be used in the connections between a platinum sensing element and the associated electronic circuitry. If the leads all experience the same thermal gradient then the emfs generated cancel each other. If, however, the temperature gradient across each lead is different then thermoelectric emfs will be generated. These will be a function of the thermal gradient and the materials concerned. For a platinum and copper combination of metals the Seebeck coefficient is approximately 7 μV/°C. The associated error in a typical resistance measurement using a current of 1 mA is about 0.02°C per degree of temperature difference in the junction temperatures. The effects of thermoelectric emfs can become particularly significant when exposed terminals are subject to heat transfer by forced convection and thermal radiation. In an ideal voltmeter the reading will be zero when the input is zero. Any error for this condition is referred to as the offset voltage and is additive at higher input potentials. For most modern voltage measurement devices the offset voltage may be between 40 μV and 0.1 V which can cause errors in the temperature measurement of up to 0.1°C. It is possible to use lower uncertainty devices but these often need to be used in temperature-controlled environments and must obviously be calibrated at regular intervals in order to ensure traceability of the measurement. A further error can be caused by the electrolytic effect of moisture connecting lead wires to earth. This can be a problem in humid environments and is hard to quantify. The combined effects of all three measurement errors can typically result in uncertainties of not better than ±0.02°C.

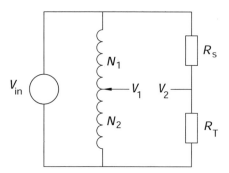

Figure 6.11 A simple form of AC resistance bridge

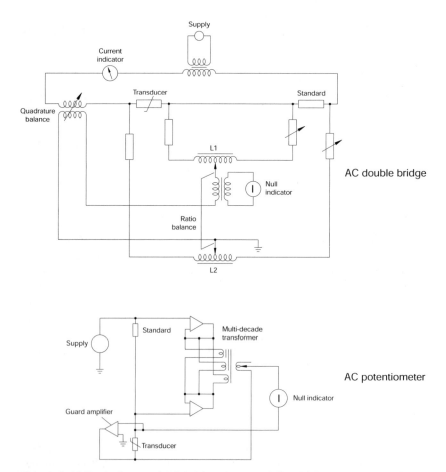

Figure 6.12 Basic forms of AC bridge designs (after ESDU 84036, 1984)

All the errors associated with the measurement of resistance using DC methods, generation of thermoelectric emf, offset voltage effects and electrolytic effects, are independent of the measuring current. By reversing the current systematically and averaging the readings these errors should disappear. This is the principle behind alternating current (AC) resistance measurement. One problem associated with the offset voltage of many voltmeters and amplifiers is that it varies erratically exhibiting 1/f noise. By using an AC current frequency above a few tens of Hertz such 1/f noise is effectively eliminated. The added benefit of running above a few tens of Hertz is that transformers can be used to establish accurate ratios of AC voltage and allow very accurate measurement of resistance. In an AC resistance bridge one arm is replaced by a variable ratio transformer as illustrated in Figures 6.11 and 6.12. In use the transformer turns-ratio is adjusted to obtain a bridge balance by a null detector. The resistance of the sensing element can be determined by

$$R_T = \frac{N_2}{N_1} R_s \qquad (6.27)$$

Alternating current bridges can be designed using multistage transformers with an effective number of turns of more than 100 million. This enables the resistance to be measured with an uncertainty corresponding to $25\,\mu K$. Generally DC instruments are suitable for uncertainties between $\pm0.001°C$ and $\pm1°C$, although they are still being improved, and AC bridges are necessary for uncertainties better than $\pm0.02°C$. Methods using both DC and AC bridges are reviewed by Wolfendale et al. (1982) and Connolly (1998a).

6.2.5 Errors

Errors inherent with a PRT include thermal disturbance errors, oxidation, corrosion and strain of the sensing element, self-heating of the sensing element due to the current and signal transmission errors. Most errors can be recognized by analysis or tests and appropriate rectifying action can be taken or allowances made in the interpretation of the measurement.

The thermal disturbance caused by a PRT can be assessed using the principles described in Chapter 2. Of particular concern is the size of the sensing element which in the case of SPRTs and some IPRTs is relatively large. As a result the probe needs to be immersed in the medium to an appropriate depth to ensure that the whole of the sensing element comes into thermal equilibrium with the medium of interest.

When a current is passed through a conductor, power will be dissipated at a rate of I^2R, which will in turn raise the temperature of the resistance element. This elevation of temperature over that of the undisturbed medium is called the self-heating error. Its magnitude ranges from negligible values to as much as 1°C. The self-heating error can be simplistically modelled as:

$$\Delta T = \frac{I^2 R}{\zeta} \qquad (6.28)$$

where: ΔT = temperature rise due to self-heating error (°C)
  I = current through sensing element (A)
  R = resistance of sensing element (Ω)
  ζ – dissipation constant (W/°C).

The rate of dissipation will depend on thermal conduction within the sensor assembly and the thermal convection boundary conditions for the sensor in the application. Table 6.4 lists a range of typical dissipation constants for unsheathed PRTs with 100 Ω elements and a 1 mA sensing current. The self heating error is a function of the square of the current and heat transfer and for these reasons the current used should be minimized and good thermal contact between the sensor and surrounding medium ensured.

Table 6.4 Range of dissipation constant for unsheathed PRTs with 100 Ω elements and a 1 mA sensing current. (After Nicholas and White, 1994)

Medium	Dissipation constant (W/K)	ΔT (mK)
Still air	0.001 to 0.01	10 to 100
Still water	0.002 to 0.4	0.25 to 50
Flowing water	0.01 to 1	0.1 to 10

Corrections for the self-heating error can be made by altering the sensing current and making a second measurement. These results can be used in equation (6.29) to calculate the zero current reading. For data T_1 and T_2 at currents of I_1 and I_2, the zero current reading, assuming the temperature of the medium of interest does not change, is given by

$$T_0 = T_1 - \frac{I_1^2}{I_1^2 - I_2^2}(T_1 - T_2) \qquad (6.29)$$

Most designs of IPRT exhibit some hysteresis in resistance on thermal cycling with the resistance indicated at a known temperature, say, on heating being different from that shown on cooling. The hysteresis is due to two effects: strain and electrical shunting. Thermal expansion of the substrate relative to the platinum sensing element will cause strain on the platinum. This will distort the lattice structure of the platinum and alter the temperature coefficient. Some substrate materials such as magnesia have the unfortunate drawback of absorbing moisture. This reduces the insulation resistance and causes the indicated temperature to be depressed below its real value on heating as the water boils off and to be high on cooling as the magnesia is yet to reabsorb moisture. As this effect is dependent on water its influence is likely only on temperatures below about 100°C. An example of hysteresis observed on cycling to relatively high temperatures for a fully supported IPRT is illustrated in Figure 6.13.

Above about 250°C platinum becomes progressively more susceptible to contamination. The effect of contamination is to increase impurities in the metal lattice and increase the resistance. The most likely cause of contamination is the migration of iron, manganese and chromium from stainless steel sheaths. For example, exposure of an unprotected ceramic element at 500°C for several hours can cause errors of several degrees. Unlike crystal defects, the impurities cannot later be removed by annealing and effectively render the thermometer useless. The migration of contaminants from a sheath can be reduced by heat treatment of the sheath in air or oxygen before the sensor is assembled as this produces a layer of oxide on the metal surface and acts as a barrier to the migration of metal atoms. Above 500°C quartz sheaths provide a relatively robust solution until the temperature limits of this material are reached. Above 800°C platinum sheaths must be used.

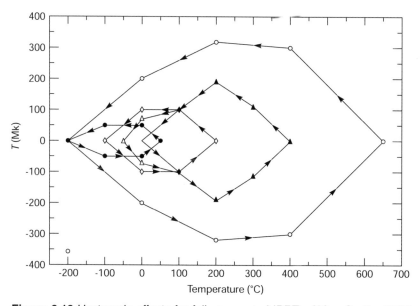

Figure 6.13 Hysteresis effects for fully supported IPRTs. (After Curtis, 1982)

Platinum, although considered relatively stable, can oxidize at elevated temperatures (Berry, 1982a,b). The combined effects of changes in the overall composition of platinum due to oxidation and thermally induced strain have been investigated by Trietley (1982). Errors equivalent to several degrees were reported as a result of repeated cycling above 500°C.

6.2.6 Calibration

Some PRTs will be calibrated prior to delivery by the manufacturer at a specific temperature such as the ice point. However, a regular testing programme should be devised in order to ensure validity of the temperature measurements made. The two common forms of calibration used for IPRTs are:

1 the fixed-point calibration method
2 the comparison method.

Fixed-point calibration involves bringing the sensor into thermal equilibrium with a substance at a known temperature such as the triple point, freezing point or melting point of a variety of purified substances. Fixed points for a

Figure 6.14 An example of a fixed-point cell for establishing the melting point of gallium to within 0.00025°C. Photograph courtesy of Isothermal Technology Ltd

variety of substances are listed in Table 1.2. This process is undertaken using a specialized item of equipment called a fixed-point cell and an example is illustrated in Figure 6.14. The fixed-point cell serves the purpose of raising the temperature to the appropriate value under controlled conditions and preventing contamination of the substance concerned. The general procedure involves preparation of the fixed-point cell, insertion of the thermometer to be calibrated, thermal stabilization of the system (this may take 15 to 30 minutes), subsequent measurement of the resistance of the thermometer and recording of the data. Fixed-point calibrations can be undertaken with very low uncertainty (within ±0.001°C) but the process is time-consuming and a fixed-point cell can normally accommodate only one sensor at a time. An SPRT used for interpolating the ITS-90 must, for example, be calibrated at the defined fixed points. Detailed procedures for conditioning the PRT and achieving fixed points are outlined by Connolly (1998b).

The comparison calibration method is normally used for IPRTs and secondary SPRTs. In this method the thermometers being calibrated are compared to the reading from a calibrated thermometer whilst all the devices are immersed in an isothermal environment where the temperature is stable as well as being uniform. Isothermal baths are available (see Figures 5.22 and 5.23) for comparison calibrations between –100°C and 600°C using a variety of liquids such as silicone oil, water, alcohol or molten salts depending on the temperature limit of the substance (Table 6.5). Isothermal baths are heated using an electrical coil and an automatic stirrer or circulation system is provided to ensure that the environment is at a uniform temperature. For temperatures above 500°C fluidized beds and metal block baths are used although the uniformity of temperature within the calibration environment is less stable.

The procedure used for a comparison calibration is itemized below:

1 Insert the standard thermometer and the thermometers to be calibrated into the isothermal bath.
2 Allow the temperature within the bath to stabilise and become uniform.
3 Measure the resistance of the standard thermometer and use this value to determine the temperature of the bath.
4 Measure the resistance of each thermometer being calibrated.
5 Change the temperature of the bath and repeat steps 2–4 for as many temperatures as appropriate.

Table 6.5 Temperature range for isothermal bath mediums

Substance	Temperature range
Water	5 to 90 °C
Silicone oil	40 to 200 °C
Hot salt	150 to 600 °C

As a simple check for an IPRT or secondary SPRT, the ice point of water (which is 0.01°C lower than its triple point) provides a reference temperature with an uncertainty of 0.002°C provided appropriate procedures are followed (Horrigan, 1998; Wise, 1976).

Once data for resistance versus temperature for a particular sensor has been obtained, the calibration coefficients defining the temperature resistance characteristic for that sensor should be determined. These enable the temperature at any intermediate value to be evaluated. The coefficients concerned may be those from the ITS-90 interpolation equations or the Callendar–Van Dusen equation described in Section 6.2.1. As a general rule of thumb if the minimum required uncertainty for the measurement is less than or equal to 0.1°C then the ITS-90 interpolation equations are probably the most appropriate. If, however, the uncertainty of measurements desired is greater than 0.1°C then simpler equations can be used.

The frequency at which a PRT should be calibrated depends mostly on the temperature of operation. For operation below 400°C drift of the calibration is not a problem and periodic calibration, compatible with the need for traceability (say, once a year) may well be adequate. At higher temperatures from about 500°C to 600°C drift becomes significant and can be as large as several degrees per year and more frequent calibration is necessary.

Calibration of an RTD requires measurement of the resistance of both the SPRT and the PRT being calibrated. It is therefore essential that the method for determining resistance has itself been calibrated and is reliable.

6.3 Copper and nickel resistance thermometers

In addition to platinum, copper, nickel, nickel–iron and rhodium–iron are used in resistance thermometry. The resistance temperature characteristic of copper is fairly linear within 0.1°C at temperatures less than 200°C and this can have advantages in terms of the cost of the associated electronic measuring or control circuitry. The disadvantage of materials such as nickel and copper is their low resistance and their susceptibility to corrosion. Nickel resistance thermometers are relatively cheap in comparison to PRTs and have high sensitivity. The resistance temperature characteristic for nickel RTDs has been standardized by DIN 43760 for the range −60 to 180°C and is defined by

$$R_T = R_0 (1 + AT + BT^2 + CT^4) \tag{6.30}$$

where: $R_0 = 100\,\Omega$
$A = 5.450 \times 10^{-3}/°C$
$B = 6.65 \times 10^{-6}/°C^2$
$C = 2.695 \times 10^{-11}/°C^4$.

The high sensitivity and resistance of nickel RTDs make them a popular choice for use in the control and monitoring of, for example, air-conditioning systems. Nickel–iron RTDs with a composition of 70% nickel and 30% iron are available under the trade name of Balco. The basis for use of these devices is their high resistance. Ice point resistances can be between $2\,k\Omega$ and $10\,k\Omega$.

6.4 Rhodium–iron, doped germanium and carbon resistors

The PRT is suitable for use at low temperatures as identified by its use for the ITS-90 down to 13.81 K. Below 20 K, however, the sensitivity drops off but there are other resistance devices with favourable characteristics at low temperatures such as rhodium–iron, doped germanium and carbon resistors. Above 30 K, rhodium iron alloy (0.5% iron in rhodium) provides a similar resistance–temperature characteristic to platinum. Below 30 K, the sensitivity drops to a minimum between 25 and 15 K and then rises again giving a thermometer with good sensitivity at low temperatures (Rusby, 1982; Schuster, 1992). Germanium resistors are commercially available with a relatively wide temperature range (0.05–325 K). Doped germanium is commonly used with commercial devices, typically consisting of a chip of the semiconductor encapsulated in a 3 mm diameter, 8.5 mm long cylinder (e.g. the GR-200A available from Lake Shore Cryotronics Inc.). Germanium RTD are particularly suitable for temperature measurements in the range 0.05–30 K giving sensitivities of approximately 102–104 Ω/K at 1 K. Certain types of commercial resistors have been identified (following the original work of Clement and Quinnell, 1952) as having a resistance–temperature characteristic at low temperatures similar to that of germanium. Examples include radio resistors from Allen Bradley (USA), Matsushita (Japan), Airco Speer Co. (USA, albeit no longer in production) and Cryocal Co. (see Quinn, 1990; Rubin, 1970, 1997). The procedure commonly adopted is to remove the insulating skin and pot the resistance element in a protective capsule to prevent moisture absorption and stability problems. In order to improve the stability one option is to use carbon glass resistance thermometers, which consist of carbon fibres trapped in a glass matrix (see Besley, 1979; Lawless, 1972, 1981).

6.5 Thermistors

If the level of uncertainty is less critical then a cheaper form of RTD than PRTs is the thermistor. Thermistors consist of a ceramic semiconductor whose resistance is sensitive to temperature and were originally named from a shortened form of the term thermally sensitive resistor (Becker *et al.*, 1946). Thermistors have very high sensitivities with values of the order of

50 mV/°C, which is more than a hundred times that of PRTs and a thousand times that of thermocouples. Modern thermistors are usually mixtures of oxides such as the oxides of nickel, manganese, iron, copper, cobalt and titanium, other metals and doped ceramics. Thermistors can have either a negative temperature coefficient (NTC), where the resistance reduces with temperature, or a positive temperature coefficient (PTC) depending on the type of materials used. NTC and PTC thermistors each have specific features that make them appropriate for particular sensor applications. NTC thermistors are the most common and can operate at temperatures between −200°C and 1000°C. For temperatures above 300°C the refractory metal oxide devices are appropriate, whilst for temperatures of the order of 700°C devices utilizing zirconia doped with rare-earth oxides can be used. For low-temperature applications, non-stoichiometric iron oxides are suitable. PTC thermistors are also available for temperature measurement and are commonly used in control applications.

6.5.1 NTC thermistors

Negative temperature coefficient thermistors have the property that their resistance reduces dramatically with increasing temperature. They are normally manufactured using metal oxides such as those of manganese, nickel, cobalt, iron, copper and titanium. In the manufacturing process, a mixture of two or more metal oxides are combined with a suitable binder and formed into the desired geometry, dried and then sintered at an elevated temperature. By careful control of the oxide type, proportions and sintering atmosphere, a wide range of resistivities and temperature coefficient characteristics can be obtained. Commercial NTC thermistors can be broadly classified into two groups depending on how the leads are connected to the thermistor body: bead thermistors and metallized surface contacts. Thermistors are usually designated by their resistance at 25°C, with resistances commonly available between 10 Ω and 5 MΩ.

Bead thermistors have platinum alloy, silver or gold lead wires, which are directly sintered into the ceramic semiconductor body. Types of bead thermistors include bare beads, glass coated beads, glass probes, glass rods and bead-in-glass enclosures (Figure 6.15). Bare beads provide little if any protection against the environment and therefore should only be used where the environment is relatively inert. The use of a hermetic seal such as glass provides a tenfold improvement in the stability of thermistors in comparison to bare beads. Glass-coated thermistors are available or the bead can be embedded in the tip of a glass rod. Figure 6.16 illustrates the typical manufacturing process where the beads are formed in a series between two wires. These wires are then cut in appropriate locations to form the desired lead-out configuration.

Metallized surface contact thermistors consist of a metallized body fired onto ceramic thermistor material. A wide range of types of metallized surface

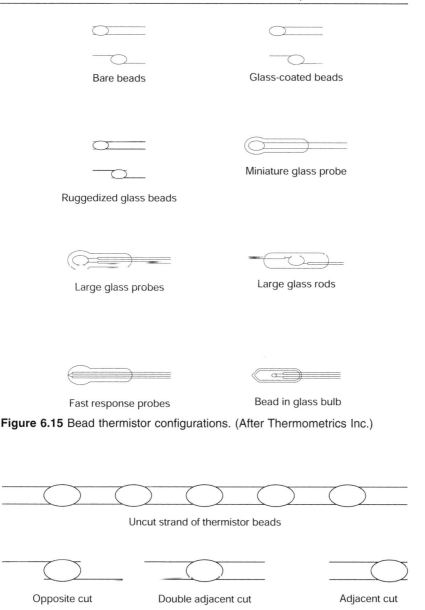

Bare beads

Glass-coated beads

Ruggedized glass beads

Miniature glass probe

Large glass probes

Large glass rods

Fast response probes

Bead in glass bulb

Figure 6.15 Bead thermistor configurations. (After Thermometrics Inc.)

Uncut strand of thermistor beads

Opposite cut

Double adjacent cut

Adjacent cut

Figure 6.16 Bead strands and lead configurations

contact thermistors are available including disks, chips or wafers, surface mounts, flakes, rods and washers, some of which are illustrated in Figure 6.17. Disk and chip thermistors tend to be larger than bead types and therefore have slower response times and larger thermal disturbance effects. Disk thermistor diameters range from 0.75 mm to 25.4 mm and are often used in low-cost

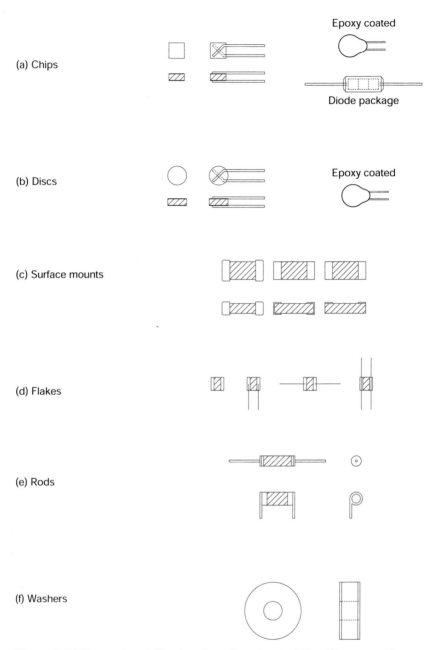

Figure 6.17 Types of metallized surface thermistors. (After Thermometrics Inc.)

thermometry applications. Typical cross-sectional areas for NTC chip thermistors range from 0.25 mm × 0.25 mm to 3 mm × 3 mm and thickness from 0.15 mm to 0.75 mm. Chip thermistors are sometimes used for precise measurements because of their low thermal disturbance, fast response and reasonable stability. Rod thermistors are formed by extruding a mixture of oxide powders and binder through a die. Their size and therefore relatively large time constants make them useful for temperature compensation or time-delay applications. Washer thermistors are similar to the disk types except that a hole is formed through the middle during the fabrication process. This can be of use in locating the device.

The typical variation of resistance with temperature for a variety of NTC thermistors is illustrated in Figure 6.18. Here the data is presented as a ratio of the resistance at a given temperature with the resistance at 25°C. The relationship of resistance for an NTC thermistor to temperature can be approximated by

$$R_T = A \exp(\beta/T) \tag{6.31}$$

where: R_T = resistance at temperature T (Ω)
A = constant for the particular thermistor under consideration (Ω)
β = constant for the particular thermistor under consideration (K)
T = absolute temperature (K).

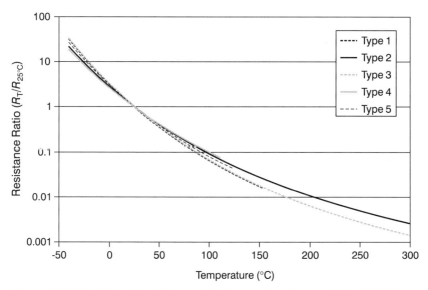

Figure 6.18 Variation of resistance with temperature for a variety of NTC thermistors. Data courtesy of Thermometrics Inc.

Equation (6.31) can be rewritten in the form

$$R_T = R_0 \exp\left[\beta\left(\frac{1}{T} - \frac{1}{T_0}\right)\right] \ (\Omega) \tag{6.32}$$

where: T_0 = a reference temperature, normally 290.15 K (25°C) or 273.15 K (0°C)

R_0 = resistance at temperature T_0 (Ω).

The temperature coefficient of resistance is given by

$$\alpha = \frac{1}{R_T}\frac{dR}{dT} = \frac{\beta}{T^2} \ (°C^{-1}) \tag{6.33}$$

Typical values for the temperature coefficient of resistance, α, are between 3%/°C and 6%/°C.

Equations (6.31) to (6.33) are only valid over small temperature spans where the slope of the $\ln(R)$ versus $1/T$ characteristic approximates a linear relationship. The error in the use of equation (6.31) for different temperature spans is listed in Table 6.6.

For more accurate results a polynomial can be used to represent the resistance temperature characteristic. The degree used for the polynomial depends on the temperature range and the type of thermistor. A third-order polynomial (equation (6.34)) is adequate for most systems:

$$\ln R_T = A_0 + \frac{A_1}{T} + \frac{A_2}{T^2} + \frac{A_3}{T^3} \tag{6.34}$$

where A_0, A_1, A_2, A_3 are constants, or

$$\frac{1}{T} = \alpha_0 + \alpha_1 R_T + \alpha_2 R_T^2 + \alpha_3 R_T^3 \tag{6.35}$$

where α_0, α_1, α_2, α_3 are constants.

Table 6.6 Approximate error inherent in equation (6.31) for specific temperature spans. Data after Thermometrics Inc.

Temperature span (°C)	Error (°C)
10	0.01
20	0.04
30	0.1
40	0.2
50	0.3

The T^2 terms in equations (6.34) and (6.35) do not contribute significantly and can sometimes be neglected:

$$\frac{1}{T} = \alpha_0 + \alpha_1 R_T + \alpha_3 R_T{}^3 \tag{6.36}$$

Seinhart and Hart (1968) compared the results between equations (6.35) and (6.36) for the range $-2°C$ to $30°C$ and reported no significant loss in uncertainty if the second-order term is neglected. In addition, Mangum (1983) reported interpolation errors of only 0.001°C over the range 0–70°C in the use of this equation. Sapoff *et al.* (1982) demonstrated that the interpolation errors in the use of equation (6.34) do not exceed the total measurement uncertainties for a wide range of glass probe thermistors. In order to determine the constants in equations (6.34) and (6.35), four calibration points are necessary. Equation (6.36) needs just three calibration points.

In order to measure the resistance of a thermistor a current must be passed through the circuit. As a result of I^2R Joule heating, power will be dissipated in the thermistor and this will heat it above its ambient temperature. The recommended practice is to limit the value of the current flow to such a value that the temperature rise due to I^2R power dissipation is less than the precision required for the temperature measurement. A typical thermistor, for example, with a resistance of $5\,k\Omega$ is capable of dissipating $1\,mW$ with a temperature rise of $1°C$. The corresponding self-heating factor, defined as the ratio of temperature rise to power dissipated, is $1°C/mW$. If the uncertainty required for the temperature measurement is 0.5°C then the power dissipated should be less than 0.5 mW. The maximum current to be used in the circuit can then be calculated for this example from

$$I = \sqrt{\text{power}/R} = \sqrt{0.0005/5000} = 0.316\,\text{mA} \tag{6.37}$$

The rate at which energy is dissipated as heat in a thermistor can be modelled by

$$\frac{dQ}{dt} = I^2R = \delta(T - T_\infty) + mc_p \frac{dT}{dt} \tag{6.38}$$

where: Q = energy supplied to thermistor (J)
t = time (s)
T = temperature (°C)
T_∞ = ambient temperature (°C)
c_p = specific heat capacity of the thermistor (J/kg·K)
m = mass of the thermistor (kg)
δ = constant determining the rate of heat loss to the surrounding environment as a function of temperature difference (J/s·K).

If the power is constant then the solution to equation (6.38) for temperature is

$$T = T_\infty + \frac{\text{power}}{\delta}\left[1 - \exp\left(-\frac{\delta}{mc_p}t\right)\right] \tag{6.39}$$

If the current supplied to the thermistor is reduced to such a value that there is negligible self-heating then equation (6.38) can be simplified to

$$\frac{dT}{dt} = -\frac{\delta}{mc_p}(T - T_\infty) \tag{6.40}$$

Integration of this equation leads to

$$T = T_\infty + (T_i - T_\infty)\exp\left(-\frac{t}{\tau}\right) \tag{6.41}$$

where: T_i = initial thermistor temperature (°C)
τ = mc_p/δ = thermal time constant (s).

Tables 6.7 and 6.8 provide dissipation and time constant data, to MIL-T-23648A (American Military Standards) procedures, for a range of glass and metallized surface thermistors respectively. The temperature increment used for determining the dissipation constant is from self-heating the thermistors to 75°C from an ambient temperature of 25°C. The time constant data relates to allowing the thermistors to cool from 75°C to an ambient temperature of 25°C.

Table 6.7 Thermal properties of selected hermetically sealed bead and probe thermistors. (After Sapoff, 1999)

Thermistor type	Diameter (mm)	Dissipation constant (mW/°C) still water	Time constant (s) still air
Glass-coated bead	0.13	0.045	0.45
Glass-coated bead	0.25	0.09	0.9
Glass-coated bead	0.36	0.10	0.98
Ruggedized bead	0.41	0.12	1.1
Glass probe	0.63	0.19	1.75
Glass-coated bead	0.89	0.35	3.8
Glass-coated bead	1.1	0.40	4.0
Ruggedized bead	1.4	0.51	4.3
Glass probe	1.5	0.72	4.4
Glass probe	2.16	0.90	4.5
Glass probe	2.5	1.0	4.5

Table 6.8 Thermal properties of selected metallized surface contact thermistors. (After Sapoff, 1999)

Thermistor type	Diameter (mm)	Dissipation constant (mW/°C)	Time constant (s)
Chip or disc in glass diode package	2	2 to 3	7 to 8
Interchangeable epoxy coated chip or disc	2.4	1	10
Disc with radial or axial leads	2.5	3 to 4	8 to 15
Disc with radial or axial leads	5.1	4 to 8	13 to 50
Disc with radial or axial leads	7.6	7 to 9	35 to 85
Disc with radial or axial leads		10.28 to 11	28 to 150
Disc with radial or axial leads		12.75 to 16	50 to 175
Disc with radial or axial leads		19.615 to 20	90 to 300
Disc with radial or axial leads		25.424 to 40	110 to 230
Disc with radial or axial leads	1.3	25 to 33	16 to 20
Disc with radial or axial leads	1.8	4 to 10	35 to 90
Disc with radial or axial leads	4.4	8 to 24	80 to 125

In order to measure temperature with a thermistor, its resistance must be determined. This can be achieved using a simple microammeter circuit or a Wheatstone bridge as illustrated in Figures 6.19 and 6.20. For the circuit shown in Figure 6.19, as thermistors have a comparatively high resistance, then as long as the voltage is constant, the current flow will be determined by change in the resistance of the thermistor. The thermistor can be mounted some distance from the ammeter and ordinary copper wire can be used for the

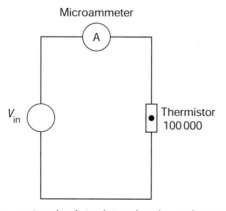

Figure 6.19 Microammeter circuit to determine the resistance of a thermistor

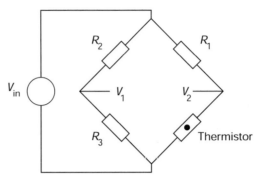

Figure 6.20 Use of a Wheatsone bridge to determine the resistance of a thermistor

leads. If the thermistor has nominal resistance of the order of $10^5 \, \Omega$, then the resistances of the lead wire can normally be neglected without impairing the uncertainty of the temperature reading significantly. For example, if the change of resistance per degree temperature change, α, is 4%/°C for the thermistor and if the resistance of lead wires is $10 \, \Omega$, then the lead wire error would be equivalent to only 0.0025°C error in temperature:

$$\Delta T_{\text{error}} = \frac{R_{\text{leads}}}{\alpha R_{\text{thermistor}}} = \frac{10}{0.04 \times 10^5} = 0.0025°C \tag{6.42}$$

Even if the nominal resistance of the thermistor was $1 \, \text{k}\Omega$, the error would still be only 0.25°C. As can be seen, the effect of lead wire resistance is not that significant and as a result three- or four-lead wire bridges are not commonly used in thermistor circuits.

The resistance characteristic of a thermistor expressed by equation (6.31) is negative and non-linear. This can be offset if desired by using two or more matched thermistors packaged in a single device so that the non-linearities of each device offset each other. The output voltage for a divider network if resistance R_1 and R_2 in series is

$$V_0 = \frac{V_s R_1}{R_1 + R_2} \text{ (V)} \tag{6.43}$$

If R_1 is a thermistor, the relationship between V_0 and temperature is non-linear and S-shaped. If R_1 is modified by the addition of another thermistor and resistor of appropriate values, the linearity of the centre section can be extended to cover a relatively wide temperature range. Over this range the temperature resistance characteristic can be approximated by a linear relationship. Packaged devices incorporating combinations of thermistors

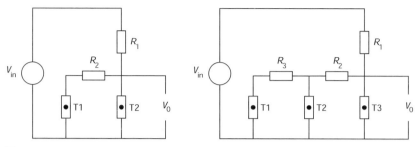

Figure 6.21 Two- and three-thermistor networks.

(Figure 6.21) in linearized circuits are commercially available and procedures for determining appropriate circuits are outlined in Beakley (1951), Sapoff and Oppenheim (1964), Sapoff (1980) and Blackburn (2001).

Thermistors can be calibrated using procedures similar to those detailed for thermocouples and PRTs. The uncertainty of thermistors can be as low as ±0.01°C to ±0.05°C, (ASTM E879–93), although commercial applications often result in an uncertainty of the order of ±1°C. The disadvantage of thermistors is their susceptibility to decalibration and drift due to changes in the semiconductor materials.

6.5.2 PTC thermistors

In PTC thermistors the resistance increases with temperature. They are manufactured using barium titanate and are not normally used in temperature measurement. These devices are selected when a significant change in the resistance is required at a specific temperature or current. Examples of applications include temperature sensing, switching, protection of windings in electric motors and transformers.

6.6 Semiconductor devices

Semiconductor-based temperature sensors are used in an increasing proportion of temperature measurement applications, mainly due to the need to monitor integrated circuit temperatures in many electronics applications. Sensors based on simple transistor circuits can be readily incorporated as part of an integrated circuit to provide on-board diagnostic or control capability. The typical temperature range for these devices is −55°C to 150°C. The temperature range for some smart integrated circuits can be extended to 185°C.

The majority of semiconductor junction sensors use a diode-connected bipolar transistor. If the base of the transistor is shorted to the collector (see Figure 6.22) then a constant current flowing in the remaining p–n junction (base

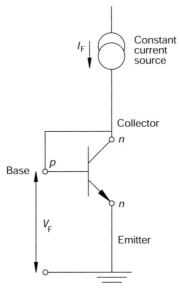

Figure 6.22 A bipolar transistor configured as a temperature sensor with the base shorted to the collector

to emitter) produces a forward voltage that is proportional to absolute temperature. A constant forward voltage supplied to an ideal p–n semiconductor junction produces a forward voltage drop that can be modelled by

$$V_F = \left(\frac{kT}{q}\right) \ln\left(\frac{I_F}{I_S}\right) \qquad (6.44)$$

where: k = Boltzmann's constant (1.38×10^{-23} J/K)
T = temperature (K)
q = charge of electron (1.6×10^{-19} C)
I_F = forward current (A)
I_S = junction's reverse saturation current (A).

In practice, the overall forward voltage drop has temperature coefficient of approximately −2 mV/°C.

The basic components of a semiconductor sensor are a bipolar transistor and a constant current supply. More sophisticated circuitry can be added to improve linearity, precision and specific features such as high and low set points. The sensor is produced in the form of a miniature integrated circuit and enclosed in a standard electronic package such as a TO99 can, a TO-92 plastic moulding or a DIP plug as illustrated in Figure 6.23.

There are a large number of types of semiconductor temperature sensor. This is because of their versatility, not only as a temperature sensor but also

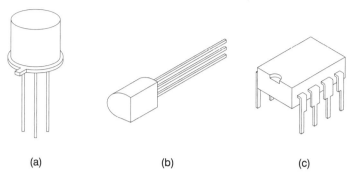

Figure 6.23 Typical semiconductor temperature sensor packaging. (a) TO99 can. (b) TO-92 plastic moulding. (c) DIP plug

as they can be integrated into a controller. For temperature sensing only, small-signal semiconductors such as the 2N2222 and 2N2904 transistors are common choices. Integrated temperature sensors such as the LM35 from National Semiconductors, provide a linear output voltage of 10 mV/°C operating in the range from –40°C to 110°C. The TMP-01 programmable temperature controller available from Analogue Devices gives a control signal from one of two outputs when the device is above or below a specific temperature that can be set by user-selected external resistors. The gate threshold voltage of a power MOSFET can be used to determine the junction temperature of a semiconductor component (Frank, 1999). Semiconductor temperature sensors can be incorporated within a larger integrated circuit to produce a smart sensor that can control regions of the circuitry on a chip at predetermined temperatures.

Semiconductor-based ICs suffer from a number of sources of uncertainty. The slope of the sensor output characteristic is only nominally adjusted to the quoted value. Using these devices without an external calibration circuit involves significant uncertainties. The sensor characteristic is non-linear, particularly at the ends of temperature ranges. This can be reduced by the addition of further electrical circuitry. A self-heating error occurs due to the forward current. This can be minimized by the use of low currents of the order of 0.1 mA and below. The uncertainty associated with one semiconductor device, the LM35, which has an output of 10 mV/°C, is ±0.8°C.

The advantages of these devices are their linearity, simple circuitry, good sensitivity, reasonable price and ready availability. As they are a high-impedance, current-based unit they can be used for remote sensing with just a twin copper cable required for connection purposes. Suitable digital voltmeters, some scaled in temperature units, are available for use with silicon transistor sensors. Such voltmeters frequently incorporate the necessary power supplies to operate the sensors. A disadvantage is the need to add extra circuitry and for calibration to attain a reasonable uncertainty.

6.6.1 Diode thermometers

The forward voltage drop across a p–n junction increases with decreasing temperature. For some semiconductors, the relationship between voltage and temperature is almost linear; in silicon this occurs between 400 K and 25 K with a corresponding sensitivity of approximately 2.5 mV/K. The two most commonly used semiconductors for thermometry are GaAs and Si. The typical voltage sensitivities for these devices are illustrated in Figure 6.24. For silicon at temperatures below about 25 K, when the forward voltage approaches 1.1 V, the characteristic function for the voltage temperature relationship changes. As can be seen in Figure 6.24, Si diodes give a lower output than GaAs diodes but have better stability and are cheaper and easily interchangeable. Generally rectifying diodes are used and these can be potted in a small container. Commercial versions are available, for example the 1.25 mm in diameter and 0.75 mm long DT-420 device, from Lake Shore Cryotronics. Zener diodes have also been used to indicate temperature (Szmyrka-Grzebyk and Lipinski, 1995).

The virtues of diode thermometers are their low price, a simple voltage–temperature relationship, a relatively large temperature range (1 to 400 K), no need for a reference bath or junction, relatively high sensitivity, an uncertainty lower than ±50 mK (Krause and Dodrill, 1986) and simplicity of operation with a constant current source and a digital voltmeter. Measurements down to 1 K do, however, require careful calibration. Errors can occur if the supply current is not a true DC but has an AC component due to, say, noise induced in the circuit from improper shielding, electrical grounds or ground loops. To minimize noise effects, all the instrumentation should be electrically shielded

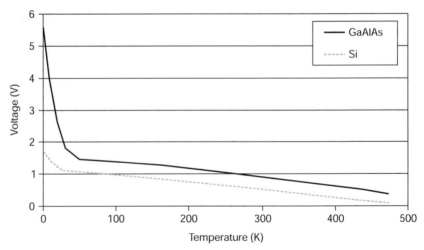

Figure 6.24 Voltage–temperature characteristic for GaAlSi and Si diode thermometers

and proper grounding techniques used (Morrison, 1998). The diode current supply should have a single ground, generally at the voltmeter, which then requires a floating current source. A current between 1 μA and 100 μA can be used but generally needs to exceed 10 μA in order to overcome noise problems. However, this can cause problems associated with self-heating at very low temperatures.

6.7 Selection

The selection process for RTDs can follow a number of approaches. The decision to use an RTD must be considered with reference to other types of temperature measurement. Is an RTD the most suitable type of temperature measurement or might another method such as using a thermocouple or an infrared method be more suitable? The relative merits of PRTs versus some competing categories of invasive instrumentation are listed in Table 6.9. The decision to use say a PRT may be based on uncertainty requirements. Generally below about 700°C the best PRTs are less uncertain and more reliable than any thermocouple (Quinn, 1990). Some of the criteria involved in the selection of a PRT are listed below:

- *Temperature range.* Different PRTs are suitable for different temperature ranges due to thermal expansion considerations and material limitations. The maximum temperature to which a PRT can be used is usually a function of what temperature the sheath and the insulation can withstand and whether the sensor will be affected by oxidation and drift. The minimum temperature capability of a sensor can be limited by the substrate used for supporting the sensor as some ceramics can shatter due to thermal shock when they undergo rapid cooling.
- *Uncertainty.* The uncertainty of PRTs is between ±1°C and ±1 mK and depends on the temperature range, type of construction and calibration. A rule of thumb is that the cost of a thermometer is inversely proportional to the required uncertainty; a ±1 mK system comprising the sensor and the sensing electronics may cost one thousand times as much as a ±1°C system.
- *Chemical compatibility.* Will the sheath material resist chemical attack in the environment of the application? Some liquids can corrode stainless steel sheaths or the insulation used for the leads.
- *Shock resistance.* Will the sensor withstand the mechanical vibration in the application or the strain imposed on the sensor?
- *Response.* The response rate of a sensor depends on the thermal mass of the sensor, its installation and the local heat transfer. Essentially the larger the sensor package, the slower the response.

Table 6.9 Comparison table for various categories of invasive temperature sensor. (After Liptak, 1993, data various sources)

Evaluation criteria	PRT 100 Ω wire wound	PRT 100 Ω thin film	Ni RTD 1000 Ω wire wound	Thermistor	Thermocouple	Semi-conductor
Cost	High	Low	Medium	Low	Low	Low
Temperature range	-240°C to 900°C	-200°C to 650°C	-212°C to 316°C	-73°C to 260°C	-268°C to 2316°C	-49°C to 125°C
Stability	Good	Good	Fair	Poor	Poor to fair Variable, Prone to ageing	Good to fair
Uncertainty (typical values)	Low (0.1 to 1.0°C)	Low	Medium	Medium	Medium (0.5°C to 5°C)	Medium
Repeatability	Excellent	Excellent	Good	Fair to Good	Fair	Good
Sensitivity	Medium	High	High	High	Low	High
Response (typical values)	Medium (1 to 50s)	Medium to fast	Medium	Medium to fast	Medium to fast (0.1 to 10s)	Medium to fast
Linearity	Good	Good	Fair	Poor	Fair	Good
Excitation	Current	Current	Current	Current	None	Current
Self-heating	Low	Medium	Medium	High	N/A	Low
Point-end sensitive	Fair	Good	Poor	Good	Excellent	Good
Lead effect	Medium	Low	Low	Very low	High	Low
Physical size	Medium to small	Small to large	Large	Small to medium	Small to large	Small to medium

References

Books and papers

Beakley, W.R. The design of thermistor thermometers with linear calibration. *Journal of Scientific Instruments*, **28**, 176–179, 1951.

Becker, J.A., Green, C.B. and Pearson, G.L. Properties and uses of thermistors. *Trans. AIEE*, **65**, 711, 1946.

Berry, R.J. The influence of crystal defects in platinum on platinum resistance thermometry. In Plumb, H.H. (Editor), *Temperature. Its Measurement and Control in Science and Industry*, Vol. 4, pp. 937–949. Instrument Society of America, 1972.

Berry, R.J. Evaluation and control of platinum oxidation errors in standard platinum resistance thermometers. In Schooley, J.F. (Editor), *Temperature. Its Measurement and Control in Science and Industry*, Vol. 5, pp. 743–752. American Institute of Physics, 1982a.

Berry, R.J. Oxidation, stability and insulation characteristics of Rosemount standard platinum resistance thermometers. In Schooley, J.F. (Editor), *Temperature. Its Measurement and Control in Science and Industry*, Vol. 5, pp. 753–762. American Institute of Physics, 1982b.

Besley, L.M. Stability characteristics of carbon glass resistance thermometers. *Rev. Sci. Instr.*, **50**, 1626–1628, 1979.

Blackburn, J.A. *Modern Instrumentation for Scientists and Engineers*. Springer, 2001.

Callendar, H.L. On the practical measurement of temperature. Experiments made at the Cavendish laboratory, Cambridge. *Phil. Trans. Royal Society, London*, **178**, 161, 1887.

Callendar, H.L. On construction of platinum thermometers. *Phil. Mag*, **34**, 104, 1891.

Callendar, H.L. Notes on platinum thermometry. *Phil. Mag*, **47**, 191, 1899.

Clement, J.R. and Quinnell, E.H. The low temperature characteristics of carbon composition thermometers. *Review of Scientific Instruments*, **23**, 213–216, 1952.

Connolly, J.J. Resistance thermometer measurement. In Bentley, R.E. (Editor), *Handbook of Temperature Measurement, 2, Resistance and Liquid in Glass Thermometry*. Springer, 1998a.

Connolly, J.J. Industrial resistance thermometers. In Bentley, R.E. (Editor), *Handbook of Temperature Measurement, 2, Resistance and Liquid in Glass Thermometry*. Springer, 1998b.

Curtis, D.J. Thermal hysteresis effects in platinum resistance thermometers. In Schooley, J.F. (Editor), *Temperature. Its Measurement and Control in Science and Industry*, Vol. 5, pp. 803–812. American Institute of Physics, 1982.

ESDU 84036. *Temperature Measurement: Resistance thermometry.* Engineering Sciences Data Unit, 1984.

Frank, R. Semiconductor junction thermometers. Section 32.5, in Webster, J.G. (Editor), *The Measurement, Instrumentation and Sensors Handbook*. CRC Press, 1999.

Hashemian, H.M. and Petersen, K.M. Achievable accuracy and stability of industrial RTDs. In Schooley, J.F. (Editor), *Temperature. Its Measurement and Control in Science and Industry*, Vol. 6, pp. 427–432. American Institute of Physics, 1992.

Horrigan, C. Calibration enclosures. In: Bentley, R.E. (Editor), *Handbook of Temperature Measurement, Vol. 2, Resistance and Liquid in Glass Thermometry.* Springer, 1998.

Johnston, J.S. Resistance thermometry. *Temperature-75*, pp. 80–90, 1975.

Jones, T.V. The thin film heat transfer gauge – a history and new developments. *Proc. 4th UK Conference on Heat Transfer*, Paper C510/150/95, pp. 1–12, 1995.

Krause, J.K. and Dodrill, B.C. Measurement system induced errors in diode thermometry. *Rev. Sci. Instrum*, **57**, 661–665, 1986.

Lawless, W.N. Thermometric properties of carbon impregnated porous glass at low temperatures. *Review of Scientific Instruments*, **43**, 1743–1747, 1972.

Lawless, W.N. Thermal properties of carbon impregnated porous glass at low temperatures. *Review of Scientific Instruments*, **52**, 727–730, 1981.

Li Xumo, Zhang Jinde, Su Jinroug and Chen Deming. A new high temperature platinum resistance thermometer. *Metrologia*, **18**, 203–208, 1982.

Liptak, B.G. (Editor). *Temperature Measurement*. Chilton Book Co., 1993.

Mangum, B.W. The triple point of succinonitrile and its use in the calibration of thermistor thermometers. *Review of Scientific Instruments*, **54**, 1687, 1983.

Mangum, B.W. and Furukawa, G.T. Guidelines for realizing the ITS-90. NIST Technical Note 1265, 1990.

Meyers, C.H. Coiled filament resistance thermometers. *National Bureau of Standards Journal of Research*, **9**, 807–813, 1932.

Morrison, R. *Grounding and Shielding Techniques*, 4th edition. Wiley, 1998.

Nicholas J.V. and White D.R. *Traceable Temperatures*. Wiley, 1994.

Quinn, T.J. *Temperature*, 2nd edition. Academic Press, 1990.

Preston-Thomas, H. The international temperature scale of 1990 (ITS-90). *Metrologia*, **27**, 3–10, 1990.

Riddle, J.L., Furukawa, G.T. and Plumb, H.H. *Platinum Resistance Thermometry.* NBS 1973.

Rubin, L.G. Cryogenic thermometry: a review of recent progress. *Cryogenics*, **10**, 14–20, 1970.

Rubin, L.G. Cryogenic thermometry: a review of progress since 1982. *Cryogenics*, **37**, 341–356, 1997.

Rusby, R.L. The rhodium–iron resistance thermometer: Ten years on. In Schooley, J.F. (Editor), *Temperature. Its Measurement and Control in Science and Industry.* Vol. 5(2), pp. 829–834, American Institute of Physics, 1982.

Sapoff, M. Thermistors: Part 4, Optimum linearity techniques. *Measurements and Control*, **14**(10), 1980.

Sapoff, M. Thermistor thermometers. Section 32.3, in Webster, J.G. (Editor). *The Measurement, Instrumentation and Sensors Handbook*. CRC Press, 1999.

Sapoff, M. and Oppenheim, R.M. Theory and application of self-heated thermistors. *Proc IEEE*, Vol. 51, p. 1292, 1964.

Sapoff, M., Siwek, W.R., Johnson, H.C., Slepian, J. and Weber, S. The exactness of fit of resistance-temperature data of thermistors with third degree polynomials. In Schooley, J.F. (Editor), *Temperature. Its Measurement and Control in Science and Industry.* Vol. 5, pp. 875–888. American Institute of Physics, 1982.

Sawada, S. and Mochizuki, T. Stability of 25 ohms platinum thermometer up to 1100°C. In Plumb, H.H. (Editor), *Temperature. Its Measurement and Control in Science and Industry.* Vol. 4, pp. 919–926, Instrument Society of America, 1972.

Schuster, G. Temperature measurement with rhodium–iron resistors below 0.5 K. In Schooley, J.F. (Editor), *Temperature. Its Measurement and Control in Science and Industry.* Vol. 6(1) pp. 449–452. American Institute of Physics, 1992.

Seinhart, J.S. and Hart, S.R. Calibration curves for thermistors. *Deep Sea Research,* **15**, 497, 1968.

Siemens, W.H. On the increase of electrical resistance in conductors with rise of temperature and its application to the measure of ordinary and furnace temperatures; also on a simple method of measuring electrical resistances. *Proc. Royal Society, London,* **19**, 443, 1871.

Sostman, H.E. and Metz, P.D. Fundamentals of thermometry Part III. The standard platinum resistance thermometer. *Isotech Journal of Thermometry,* **2**, No. 1, 1991.

Szmyrka-Grzebyk, A. and Lipinski, L. Linear diode thermometer in the 4–300 K temperature range. *Cryogenics,* **35**, 281–284, 1995.

Trietley, H.L. Avoiding error sources in platinum resistance temperature measurement. *Instrumentation Technology,* **29**(2), 57–60, 1982.

Wise, J.A. *Liquid in Glass Thermometry,* NBS Monograph 150, 1976.

Wolfendale, P.C.F., Yewen, J.D. and Daykin, C.I. A new range of high precision resistance bridges for resistance thermometry. In Schooley, J.F. (Editor), *Temperature. Its Measurement and Control in Science and Industry,* Vol. 5(2), pp. 729–732. American Institute of Physics, 1982.

Wood, S.D., Mangum, B.W., Filliben, J.J. and Tillett, S.B. An investigation of the stability of thermistors. *J. Research of the NBS,* 247–263, 1978.

Standards

ASTM E1137. Standard specification for industrial platinum resistance thermometers (1997).

ASTM E879–93. Standard specification for thermistor sensors for clinical laboratory temperature measurements (1993).

BS 1041: Part 3. Temperature measurement. Guide to the selection and use of industrial resistance thermometers (1989).

EN 60751. Industrial platinum resistance thermometer sensors (1996).

DIN 43760

IEC-751

Web sites

At the time of going to press the world wide web contained useful information relating to this chapter at the following sites.

http://ametherm.com/ntc_desription.htm
http://analog.digital.vmic.com/products/hw_damc_vme_3220.html
http://content.honeywell.com/building/components/Hycal_Html/HYT8.asp
http://edl-inc.com/RTD%20Sensors.htm
http://marlinmfg.com/mCata.htm
http://sensycon.com/txt/txt_21be.htm
http://thermometricscorp.com/index.html
http://thermoprobe-inc.com/html/why_rtds.htm
http://users.aol.com/rhines81/cgs.htm

http://www.advmnc.com/ASL/frbridge.htm
http://www.ari.co.uk/Product%20Page.htm#RTD
http://www.aslinc.com/thermometer.htm
http://www.aslltd.co.uk/prt.htm
http://www.bbrown.com/products/XTR105/
http://www.conaxbuffalo.com/products/temp_sensors.html
http://www.crouse-hinds.com/products/intrinsically_safe/techref/article2.htm
http://www.enercorp.com/temp/temp.htm
http://www.engberg.dk/prices/cbi/pri_acc.htm
http://www.engelhard.com/
http://www.execulink.com/~elkor/etrtd.htm
http://www.fenwal.com/
http://www.hartscientific.com/products/bridge.htm
http://www.hartscientific.com/products/othersprt.htm
http://www.heraeus.de/e/sensors.htm
http://www.industrialtechnology.co.uk/labfac.htm
http://www.iotech.com/da/adthermo_g.html
http://www.isotech.co.uk/calibration.html
http://www.jms-se.com/section3.html
http://www.kscorp.com/www/camac/3500/3565.html
http://www.labfacility.co.uk/frames.html
http://www.minco.com/sensors.htm
http://www.mtisensors.com/rtds.html
http://www.nist.gov/cstl/div836/836.05/greenbal/sprtc.html
http://www.okazaki-mfg.com/tec_info/tec_info.html
http://www.omega.com/
http://www.picotech.com/applications/pt100.html
http://www.pondengineering.com/
http://www.pyromation.com/
http://www.pyromation.com/products/rtd.html
http://www.rdfcorp.com/
http://www.rosemount.com/products/temperature/accessories.html
http://www.rtdco.com/producto.htm
http://www.scientificinstruments.com/sensors/rtd.htm
http://www.sensoray.com/html/7429data.htm
http://www.sensorsci.com/thermistors_rtd_thermocouples.htm
http://www.sisweb.com/ms/sis/prt.htm
http://www.tc.co.uk/index2.htm
http://www.temperatures.com/rtdhow.html
http://www.temp-pro.com/tpprdrtd.htm
http://www.thermo-kinetics.com/manufact.htm
http://www.thermometrics.com/htmldocs/ntcres.htm
http://www.thomasregister.com/olc/rtielectronics/surfmnt.htm
http://www.tinsley.co.uk/index.htm
http://www.weedinstrument.com/indhmpg.htm

http://www.wici.com/technote/tmprmch1.htm#rtds
http://www.wuntronic.de/sensors/therm_cal.htm
http://www.ysi.com/ysi/medweb.nsf

Nomenclature

A = cross-sectional area (m^2) or a constant
c_p = specific heat capacity of the thermistor (J/kg·K)
I = current (A)
I_F = forward current (A)
I_S = junction's reverse saturation current (A)
k = Boltzmann's constant (1.38×10^{-23} J/K)
L = length (m)
m = mass of the thermistor (kg)
N = number of turns
q = charge of electron (1.6×10^{-19} C)
Q = energy supplied to thermistor (J)
R = resistance (Ω)
R_L = lead resistance (Ω)
R_S = resistance of standard resistor (Ω)
R_T = resistance at temperature T (Ω)
R_{TP} = resistance at the water triple point (Ω)
R_T' = resistance in presence of impurities (Ω)
R_0 = resistance at 0°C (Ω)
R_0' = ice point resistance in presence of impurities (Ω)
R_{100} = resistance at 100°C (Ω)
t = time (s)
T = temperature (°C or K)
T_i = initial temperature (°C)
T_∞ = ambient temperature (°C)
V = voltage (V)
W = ratio of resistance
α = temperature coefficient of resistance ($\Omega\Omega^{-1}$ °C^{-1} or °C^{-1})
β = constant for the particular thermistor under consideration (K)
δ = constant determining the rate of heat loss to the surrounding environment as a function of temperature difference (J/s·K)
ρ = resistivity (Ω·m)
ρ_t = component of resistivity due to thermal effects (Ω·m)
ρ_i = component of resistivity due to the presence of impurities (Ω·m)
ρ_d = component of resistivity due to lattice deformation arising from mechanical stress (Ω·m)
τ = thermal time constant (s)
ζ = dissipation constant (W/°C)
ΔT = temperature rise (°C)
ΔT_{error} = error in temperature (°C)

7

Manometric thermometry

Manometric thermometry refers to devices using the measurement of pressure to determine the temperature. This principle is used in gas thermometers and vapour pressure thermometers. Manometric thermometry can be used to produce measurements with very low uncertainty and is employed as a principal technique in the formulation of temperature scales and cryogenic thermometry. In addition, gas thermometry can be used in relatively simple devices to provide an indication of temperature. This chapter reviews the various techniques of manometric thermometry.

7.1 Introduction

Manometric thermometry is used here to refer to devices that utilize the measurement of the pressure of a gas in order to determine the temperature. There are two principal classes of manometric thermometers: gas and vapour pressure thermometers. Gas thermometry is based on the variation of temperature according to the ideal gas law and is applicable across a wide temperature range from a fraction of a kelvin to about 1000 K. The method is capable of very low uncertainty and is used in the ITS-90 (Preston Thomas, 1990). Gas thermometry is described in Section 7.2. Vapour pressure thermometry is based on the observation that the saturation vapour pressure of a pure substance above its liquid phase varies only with temperature. This technique can offer even lower uncertainty measurements than gas thermometry in some applications and is described in Section 7.3.

7.2 Gas thermometry

The ideal gas law is given by

$$pV = nRT \tag{7.1}$$

where: p = pressure (N/m^2)
V = volume (m^3)
n = number of moles of the gas ($n = m/M$ (m = mass, M = molar mass))
R = the universal gas constant (= 8.314510 J/mol K (Cohen and Taylor, 1999))
T = temperature (K).

Examination of equation (7.1) shows that the pressure of an ideal gas at constant volume is linearly proportional to the absolute temperature. Therefore by containing a quantity of gas in an enclosure of fixed volume and, for example, measuring the pressure the absolute temperature can be evaluated. An alternative is to maintain a constant pressure and measure the volume.

The basic components of a gas thermometer are an enclosure to contain the gas sample of interest, under carefully controlled conditions and a flow circuit to allow the pressure to be measured. Although gas thermometers can be used at temperatures up to 1000 K, the principal application has been in cryogenics and the gas enclosure, commonly known as the bulb, is usually located within a cryostat.

If a measurement of temperature is based on the ideal gas equation in the form of equation (7.1), the uncertainty depends on, among other things, the value of the gas constant. This has been modified a number of times in the light of lower uncertainty measurements (e.g. see Cohen and Taylor, 1973, 1987). As a result, a number of methods have been devised that eliminate the need for knowing the gas constant operating on the principle of maintaining either a constant pressure or a constant volume and/or a constant bulb temperature. Gas thermometry techniques include:

- absolute PV isotherm thermometry
- constant volume gas thermometry
- constant pressure gas thermometry
- constant bulb temperature gas thermometry
- two bulb gas thermometry.

The five techniques are illustrated in Figure 7.1. In this figure, T_1 refers to a known reference temperature, as defined by, say, a triple point, and T_2 is the temperature to be measured. The diagrams on the left side of the figure show the schematic arrangement and thermodynamic properties at a reference condition; those on the right side, at the measuring condition.

In pV isotherm thermometry a bulb of known volume at an unknown temperature is filled with a series of increasing amounts of gas, nR, and the corresponding pressure measured. By assuming a value for the gas constant pV/nR can be plotted as a function of n/V (equation 7.2) and the intercept of the resulting isotherm gives the unknown temperature T_1.

$$\frac{pV}{nR} = T \left[1 + B(T) \left(\frac{n}{V} \right) + C(T) \left(\frac{n}{V} \right)^2 + \ldots \right] \tag{7.2}$$

The quantity of gas can be determined by a weighing process. For instance, the bulb can be weighed when evacuated and then again with a charge of gas. This method for determining the quantity of gas has now been superseded for low uncertainty measurements by a method based on a knowledge of

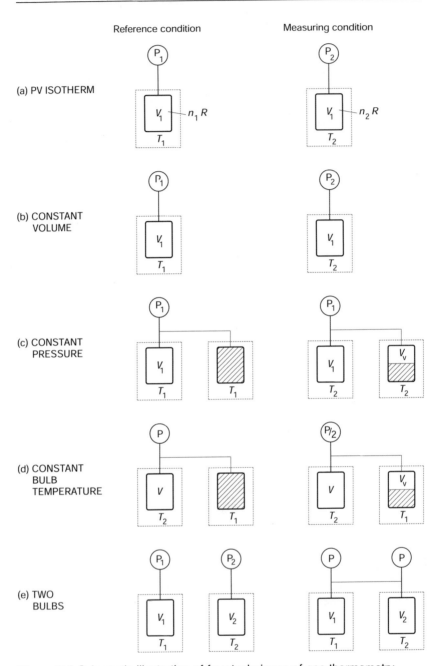

Figure 7.1 Schematic illustration of four techniques of gas thermometry.
(a) pV isotherm thermometry. (b) Constant volume gas thermometry.
(c) Constant pressure gas thermometry. (d) Constant bulb temperature gas
thermometry. (e) Two bulb gas thermometry. Figure adapted from Guildner
and Thomas (1982)

the virial coefficients of the gas at the temperature of a reference volume (see Quinn, 1990).

The most common method is constant volume gas thermometry. For this method, a bulb containing the gas is immersed in a fluid at the known reference temperature T_1. The pressure p_1 is measured under conditions of thermal equilibrium and again at temperature T_2, from which:

$$T_2 = T_1 \frac{p_2}{p_1} \tag{7.3}$$

Constant volume gas thermometers can take many forms ranging from monitoring devices to very low uncertainty instruments for the measurement of temperature for standards and scientific and industrial applications. Figure 7.2 illustrates the principal components of a constant volume gas thermometer. A system of pipes and valves allows the thermometer to be evacuated before a charge of the filling gas is admitted. The filling gas should be as close to an ideal gas as possible and must not condense over the temperature range of application. Helium, nitrogen and argon are the more common choices and air is sometimes used at higher temperatures. For low uncertainty measurements the bulb volume should have a value much greater than that of the wetted volume of the pressure-measuring device and the connection tubes. This is often not possible due to the constraints on cryostats. Bulb volumes in practice vary from approximately $1000 \, \text{cm}^3$ for low uncertainty work to as little as $50 \, \text{cm}^3$ when high uncertainty can be tolerated or practical constraints on volume cannot be overcome. For monitoring applications a Bourdon-type pressure gauge-based device as available from a number of commercial

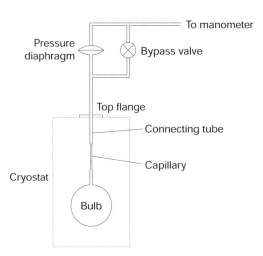

Figure 7.2 The principal components of a constant volume gas thermometer (after Pavese and Molinar, 1992)

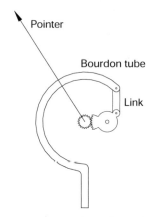

Figure 7.3 A Bourdon pressure gauge-based gas thermometer

suppliers may be suitable. The principle of operation for these devices is illustrated in Figure 7.3.

In the constant pressure gas thermometer, the mass and pressure are kept constant and the volume change resulting from a change in temperature is monitored. This can be achieved by the use of two bulbs as illustrated in Figure 7.1(b). From the ideal gas law the temperature ratio will be equal to the volume ratio which is given, for $T_2>T_1$, by:

$$T_2 = \frac{(V_1 + V_v)T_1}{V_1} \qquad (7.4)$$

The constant pressure method has not received as much attention as the constant volume technique because of the complexity of having to use an expandable enclosure.

In a constant bulb-temperature device, one bulb is always immersed at T_1 and the other always at T_2. At the reference condition the variable volume $V_v = 0$; at the measuring condition V_v expands until the pressure reaches half that at the reference condition and so:

$$T_2 = \frac{VT_1}{V_v} \qquad (7.5)$$

In the two-bulb device, the separate pressures, p_1 and p_2 are measured at the reference condition and the common pressure, p, at the measuring condition from which:

$$T_2 = \frac{V_2 T_1 (p - p_2)}{V_1 (p_1 - p)} \qquad (7.6)$$

In practice real gases do not behave exactly according to the ideal gas equation. The non-ideal nature can be modeled using the virial equation

$$p = \frac{nRT}{v}\left(1 + B(T)\left(\frac{n}{v}\right) + C(T)\left(\frac{n}{v}\right)^2 + D(T)\left(\frac{n}{v}\right)^3 + \ldots\right) \qquad (7.7)$$

where $B(T)$, $C(T)$ and $D(T)$, etc. are the second, third and fourth virial coefficients. Some of these coefficients have been evaluated for a number of gases at various temperatures (see, for example, Aziz et al., 1979; Berry, 1979; Kemp et al., 1986; Matacotta et al., 1987; Steur et al., 1987; Steur and

Table 7.1 Data for the second and third virial coefficients for ^{4}He (Quinn, 1990)

T (K)	$B(T)$ cm$^3\cdot$mol^{-1}	T (K)	$C(T)$ cm$^6\cdot$mol^{-2}
2.6	−142.44	2.6	−588.02
3	−120.34	3	715.27
3.5	−100.21	3.2	944.75
4.2221	−79.77	3.4	1057.52
5	−64.45	3.6	1105.73
10	−23.12	3.7	1115.06
13.8033	−11.82	3.8	1117.73
15	−9.47	3.9	1115.42
17.0357	−6.24	4.2	1089.74
20.2711	−2.48	5	976.34
24.5561	0.93	10	546.91
30	3.79	13.8033	421.93
40	6.90	17.0357	359.51
60	9.74	20.2711	316.96
80	10.97	24.5561	277.86
100	11.58	40	206.45
150	12.12	60	168.6
200	12.17	80	149.67
250	12.06	100	138.32
273.15	11.99	150	123.18
298.15	11.89	200	115.61
323.15	11.77	273.15	109.53
348.15	11.67		
373.15	11.56		
423.15	11.36		
573.15	10.76		
673.15	10.45		
773.15	10.14		
873.15	9.82		

Durieux, 1986; Astrov *et al.*, 1989; Luther *et al.*, 1996). In the case of helium 3, for example, C(T) and the higher-order coefficients are very small and can be neglected. Data for the second and third virial coefficients for ^{4}He and the second virial coefficient for ^{3}He are listed in Tables 7.1 and 7.2.

In order to provide low uncertainty measurements a number of corrections to account for the physical behaviour of a gas thermometer can be made including:

- the dead space in connection tubes
- for thermal expansion of the gas bulb
- the difference in density of the gas at different levels in the pressure-sensing tubes
- a thermomolecular pressure correction to account for temperature differences along the pressure-sensing tube
- for the absorption of impurities in the gas.

Practical constraints commonly dictate that the sensing bulb and the pressure sensor are not at the same temperature but are connected by means of tubes as indicated in Figure 7.2. The connection can take the form of capillary tubes chosen to minimize the volume of fluid that is not at the temperature of the bulb. A correction known as the deadspace correction can be made to account for the volume of fluid in the connection tubes and associated valves. The dead space correction can be determined by a gas expansion technique (see Berry, 1979). In this technique, the bulb is disconnected, the connection end closed off and the connection tube

Table 7.2 Data for the second virial coefficients for ^{3}He (Quinn, 1990)

T (K)	B(T) $cm^3 \cdot mol^{-1}$
1.5	−171.6
2.1768	−120.24
3	−86.03
3.5	−72.48
4.2221	−58.2
5	−47.17
7	−29.63
10	−16.11
13.8033	−7.25
15	−5.37
17.0357	−2.78
20.2711	0.29

evacuated. A gas from a known reference volume is then allowed to expand into the evacuated volume and the resulting pressure P_2 measured. The deadspace correction is then given by

$$D_R = \frac{V_R}{T_R} \left(\frac{p_1 - p_2}{p_2 - p} \right) \tag{7.8}$$

where: T_R = temperature of the gas in the reference bulb (K)
        V_R = volume of the reference bulb (m^3)
        p_1 = pressure on the reference bulb before expansion (N/m^2)
        p_2 = pressure in enclosure after expansion (N/m^2)
        p = the residual gas pressure in the deadspace volume prior to the expansion (N/m^2).

The component parts of a gas thermometer such as the bulb and connection tubes will expand with temperature. For a cubic enclosure the expansion can be determined from the relationship:

$$\frac{V_2}{V_1} = \frac{(L_1 + L_1 \, \alpha \Delta T)^3}{L_1^3} = (1 + \alpha \Delta T)^3 \tag{7.9}$$

where: V_1 = original volume (m^3)
        V_2 = final volume (m^3)
        L_1 = original length (m)
        α = coefficient of linear expansion (K^{-1})
       ΔT = temperature rise (K).

Values for the coefficient of linear expansion for a variety of materials used for gas thermometer bulbs are listed in Table 7.3.

Table 7.3 Linear coefficients of thermal expansion for a variety of materials used for gas thermometer bulbs (data after Goodfellow Cambridge Ltd (catalogue 2000/2001))

Material	α (K^{-1})
Al	23.5×10^{-6}
Graphite	0.6 to 4.3×10^{-6}
Cu	17×10^{-6}
Pt	9×10^{-6}
Invar	1.7 to 2.0×10^{-6}
Al$_2$O$_3$	8×10^{-6}

For low uncertainty measurements it is necessary to account for the variation of pressure with depth in a fluid which is given by the relation

$$\Delta p = \int_0^h \rho(h)g\,dh = \frac{Mpg}{R} \int_0^h \frac{dh}{T(h)} \tag{7.10}$$

where: h = height (m)
         $\rho(h)$ = density at height h of pressure line (kg/m^3)
         g = acceleration due to gravity (m^2/s)
         $T(h)$ = temperature at height h of pressure line (K)
         M = molar mass (kg/mol)
         p = pressure (N/m^2)
         R = gas constant (J/mol·K).

Hydrostatic pressure variation is of particular significance in low-temperature gas thermometry (see, for example, Berry, 1979; Pavese and Steur, 1987; and Quinn, 1990).

If a capillary tube has a diameter that is smaller than the mean-free path of the gas then corrections should be made for the thermomolecular pressure difference arising when there is a temperature gradient along the tube. For a capillary of a diameter less than the mean-free path of the gas the relationship between the pressures at the two ends is given by

$$p_1 = p_2 \sqrt{\frac{T_1}{T_2}} \tag{7.11}$$

Guildner and Edsinger (1982) reported pressure corrections exceeding 400 parts per million for a 0.8 mm diameter Pt 10% Rh capillary held at 10 kPa with end temperatures of 273 and 1000 K. In practical situations the thermomolecular pressure is critically dependent on the surface condition of the tube and the validity of the simple relationship given in equation (7.11) is limited. Methods for determining it are reported by McConville (1972), Weber and Schmidt (1936), Guildner and Edsinger (1976) and Berry (1979).

Constant volume gas thermometry normally assumes that the quantity of gas within the system is fixed. This is compromised by two effects: adsorption of molecules on the surface of the bulb and absorption of molecules into the body of the bulb walls. Sorption phenomena are dependent on the attractive forces between gas molecules and surface atoms of an adjacent surface. The magnitude of this effect can be of the order of ± 0.1 mK in bulk ^{4}He (Pavese and Steur, 1987; see also Berry, 1979).

The uncertainty of gas thermometry measurements depends on the care taken and the temperature range. For example, Pavese and Steur (1987) report an uncertainty of 0.5 mK for temperatures between 0.5 K and 30 K. At the lower temperatures, better uncertainty may be obtained by measuring the vapour pressure of a cryogenic liquid, (Section 7.3). The use of such vapour pressure thermometry over gas thermometry at temperatures below 4.2 K is recommended by Pavese and Steur (1987).

With the exception of monitoring-type devices, gas thermometers are not generally commercially available and component parts are usually assembled from a variety of manufacturers and a system assembled to produce a gas thermometer. Some companies such as those specializing in cryostats will assemble a gas thermometry system for a client. Gas thermometry tends to be a specialist activity and is usually confined to standards laboratories and cryogenic applications. Use of gas thermometry in low uncertainty experiments is reported by Steur (1999), Steur and Pavese (1989) and Edsinger and Schooley (1989).

7.3 Vapour pressure thermometry

The saturation vapour pressure of a pure substance above its liquid phase varies only with temperature. The relationship between vapour pressure and temperature is known with low uncertainty for a number of cryogenic liquids such as helium 3 and 4, hydrogen, neon, nitrogen, oxygen, argon, methane and carbon dioxide. Measurements of the vapour pressure can therefore be exploited to determine the temperature. Vapour pressure thermometers have good sensitivity, require relatively simple measuring equipment in comparison to, say, noise thermometry, and are inherently reproducible. Equations quantifying the relationship between pressure and temperature for a variety of cryogenic liquids are listed in equations (7.12) to (7.16) taken from Pavese (1999).

For liquid–vapour phases of helium-4,

$$T = A_0 + \sum_{i=1}^{9} A_i \left[(\ln p - B)/C \right]^i \tag{7.12}$$

where: $A_0 = 1.392408$, $A_1 = 0.527153$, $A_2 = 0.166756$, $A_3 = 0.050988$, $A_4 = 0.026514$, $A_5 = 0.001975$, $A_6 = -0.017976$, $A_7 = 0.005409$, $A_8 = 0.013259$, $B = 5.6$, $C = 2.9$ for the temperature range 1.25 to 2.1768 K, and

$A_0 = 3.146631$, $A_1 = 1.357655$, $A_2 = 0.413923$, $A_3 = 0.091159$, $A_4 = 0.016349$, $A_5 = 0.001826$, $A_6 = -0.004325$, $A_7 = -0.004973$, $B = 10.3$, $C = 1.9$ for the temperature range 2.1768 to 5 K.

For liquid–vapour phases of equilibrium hydrogen,

$$p = p_0 \exp \left(A + \frac{B}{T} + CT \right) + \sum_{i=0}^{5} b_i T^i \tag{7.13}$$

where: $A = 4.037592968$, $B = -101.2775246$, $C = 0.0478333313$, $b_0 = 1902.885683$, $b_1 = -331.2282212$, $b_2 = 32.25341774$, $b_3 = -2.106674684$, $b_4 = 0.060293573$, $b_5 = -0.000645154$ the temperature range 13.8 to 20.3 K.

For liquid–vapour phases of nitrogen,

$$\ln\left(\frac{p}{p_c}\right) = \frac{T_c}{T}\left(A\tau + B\tau^{0.5} + C\tau^3 + D\tau^6\right) \tag{7.14}$$

where: $\tau = 1 - T/T_c$, $A = -6.10273365$, $B = 1.153844492$, $C = -1.087106903$, $D = -1.759094154$, $T_c = 126.2124\,K$, $p_c = 3.39997\,MPa$ for the temperature range 63.2 to 125 K.

For liquid–vapour phases of oxygen,

$$\ln\left(\frac{p}{p_c}\right) = \frac{T_c}{T}\left(A\tau + B\tau^{1.5} + C\tau^3 + D\tau^7 + E\tau^9\right) \tag{7.15}$$

where: $\tau = 1 - T/T_c$, $A = -6.044437278$, $B = 1.176127337$, $C = -0.994073392$, $D = -3.449554987$, $E = 3.343141113$, $T_c = 154.5947\,K$, $p_c = 5.0430\,MPa$ for the temperature range 54 to 154 K.

For liquid–vapour phases of carbon dioxide,

$$\ln\left(\frac{p}{p_c}\right) = A_0\left(1 - \frac{T}{T_c}\right)^{1.935} + \sum_{i=1}^{4} A_i\left(\frac{T_c}{T} - 1\right)^i \tag{7.16}$$

where: $p_c = 7.3825\,MPa$, $T_c = 304.2022\,K$, $A_0 = 11.37453929$, $A_1 = -6.886475614$, $A_2 = -9.589976746$, $A_3 = 13.6748941$, $A_4 = -8.601763027$ for the temperature range 216.6 to 304 K.

The basic equipment required for vapour pressure thermometry is illustrated in Figure 7.4. The bulb material should have a high thermal conductivity in

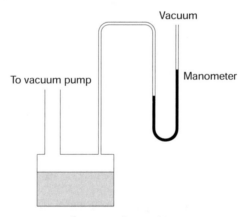

Figure 7.4 Vapour pressure thermometry system

order to ensure that the liquid takes up the temperature of the application. The choice of pressure-measuring device depends on the uncertainty requirements and includes dial manometers, diaphragm or bellows manometers, and Bourdon gauges. Dial manometers are suitable for uncertainties of the order of 1% whilst capacitive bellows pressure transducers can provide a reproduceability of better than ±0.1 Pa. The design criteria for vapour pressure thermometry are summarized by Pavese (1999) and described in detail by Pavese and Molinar (1992) and its use for the range 0.65 to 5 K is reported by Meyer and Reilly (1996).

Vapour pressure thermometry can be time-consuming as it is important to allow time for the liquid and vapour to reach equilibrium at the same temperature. In order to achieve low uncertainty measurements it is necessary to account for a number of factors such as the bulb being at a higher temperature than the cryogenic liquid, the thermomolecular pressure difference between two volumes of gas at different temperature and hydrostatic pressure effects. Methods for determining thermomolecular pressure corrections have been produced by Weber and Schmidt (1936), McConville (1972), Guildner and Edsinger (1976), Berry (1979) and Roberts and Sydoriak (1957). The pressure within a liquid will vary with depth and a higher vapour pressure is necessary in order to form bubbles below the surface of a boiling liquid. The temperature of the liquid boiling below the surface is therefore at a slightly higher value than that at the surface where the vapour pressure is measured. Cold spots caused by condensation along connection tubes can be avoided by encasing the connection tubes in a vacuum jacket or by using a heater.

7.4 Conclusions

Manometric thermometry involves measurement of pressure to determine temperature. Gas thermometers based on Bourdon-type pressure gauges are commercially available providing an indicator capability. Gas thermometers can also be assembled that are capable of very low uncertainty measurements for standards and cryogenic applications. An alternative technique based on the variation of temperature with the saturation vapour pressure of a pure substance above its liquid phase is also capable of low uncertainty measurements at low temperatures. Although the basic principles of manometric techniques are relatively simple, the undertaking of low uncertainty measurements is a complex task requiring consideration of the detailed physics of the associated components. For further information on manometric thermometry the reader is referred to the specialist texts by Pavese and Molinar (1992) and White (1987) and the more general texts by Schooley (1986) and Quinn (1990).

References

Astrov, D.A., Beliansky, L.B., Dedikov, Y.A., Polunin S.P. and Zakharov, A.A. Precision gas thermometry between 2.5 K and 308 K. *Metrologia*, **26**, 151–166, 1989.

Aziz, R.A., Nain, V.P.S., Carley, J.S., Taylor, W.L. and McConville, G.T. An accurate intermolecular potential for helium. *J. Chem. Phys.*, **70**, 4330–4342, 1979.

Berry, K.H. NPL-75, A low temperature gas thermometer scale from 2.6 K to 27.1 K. *Metrologia*, **15**, 89–115, 1979.

Cohen, E.R. and Taylor, B.N. Recommended consistent values of the fundamental physical constants. *Journal Phys. Chem. Ref. Data 2*, 663, 1973.

Cohen, E.R. and Taylor, B.N. The 1986 adjustment of the fundamental physical constants. *Reviews of Modern Physics*, **59**, 1121–1148, 1987.

Cohen, E.R. and Taylor, B.N. The fundamental physical constants. *Physics Today*, BG5-BG9, 1999.

Edsinger, R.E. and Schooley, J.F. Differences between thermodynamics temperature and (IPTS-68) in the range 230°C to 660°C. *Metrologia*, **26**, 95–106, 1989.

Guildner, L.A. and Edsinger, R.E. Deviation of international practical temperatures from thermodynamic temperatures in the temperature range from 273.16 K to 730 K. *Journal of Research of the National Bureau of Standards*, **80A**, 703–738, 1976.

Guildner, L.A. and Edsinger, R.E. Progress in NBS gas thermometry above 500°C. In Schooley, J.F. (Editor), *Temperature. Its Measurement and Control in Science and Industry*, Vol. 5(1), pp. 43–48, American Institute of Physics, 1982.

Guildner L.A. and Thomas, W. The measurement of thermodynamic temperature. In Schooley, J.F. (Editor), *Temperature. Its Measurement and Control in Science and Industry*, Vol. 5(1), pp. 9–19. American Institute of Physics, 1982.

Kemp, R.C., Kemp W.R.C. and Besley, L.M. A determination of thermodynamic temperatures and measurements of the second virial coefficient of ^{4}He between 13.81 K and 287 K using a constant volume gas thermometer. *Metrologia*, **23**, 61–86, 1986.

Luther, H., Grohmann, K. and Fellmuth, B. Determination of thermodynamic temperature and ^{4}He virial coefficients between 4.2 K and 27.0 K by dielectric constant gas thermometry. *Metrologia*, **33**, 341–352, 1996.

Matacotta, F.C., McConville, G.T., Steur, P.P.M. and Durieux, M. Measurements and calculations of the ^{3}He second virial coefficient between 1.5 K and 20.3 K. *Metrologia*, **24**, 61–67, 1987.

McConville, G.T. The effect of the measuring tube surface on thermomolecular pressure corrections in vapor pressure thermometry. In Plumb, H.H. (Editor), *Temperature. Its Measurement and Control in Science and Industry*, Vol. 4, pp. 159–165. Instrument Society of America, 1972.

Meyer, C. and Reilly, M. Realization of the ITS-90 at the NIST in the range 0.65 K to 5.0 K using ^{3}He and ^{4}He vapour pressure thermometry. *Metrologia*, **33**, 383–389, 1996.

Pavese, F. Manometric thermometers. Section 32.9 in Webster, J.G. (Editor), *The Measurement Instrumentation and Sensors Handbook*. CRC Press, 1999.

Pavese, F., and Molinar, G. *Modern Gas-based Temperature and Pressure Measurements*. Plenum Press, 1992.

Pavese, F. and Steur, P.P.M. ^{3}He constant-volume gas thermometry: calculations for a temperature scale between 0.8 K and 25 K. *J. Low Temp. Phys.*, 69(1–2), 91–117, 1987.

Preston-Thomas, H. The international temperature scale of 1990 (ITS-90). *Metrologia*, **27**, 3–10, 1990.

Quinn, T.J. *Temperature*, 2nd edition. Academic Press, 1990.

Roberts, T.R. and Sydoriak, S.G. Thermomolecular pressure ratios for ^{3}He and ^{4}He. *Phys. Rev.*, **102**, 304–308, 1957.

Schooley, J.F. *Thermometry*. CRC Press, 1986.

Steur, P.P.M. The interpolating constant-volume gas thermometer and thermal anchoring. *Metrologia*, **36**, 33–39, 1999.

Steur, P.P.M. and Durieux, M. Constant volume gas thermometry between 4 K and 100 K. *Metrologia*, **23**, 1–18, 1986.

Steur, P.P.M., Durieux, M. and McConville, G.T. Analytic expressions for the virial coefficients B(T) and C(T) of ^{4}He between 2.6 K and 300 K. *Metrologia*, **24**, 69–77, 1987.

Steur, P.P.M. and Pavese, F. He-3 constant volume gas thermometer as interpolating instrument: calculations of the accuracy limit versus temperature range and design parameters, *Cryogenics*, **29**, 135–138, 1989.

Weber, S. and Schmidt, G. Experimentelle untersuchungen uber die thermomlekulare druckdifferenz in der nahe der grenzbedingung p1/p2 = √T1/T2 und vergleichung mit der theorie. *Leiden Communications*, **246c**, 1936.

White, G.K. *Experimental Techniques in Low Temperature Physics*, 3rd edition. Oxford University Press, 1987.

Web sites

At the time of going to press the world wide web contained useful information relating to this chapter at the following sites.

http://www.dresscrinstruments.com/products/S5500.html
http://www.dresserinstruments.com/weksler/weks_catalog/weks_index.html

Nomenclature

$B(T)$ = second virial coefficient ($cm^3 \cdot mol^{-1}$)
$C(T)$ = third virial coefficient ($cm^6 \cdot mol^{-2}$)
$D(T)$ = fourth virial coefficient ($cm^9 \cdot mol^{-3}$)
D_R = deadspace correction (m^3/K)
g = acceleration due to gravity (m^2/s)
h = height (m)
L_1 = original length (m)
m = mass (kg)
M = molar mass (kg/mol)
n = number of moles of the gas
p = pressure (N/m^2)
R = the universal gas constant (= 8.314510 J/mol·K (Cohen and Taylor, 1999))

T = temperature (K)
T_R = temperature of the gas in the reference bulb (K)
V = volume (m^3)
V_v = variable volume (m^3)
V_R = volume of the reference bulb (m^3)
α = coefficient of linear expansion (K^{-1})
ΔT = temperature rise (K)
ρ = density (kg/m^3)

8

Semi-invasive temperature measurement

Semi-invasive temperature measurement techniques are classified here as those involving some form of treatment of the surface of interest such as the application of a temperature-sensitive paint and remote observation of the temperature-dependent properties of the surface application. The aims of this chapter are to introduce the various techniques and outline their use and merits.

8.1 Introduction

Some temperature measurement scenarios permit the application of a temperature-sensitive material or component on a surface. The variations, for example, of optical properties in the case of a surface coating can then be observed remotely. Surface coating methods are classed as semi-invasive here as the technique involves modification of the component of interest and therefore some disturbance to the temperature field. This is in most cases minimal and semi-invasive techniques are increasing in popularity in both industrial and scientific applications. A range of techniques have now been developed including thermochromic liquid crystals, heat-sensitive crystalline solids and paints, thermographic phosphors and pyrometric cones. Thermochromic liquid crystals provide a vivid visual indication of temperature by changing colour through the visual spectrum across a relatively narrow temperature band. They have now been widely used in scientific research and found commercial application in the form of adhesive temperature-indicating strips and forehead thermometers. Temperature-indicating paints are described in Section 8.2, crayons, pellets and labels in Section 8.3 and pyrometric cones and related deforming devices in Section 8.4. Most temperature-sensitive crystalline solids and paints melt on reaching a certain temperature and are used in the form of labels or pellets to indicate whether a particular process temperature has been attained. Thermographic phosphors and temperature-sensitive paints are materials that can be excited by the absorption of energy and subsequently emit light in a fashion inversely proportional to their temperature. A surface of interest can be coated with a thermographic phosphor and energized using, say, a laser and observed with an optical system. Pyrometric cones and bars have been developed for the ceramics and glass industries to provide an indication of both firing time and temperature. These devices typically comprise a ceramic compound that softens in a predictable fashion when heated through a specified cycle. Pyrometric cones deform by bending and can be observed through a spy hole in a kiln or high-temperature oven.

8.2 Paints

A range of temperature sensitive materials exist that can be incorporated into a binder to form a paint and provide the possibility for the remote observation of temperature. These include thermochromic liquid crystals, temperature-indicating paints, thermographic phosphors and temperature-sensitive paints and are considered in Sections 8.2.1 to 8.2.4 respectively.

8.2.1 Thermochromic liquid crystals

Thermochromic liquid crystals are commercially available materials that exhibit brilliant changes in colour over discrete temperature bands. At a particular temperature the liquid crystal material selectively reflects incident light within a certain band of wavelengths. As temperature rises a thermochromic liquid crystal mixture will turn from colourless, black against a black background, to red at a particular temperature and then pass through the visible spectrum (red to orange to yellow to green to blue to violet) before turning colourless again. The process of colour change is reversible and on cooling the colour change process is reversed (violet to blue to green to yellow to orange to red) before the mixture appears colourless again. By calibration the colours displayed can be related to the temperature of the crystals. They can therefore be used as a temperature sensor, provided a reliable means of monitoring colour change is adopted.

Liquid crystals can be classified, according to the molecular structure, into three categories: smectic, nematic and cholesteric (Figure 8.1). The molecules are arranged in layers with the long axes of the molecules parallel to each other in the plane of the layer (Fergason, 1964). Each layer is slightly rotated compared to its neighbour forming a helical pattern. The distance between layers is sensitive to a number of external influences such as temperature, electric fields, magnetic fields and shear stress. Types of thermochromic liquid crystals include cholesteric and chiral nematic. Although cholesterol itself does not exhibit liquid crystalline properties, the majority of thermochromic liquid crystals are formed from estors of cholesterol (Woodmansee, 1966). As the temperature rises from the lower limit of the temperature range the twist of the structure alters and the dominant reflected wavelength shifts and this provides the mechanism by which a coated surface appears to change colour from red to blue.

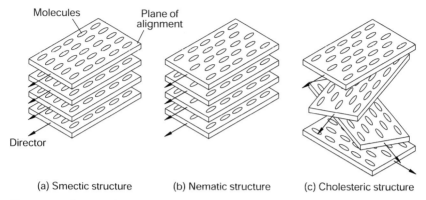

(a) Smectic structure   (b) Nematic structure   (c) Cholesteric structure

Figure 8.1 Classification of liquid crystals according to molecular structure.
(a) Smectic. (b) Nematic. (c) Cholesteric

Pure crystals deteriorate rapidly with age, allowing only a few hours of use following manufacture. They are prone to contamination from exposure to atmospheric compounds as well as ultraviolet light. These problems can be reduced by encapsulation of the crystals in a polymer coating. The coating process results in the formation of discrete spherical microcapsules with diameters in the range of a few micrometres to millimetres. Thermochromic liquid crystals are available as a water-based slurry or as a pre-formed layer on a blackened substrate of mylar or paper. Typical surface coatings are illustrated in Figure 8.2. In their water-based slurry form, liquid crystals can be applied to a surface by a number of means including screen printing, brushing, dipping, rolling and air brushing. Air brushing is favoured by a

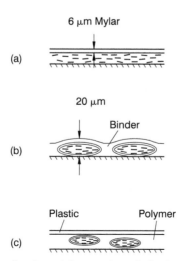

Figure 8.2 Surface application of thermochromic liquid crystals

number of heat transfer researchers for both flat and curved surfaces (e.g. see Baughn, 1995; Farina *et al.*, 1994; Roberts and East, 1996; Ireland and Jones, 2000). If the material is spread too thinly colours can appear in an irregular, jumbled pattern. If it is too thick then the surface colours have a milky appearance and slow response (Baughn, 1995).

The temperature at which a liquid crystal formulation begins to display colour is often referred to as its red-start temperature. The range of temperature over which the crystals display colours is referred to as either the colour play bandwidth or the temperature event range. The range of temperature over which a liquid crystal displays a single colour is referred to as an isochrome bandwidth and this can be as small as 0.1°C. The value of the red-start temperature and the colour play bandwidth can be controlled by selecting appropriate cholesteric estors and their proportions. Table 8.1 illustrates the red-start temperatures and colour play bandwidths for a selection of the materials available from one commercial supplier.

In order to visualize the colours it is necessary to paint thermochromic liquid crystals on a dark or blackened background. The black substrate serves the purpose of ensuring that all the incident light transmitted through the liquid crystals is absorbed and not reflected to compete with the desired signal.

Colour can be defined as a psychophysical property of light (Camci *et al.*, 1992, 1993, 1996). A number of approaches can be adopted in the use of thermochromic liquid crystals. Human observer judgement can be used by comparison of the displayed temperature with a look-up chart as in the use of forehead clinical thermometers (Figure 8.3). In experimental heat transfer measurements two approaches have been pioneered. One is based on the red, green and blue (RGB) colour system used in domestic video recorders and the other is based on a hue system.

Liquid crystals have now been widely used in the characterization of heat transfer on applications as diverse as turbine blades (Campbell and Molezzi,

Table 8.1 Red-start temperatures and colour play bandwidths for readily available thermochromic liquid crystals from Hallcrest Inc. Data courtesy of Hallcrest Ltd

Red-start temperature (°C)	Minimum colour play bandwidth (°C)	Maximum colour play bandwidth (°C)
−30	2	30
0	1	25
30	0.5	25
60	1	20
90	1	20
120	1	20

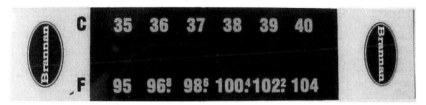

Figure 8.3 Forehead thermometers

1996; Ou *et al.*, 2000; Hoffs and Bolcs, 1995; Wang *et al.*, 1998) and jet engine nacelles to vehicle interiors (Lee and Yoon, 1998). In a typical transient heat transfer experiment the images produced by a surface coated with thermochromic liquid crystals can be recorded using a commercial domestic colour video camera (Figure 8.4). Here the signal can be considered to be made up of red, green and blue components. These are defined by:

$$R = \int_{-\infty}^{\infty} E(\lambda)R(\lambda)r(\lambda)\,d\lambda \tag{8.1}$$

$$G = \int_{-\infty}^{\infty} E(\lambda)R(\lambda)g(\lambda)\,d\lambda \tag{8.2}$$

$$B = \int_{-\infty}^{\infty} E(\lambda)R(\lambda)b(\lambda)\,d\lambda \tag{8.3}$$

where: $r(\lambda)$ = filter transmissivity
$g(\lambda)$ = filter transmissivity

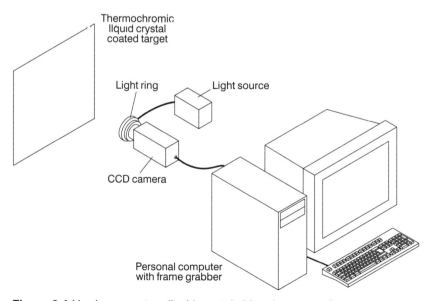

Figure 8.4 Hardware system liquid crystal video thermography

$$b(\lambda) = \text{filter transmissivity}$$
$$E(\lambda) = \text{lighting spectral distribution}$$
$$R(\lambda) = \text{surface reflectance}$$
$$\lambda \quad = \text{wavelength.}$$

The colour signal measured by the camera is encoded into a composite video signal (PAL (Phase Alternation Line) or NTSC (National Television System Committee)) format and recorded onto video tape or computer memory. The RGB components can be recorded by a signal decoder.

Although the values of the RGB signals can be used to fully specify the temperature of a thermochromic liquid crystal, a colour index approach based on the RGB signals is often used, (e.g. Wang *et al.*, 1994). The hue signal in the hue, saturation, intensity (HSI) colour definition is a simple monotonic function of the crystal temperature and is independent of local illumination strength. Hue, saturation and intensity are defined by equations (8.4)–(8.8) (Camci, 1996):

$$H = \frac{1}{360}\left[90 - \tan^{-1}\left(\frac{F}{\sqrt{3}}\right) + \begin{matrix} 0, G > B \\ 180, G < B \end{matrix}\right] \tag{8.4}$$

where

$$F = \frac{2R - G - B}{G - B} \text{ for } G \neq B \tag{8.5}$$

$$F = R \text{ for } G = B \tag{8.6}$$

Saturation and intensity are defined by

$$S = 1 - \left[\frac{\min(R, G, B)}{I}\right] \tag{8.7}$$

$$I = \frac{R + G + B}{3} \tag{8.8}$$

The frames can later be digitally analysed using a computer and frame grabber and the data converted from RGB to values of hue, saturation and intensity (HSI). Comparison of the hue values with calibration results gives the surface temperature for each pixel location (see Camci *et al.*, 1992; Farina *et al.*, 1994; Hay and Hollingsworth, 1996).

Experimental uncertainty depends on both the experimental conditions and the image processing system. Baughn *et al.* (1988) and Camci *et al.* (1993) report an uncertainty of approximately 6% in the measurement of heat transfer coefficients using thermochromic liquid crystals. The uncertainty of the surface temperature measurement should be significantly better than this and Simonich and Moffatt (1984) report a calibration uncertainty of ±0.25°C using mercury

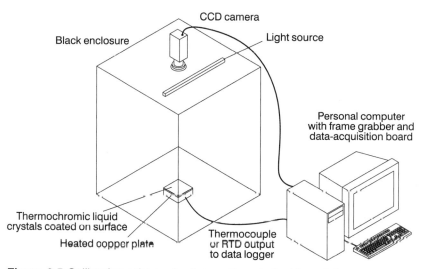

Figure 8.5 Calibration scheme for thermochromic liquid crystals

vapour lamp illumination. The wavelengths reflected by a surface coated with thermochromic liquid crystals are a function of illumination and viewing angle. Careful calibration of the image-processing system is therefore necessary. The HSI approach negates the problem of illumination but calibration is still necessary to determine the relationship between hue and the crystal output. One strategy for calibration is to construct a system comprising a test piece coated with the crystals, a light source, CCD camera and a PC with a frame grabber, (see Figure 8.5). The test piece can comprise a copper plate on top of which the thermochromic liquid crystal layer can be applied. The temperature of the copper plate can be varied by means of a heater and its temperature monitored by thermistors, thermocouples or PRTs. The heating rate and temperature sensor signals can be controlled and monitored by the data control and acquisition system on the PC. It is important that the illumination and viewing angles should be comparable to those in use in the application. Using a scheme similar to that shown in Figure 8.5, Baughn *et al.* (1999) interestingly revealed that thermochromic liquid crystals can demonstrate hysteresis in their output depending on whether they are heating up or cooling.

The speed of response of thermochromic liquid crystals depends on the viscosity of the crystals. Ireland and Jones (1987) demonstrated that liquid crystal formulations typically have a response time constant of 0.003 s.

8.2.2 Temperature-indicating paints

In addition to thermochromic liquid crystals, temperature-indicating paint is available in a number of forms. The Tempilaq paint (manufactured by Tempil), for example, is available for a wide number of ratings as listed in

Table 8.2. The paint uses the same material as used in the pellets and crayons described in Section 8.3. The paint provides a means of identifying that a peak temperature has been attained or exceeded and undergoes a non-reversible phase change from an opaque solid to a clear liquid at the rated temperature. A component can be coated with a number of patches of different rating paints in order to determine temperature distribution. Alternative paints are available that also exhibit a non-reversible but very perceptible colour change when heated. The colour tranformations for one range of paints are listed in Table 8.3. The paints can be applied by brush or spray to a grease-free base metal or over a primer coat. Limitations of temperature paints include their poor performance in regions of high thermal gradient, their inability to indicate the peak temperature reached above the rated temperature and the labour intensity required in their application and observation. Paints have been used to indicate peak temperatures and heat transfer distributions on turbine blades (Bird *et al.*, 1998; Neumann, 1989).

Table 8.2 Temperature ratings available from Tempil for their paints, crayons, pellets and temperature-sensitive labels

100–244°F	250–463°F	475–1350°F	1400–2500°F
100°F/38°C	250°F/121°C	475°F/246°C	1400°F/760°C
103°F/39°C	256°F/124°C	488°F/253°C	1425°F/774°C
106°F/41°C	263°F/128°C	500°F/260°C	1450°F/778°C
109°F/43°C	269°F/132°C	525°F/274°C	1480°F/804°C
113°F/45°C	275°F/135°C	550°F/288°C	1500°F/816°C
119°F/48°C	282°F/139°C	575°F/302°C	1550°F/843°C
125°F/52°C	288°F/142°C	600°F/316°C	1600°F/871°C
131°F/55°C	294°F/146°C	625°F/329°C	1650°F/899°C
138°F/59°C	300°F/149°C	650°F/343°C	1700°F/927°C
144°F/62°C	306°F/152°C	700°F/371°C	1750°F/954°C
150°F/66°C	313°F/156°C	750°F/399°C	1800°F/982°C
156°F/69°C	319°F/159°C	800°F/427°C	1850°F/1010°C
163°F/73°C	325°F/163°C	850°F/454°C	1900°F/1038°C
169°F/76°C	331°F/166°C	900°F/482°C	1950°F/1066°C
175°F/79°C	338°F/170°C	932°F/500°C	2000°F/1093°C
182°F/83°C	344°F/173°C	950°F/510°C	2050°F/1121°C
188°F/87°C	350°F/177°C	977°F/525°C	2100°F/1149°C
194°F/90°C	363°F/184°C	1000°F/538°C	2150°F/1177°C
200°F/93°C	375°F/191°C	1022°F/550°C	2200°F/1024°C
206°F/97°C	388°F/198°C	1050°F/566°C	2250°F/1232°C
213°F/101°C	400°F/204°C	1100°F/593°C	2300°F/1260°C
219°F/104°C	413°F/212°C	1150°F/621°C	2350°F/1288°C
225°F/107°C	425°F/218°C	1200°F/649°C	2400°F/1316°C
231°F/111°C	438°F/226°C	1250°F/677°C	2450°F/1343°C
238°F/114°C	450°F/232°C	1300°F/704°C	2500°F/1371°C
244°F/118°C	463°F/239°C	1350°F/732°C	

Table 8.3 Colour change temperatures for the Therm-O-Signal range of temperature-indicating paints

Type	Original colour	Transition temperature after 10 minutes (°C)	Colour after transition
7H10	Red	65	Black
7H12	Pink	80	Lavender
7H14	Pink	130	Blue
		300	Grey
7H16	Pink	145	Blue
		320	Grey
7H20	Ochre yellow	155	Green
		235	Dark brown
		300	Indian red
7H22	Sulphur yellow	175	Black
7H23	Orange red	240	Dark grey
		255	Grey
		335	Grey white
7H28	Green blue	270	Beige
7H30	Orange red	400	Grey
		455	Yellow
		570	Orange
7H34	Violet	400	White
7H36	Red	350	Brown grey
		430	Yellow
		550	Orange
		670	Green yellow
		815	Grey
7H38	Violet blue	100	Bright green
		210	Olive green
		250	Dark green
		260	Beige
7H40	Green	430	Salmon pink
7H42	Dark violet	420	Light violet
		500	Yellow brown
		610	Blue
		745	White
		830	Dark blue
		860	Matt black
		930	Glossy black

Some paint formulations are available that provide a reversible indication of temperature. One principle exploited is to drive water off a colourful salt, thus changing its colour. On cooling the salt can re-absorb water vapour from the atmosphere and revert to its initial colour state. One form of commercial paint providing reversible indication of temperature is the Chromonitor brand. This changes colour from red to deep maroon in the temperature range 65–71°C.

8.2.3 Thermographic phosphors

Thermographic phosphors are materials that can be excited by the absorption of energy and subsequently emit light in a temperature-dependent fashion. There are two classes of phosphors: organic and inorganic. It is the inorganic types such as Y_2O_3:Eu and La_2O_2S:Eu that tend to be used in thermometric applications. Phosphors are usually fine white or slightly coloured powders. Prior to excitation the material's electronic levels are populated in their ground state. They can be excited by the absorption of energy in some way such as electromagnetic radiation (visible or UV light, X- or γ-rays), particle beams (electrons, neutrons or ions) or by an electric field. After the absorption of energy the atomic configuration of the material may not remain excited but return to its initial or some intermediate state as illustrated schematically in Figure 8.6.

When considering the behaviour of phosphors a number of definitions of terms are helpful. Luminescence is the absorption of energy by a material and the subsequent emission of light. Fluorescence is the same as luminescence but the emission is usually in the visible band and has a duration of typically 10^{-9} to 10^{-3} s. Phosphorescence is a type of luminescence with a duration of about 10^{-3} to 10^{3} s (Allison and Gillies, 1997).

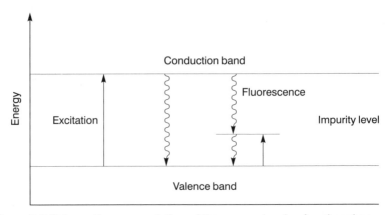

Figure 8.6 Schematic representation of the energy levels of a phosphor

Thermographic phosphors can be used to indicate temperatures from cryogenic levels (Cates *et al.*, 1997; Simons *et al.*, 1996) to 2000°C. A phosphor-based thermometry system will generally comprise:

- a source of excitation energy
- a means of delivering the energy to the target
- a fluorescing medium bonded to the target

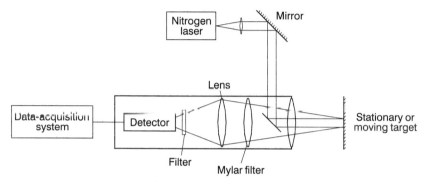

Figure 8.7 Typical measurement system for the use of thermographic phosphors. (After Allison and Gillies, 1997)

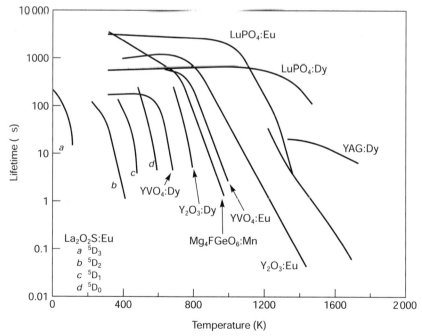

Figure 8.8 Lifetime versus temperature for a variety of phosphors. Reproduced from Allison and Gillies (1997)

- an optical system to collect the fluorescence
- a detector
- a data-acquisition and analysis system.

A typical layout for a thermographic phosphor thermometry system is illustrated in Figure 8.7. When selecting the type of phosphor consideration should be given to the temperature range of interest. The temperature range capability and lifetimes for various thermographic phosphors are illustrated in Figure 8.8. In addition, the chemical compatibility of the phosphors with the surface of interest and the surrounding atmosphere should be considered. Fortunately most thermographic phosphors are ceramics and therefore relatively inert. The bonding of the phosphor can be achieved by mixing the phosphor slurry with epoxy, paint or glue and then brushing or spraying the mixture onto the surface of interest. For harsh environments (high mechanical shock, large thermal gradients, etc.) chemical bonding by vapour deposition, RF sputtering and laser ablation can also be considered.

The intensity of emitted light is an inverse function of temperature and in the case of European based phosphors experiencing continuous illumination is given by (Fonger and Struck, 1970).

$$I_r = \left[a_j + a_j \, A e^{-\Delta E / kT} \right]^{-1} \tag{8.9}$$

where: I_T = intensity
a_j = probability rate
A = factor related to a_j
E = energy
T = temperature
k = Boltzmann's constant
a_{CTS} = charge transfer state rate.

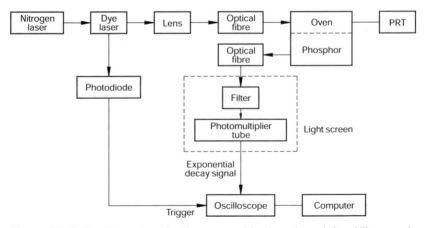

Figure 8.9 Calibration set-up for thermographic phosphors (after Allison and Gillies, 1979)

Calibration of thermographic phosphors generally involves the adoption of a scheme similar to that illustrated in Figure 8.7 but with some independent means of monitoring the phosphor temperature such as a PRT as indicated in Figure 8.9. Here the sample of phosphor to be calibrated can be bonded to a substrate or spread out in a ceramic boat. The sample can be placed in an oven and a ceramic-clad optical fibre used to convey the signals into and out of the high-temperature cavity.

Applications have included temperature measurements of flat plates in supersonic flows (Bradley, 1953), wind tunnel models (Czysz and Dixon, 1968, 1969), turbine blades and components (Tobin *et al.*, 1990; Noel *et al.*, 1991, 1992; Alaruri *et al.*, 1995, 1999; Feist and Heyes, 2000), curved surfaces (Ervin *et al.*, 1995), surface temperature fields (Edge *et al.*, 2000), colour TV screens (Kusama, 1976), textiles during microwave drying (Dever *et al.*, 1990) and tumours (Sholes and Small, 1980). The thermographic phosphor thermometry technique can offer sensitivities of 0.05°C and an uncertainty of 0.1–5% of the Celsius temperature reading (Allison and Gillies, 1997). One of the merits of this technique is its independence of emissivity, a problem associated with many of the infrared-based techniques described in Chapter 9.

8.2.4 Temperature-sensitive paints

Another class of temperature-indicating paints exist that are known as temperature-sensitive paints (TSPs). The paint comprises luminescent molecules and a polymer binder material that can be dissolved in a solvent and has similarities to the thermographic phosphors described in Section 8.2.3. TSPs can be applied by brush or spray and as the paint dries the solvent evaporates to leave a polymer matrix with luminescent molecules embedded in it. The principle exploited in these paints is photoluminescence whereby a probe molecule is promoted to an excited electronic state by the absorption of a photon of appropriate energy in a fashion comparable to thermographic phosphors. The difference is that the intensity of luminescence is related to temperature by photophysical processes known as thermal and oxygen quenching (Gallery *et al.*, 1994). The intensity of luminescence is inversely proportional to temperature (Donovan *et al.*, 1993; Cattafesta, 1998). Examples of TSPs include rhodamine dyes and europium thenoyltrifluoroacetonate. The technique has been used in turbomachinery testing and heat transfer experiments (Hubner *et al.*, 1999; Crafton *et al.*, 1999; Hamner *et al.*, 1997). Cattafesta *et al.*, (1998) report an average uncertainty of ±0.3°C for several sample applications of rhodamine dyes operating in the temperature range 0–95°C and using an industrial-grade CCD camera.

8.3 Temperature-sensitive crayons, pellets and labels

A number of products are available that provide an indication that a particular temperature has been attained or exceeded. Examples include a variety of heat-sensitive crayons, pellets, labels and the paints described in the previous section. On heating above some critical temperature the indicator material melts, fuses or changes composition providing, in most cases, a permanent record that a process temperature has been reached or exceeded.

One of the simplest devices is the heat-sensitive crayon stick (Figure 8.10). These were originally developed to allow temperatures during welding and metal fabrication to be monitored. For applications below about 371°C they can be used to mark the surface before heating begins. Once dry, the opaque crayon mark will change to a distinct melted residue when the prescribed temperature rating for that crayon has been attained. At temperatures above 371°C under prolonged heating the crayon marks tend to evaporate or become absorbed. Under these conditions, standard practice is to stroke the workpiece with the crayon periodically during the heating process and observe whether the crayon mark melts and leaves a liquid smear. It should be noted that colour changes during heating should be disregarded. It is the melting of the formulation that provides the indication of the rated temperature. Typical uses include monitoring of preheat, interpass and postweld heat treatment, annealing and stress releaving. The crayons are available in a wide number of temperature ratings as indicated in Table 8.2. In addition to the crayons,

Figure 8.10 A temperature-sensitive crayon. Once applied to a surface, the formulation will melt at a particular temperature providing a permanent indication that a process temperature has been attained

pellets are also available, and at the same ratings listed in Table 8.2 for those manufactured by Tempil. Again these melt on attaining the rated temperature. An alternative formulation of pellet material is available for use in reducing atmospheres.

Self-adhesive labels consisting of a temperature-sensitive indicator sealed under a transparent window are also available (see Figure 8.11). The indicating material permanently changes colour from say light grey to black at the rated temperature. Labels are available in a wide number of configurations as illustrated in Figure 8.12. Applications include use on printed circuit boards and electronic components to monitor safe operating temperatures and to verify, for example, that a device has not been exposed to too high environmental temperatures thereby invalidating, say, warranty conditions. The typical speed of response for temperature-sensitive labels is between one and several seconds.

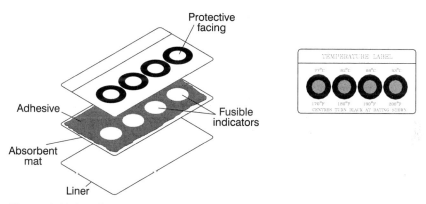

Figure 8.11 A self-adhesive temperature-indicating label consisting of a temperature-sensitive indicator sealed under a transparent window

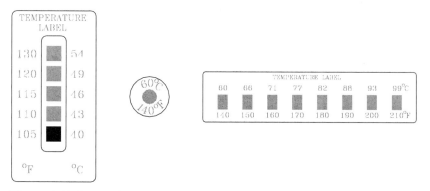

Figure 8.12 Some of the configurations available for self-adhesive temperature-indicating labels

Table 8.4 Temperature equivalents (°C) for Orton pyrometric cones

	Self-supporting cones						Large cones				Small
	Regular			Iron-free			Regular		Iron-free		Regular
	Heating rate °C/h (last 90–120 minutes of firing)										
Cone	15	60	150	15	60	150	60	150	60	150	300
022		586	590				N/A	N/A			630
021		600	617				N/A	N/A			643
020		626	638				N/A	N/A			666
019	656	678	695				676	693			723
018	686	715	734				712	732			752
017	705	738	763				736	761			784
016	742	772	796				769	794			825
015	750	791	818				788	816			843
014	757	807	838				807	836			870
013	807	837	861				837	859			880
012	843	861	882				858	880			900
011	857	875	894				873	892			915
010	891	903	915	871	886	893	898	913	884	891	919
09	907	920	930	899	919	928	917	928	917	926	955
08	922	942	956	924	946	957	942	954	945	955	983

	C1	C2	C3	C4	C5	C6	C7	C8	C9	C10	C11
07	1008	980	970	985	973	982	971	953	987	976	962
06	1023	996	991	1011	995	998	991	969	1013	998	981
05½	1043	1020	1011	1023	1012	1021	1012	990	1025	1015	1004
05	1062	1044	1032	1046	1030	1046	1037	1013	1044	1031	1021
04	1098	1067	1060	1070	1060	1069	1061	1043	1077	1063	1046
03	1131	1091	1087	1101	1086	1093	1088	1066	1104	1086	1071
02	1148	1113	1102	1120	1101	1115	1105	1084	1122	1102	1078
01	1178	1132	1122	1137	1117	1134	1123	1101	1138	1119	1093
1	1184	1146	1137	1154	1136	1148	1139	1119	1154	1137	1109
2	1190			1162	1142				1164	1142	1112
3	1196	1160	1151	1168	1152	1162	1154	1130	1170	1152	1115
4	1209			1181	1160				1183	1162	1141
5	1221			1205	1184				1207	1186	1159
5½	N/A			1223	1201				1225	1203	1167
6	1255			1241	1220				1243	1222	1185
7	1264			1255	1237				1257	1239	1201
8	1300			1269	1247				1271	1249	1211
9	1317			1278	1257				1280	1260	1224
10	1330			1303	1282				1305	1285	1251
11	1336			1312	1293				1315	1294	1272
12	1355			1326	1304				1326	1306	1285
13				1346	1321				1348	1331	
13½									1367	1352	

| | Self-supporting cones | | | | | | Large cones | | | | Small |
| | Regular | | | Iron-free | | | Regular | | Iron-free | | Regular |
Cone	15	60	150	15	60	150	60	150	60	150	300
14		1365	1384				1388	1366			
14½		1386	1409								
15		1417	1428				1424	1431			
15½		1436	1445								
16		1457	1475				1455	1473			
17		1479	1487				1477	1485			
18		1502	1508				1500	1506			
19		1522	1530				1520	1528			
20		1544	1551				1542	1549			
21		1566	1571				1564	1569			
22											
23		1588	1592				1586	1590			
24											
25											
26							1589	1605			
27							1614	1627			

Heating rate °C/h (last 90–120 minutes of firing)

Cone		
28	1619	1633
29	1624	1645
30	1636	1654
31	1661	1679
31½	1684	1698
32	1706	1717
32½	1718	1730
33	1732	1741
34	1757	1759
35	1784	1784
36	1798	1796
37		1820
38		1850
39		1865
40		1885
41		1970
42		2015

Notes
Temperatures shown are for specific mounted height above base. For self-supporting – 1¾″. For large 2″. For small cones, 15/16″. For large cones mounted at 1¾″ height, use self-supporting temperatures.

8.4 Pyrometric cones, thermoscope bars and Bullers rings

Although it is possible to measure the temperature within a kiln or oven using sensors such as thermocouples and infrared thermometers, the ceramics and glass industries routinely make use of devices known as pyrometric cones, bars and rings. Pyrometric cones, also known as Harrison, Seger or Orton cones, are slender pyramids as shown in Figure 8.13 and are manufactured from ceramic composites. On heating the effect of gravity

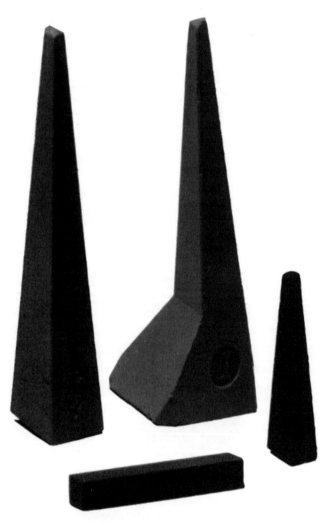

Figure 8.13 A selection of pyrometric cones. Photograph courtesy of Orton Ceramics (2000)

causes the cone to progressively bend as the material softens over a particular range of temperatures. The progressive deformation of a cone during firing is illustrated in Figure 8.14.

Pyrometric cones do not provide an exact indication of temperature. Their deformation is a function of both the rate of heating and the temperature cycle. They are designed to provide an indication of process completion, for instance that firing is complete. A pyrometric cone will be selected that will bend when heated at a prescribed rate and then held at an elevated temperature for a set period.

Cones vary in shape, size, method of use and performance according to the manufacturer. Pyrometric cones traditionally include identification numbers on their base or side, ranging in the case of Orton cones from 022 to 42. The temperature ranges, dependent on the rate of heating, for the various cones are listed in Table 8.4. Cones 022 to 011 are designed for use in the range of 590–860°C and tend to be used for overglaze decorations, lustres, enamels, glass fusing and annealing. Cone numbers 010 to 3 cover the temperature range 890–1170°C and are used to fire clay bodies, wall tiles, glazes and some structural clay products. Cones 4 to 12 cover the temperature range 1180–1330°C and are used in firing porcelain, floor tiles, china, stoneware and refractory materials. Cones 13 to 42 are used to fire products at

Figure 8.14 As temperature is maintained or increased above the rating for a particular cone, the cone will progressively bend and deform due to the effect of gravity on the softened material. Photograph courtesy of Orton Ceramics (2000)

temperatures up to 2015°C. Although the cones do not measure temperature directly, the cone-bending behaviour is related to temperature. Generally the faster the firing, the higher the temperature required to bend the cone and the slower the firing, the lower the temperature required to bend the cone. Pyrometric cones have been manufactured by a number of companies. Determination of performance and equivalents has been standardized in ANSI/ASTM C24-79 (1984) and reported by Fairchild and Peters (1926) and Beerman (1956).

A range of cones and bars are available. Small cones and bars tend to be used in conjunction with a kiln-sitter, a device which cuts power to a kiln when the cone or bar has deformed. Self-supporting cones are the easiest to use as there is no need to use or form a special supporting base. They can simply be placed in the region of interest in the kiln. Large cones must be supported in use. This can be achieved by use of a cone holder, clay pat (Figure 8.15) or a special wire holder. When using large cones it is important to mount them at the correct angle and height to ensure consistent bending. For instance, for the large Orten cones, recommended practice is to ensure that 50 mm of the cone is exposed above the cone plaque or clay pat and that the mounting angle is at 8°. Cones are sensitive to reducing atmospheres and these can cause them to bloat and bend backwards giving a false indication of temperature. In order to overcome this problem large cones manufactured from an iron-free composition are also available.

Figure 8.15 Use of a clay pat to support pyrometric cones. The pat can be been pierced when soft to allow the escape of water vapour. Cones courtesy of Lizzy Mellen

Common practice is to use a number of different cones, each with a higher bending temperature, simultaneously. One option is to use three cones: a guide cone, a firing cone and a guard cone. The guide cone is selected so that it will deform as the firing process is approaching maturity. The firing cone is selected so that its deformation is complete when the firing process has ended and the guard cone provided should not deform unless the firing process has exceeded the time–temperature specifications.

The temperature indicated by a cone can be determined by comparison of the bending angle with a chart supplied by the manufacturer. The bending angle can be judged by eye using a clock based approach as indicated in Figure 8.16 or a template approach as illustrated in Figure 8.17. It should be noted that the divisions of the clock-based approach do not indicate constant temperature intervals. This is illustrated by example in Figure 8.18 where at

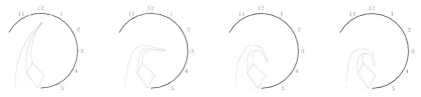

Figure 8.16 Clock-based model for the interpretation of cone bend angle. (After Orton Ceramics)

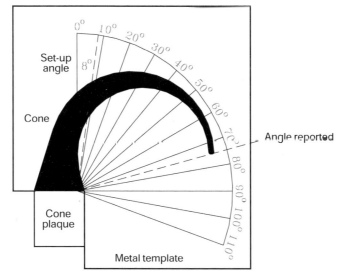

Figure 8.17 Use of a template supplied by the cone manufacturer to determine cone bend angle. (After Orton Ceramics)

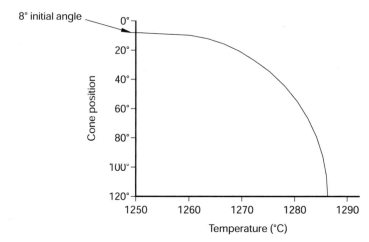

Figure 8.18 Variation of temperature with cone angle. (After Orton Ceramics (2000)

1270°C an angular deflection of 10° represents a temperature change of 5°C whilst at 1285°C an angular deflection of 10° represents a temperature rise of just 1°C.

General guidance in the use of cones is given below (after Smith, 1995):

- Face the cones to the source of heat i.e. towards the burners. If in the case of a cross-draught kiln they are faced towards the flues they will bend backwards towards the flame and cannot be considered to give an accurate indication of the firing process.
- Follow the manufacturer's guidance on the mounting height and angle.
- Rather than use cone stands use a small volume of wadding clay to form a clay pat (50% ball clay and 50% sand by volume) and use a match or pencil to spike it all over, to prevent it from exploding when the water content vaporizes.
- Ensure that the cones are visible through the spy hole.
- Keep a record of the type of cone and the cone location within the kiln along with the qualitative results of the firing. This will provide a valuable record of successful and unsuccessful practice.
- If possible always place the cones in the same position within the kiln.
- Use gas welding glasses to view cones at temperatures above 1150°C as the thermal radiation at these temperatures can damage the eyes.

Bars and rings known as thermoscope bars and Bullers rings respectively are also sometimes used to indicate temperature in the firing process. Thermoscope bars are used by mounting them in a special stand (Figure 8.19). They soften and deform with temperature and this deformation can be

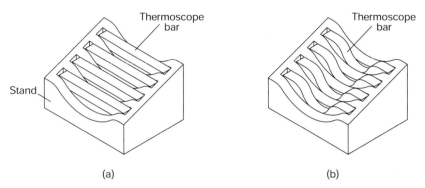

Figure 8.19 Thermoscope bars mounted on a stand. (a) Before firing. (b) After firing. (After BS 1041: Part 7)

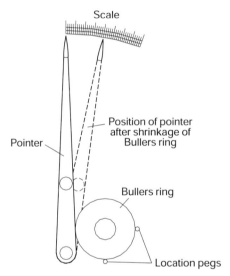

Figure 8.20 Measurement of the contraction of a Bullers ring. (After BS 1041: Part 7)

measured and compared with a look-up table supplied by the manufacturer to provide an indication of temperature. The range of operation for temperature measurement for thermoscope bars is between 590°C to 1525°C in intervals of about 15°C to 30°C. Bullers rings are precision dust pressed devices that are designed to shrink linearly with temperature rise. The contraction can be measured using a special gauge as illustrated in Figure 8.20. The temperature range capability for Bullers rings is between approximately 960°C and 1400°C. The usage of thermoscope bars and Bullers rings is outlined in BS 1041: Part 7.

References

Books and papers

Alaruri, S., Bonsett, T., Brewington, A., McPheeters, E. and Wilson, M. Mapping the surface temperature of ceramic and superalloy turbine engine components using laser-induced fluorescence of thermographic phosphor. *Optics and Lasers in Engineering*, **31**, 345–351, 1999.

Alaruri, S., Mcfarland, D., Brewington, A., Thomas, M. and Sallee, N. Development of a fiberoptic probe for thermographic phosphor measurements in turbine-engines. *Optics and Lasers in Engineering*, **22**, 17–31, 1995.

Allison, S.W. and Gillies G.T. Remote thermometry with thermographic phosphors: Instrumentation and applications. *Review of Scientific Instruments*, **68**(7), 2615–2650, 1997.

Baughn, J.W. Review – Liquid crystal methods for studying turbulent heat transfer. *International Journal of Heat and Fluid Flow*, **16**, 365–375, 1995.

Baughn, J.W., Anderson, M.R., Mayhew, J.E. and Wolf, J.D. Hysteresis of thermochromic liquid crystal temperature measurement based on hue. *Journal of Heat Transfer*, **121**, 1067–1072, 1999.

Baughn, J.W., Ireland, P.T., Jones, T.V. and Saniei, N. A comparison of the transient and heated coating methods for the measurement of local heat transfer coefficients on a pin fin. ASME Paper 88-GT-180, 1988.

Beerman, H.P. Calibration of pyrometric cones. *Journal of the American Ceramic Society*, **39**, 47–54, 1956.

Bird, C., Mutton, J.E., Shepherd, R., Smith, M.D.W. and Watson, H.M.L. Surface temperature measurement in turbines. *AGARD CP 598*, 21.1–21.10, 1998.

Bradley, L.C. *Review of Scientific Instruments*, **24**, 219, 1953.

Camci, C. Liquid crystal thermography. In *Temperature Measurements*, Von Karman Institute for Fluid Dynamics, Lecture Series 1996–07, 1996.

Camci, C., Kim, K. and Hippensteele, S.A. A new hue capturing technique for the quantitative interpretation of liquid crystal images used in convective heat transfer studies. *Journal of Turbomachinery*, **114**, 765–775, 1992.

Camci, C., Kim, K., Hippensteele, S.A. and Poinsatte, P.E. Evaluation of a hue capturing based transient liquid crystal method for high-resolution mapping of convective heat transfer on curved surfaces. *Journal of Heat Transfer*, **115**, 311–318, 1993.

Campbell, R.P. and Molezzi, M.J. Applications of advanced liquid crystal video thermography to turbine cooling passage heat transfer measurement. ASME Paper 96-GT-225, 1996.

Cates, M.R., Beshears, D.L., Allison, S.W. and Simmons, C.M. Phosphor thermometry at cryogenic temperatures. *Review of Scientific Instruments*, **68**, 2412–2417, 1997.

Cattafesta, L.N., Liu, T. and Sullivan, J.P. Uncertainty estimates for temperature-sensitive paint measurements with charge-coupled device cameras. *AIAA Journal*, **36**, No. 11, 2102–2108, 1998.

Crafton, B.T., Lachendro, J., Guille, N., and Sullivan, J.P. Application of temperature pressure sensitive paint to an obliquely impinging jet. AIAA Paper 99–0387, 1999.

Czysz, P. and Dixon, W.P. Thermographic heat transfer measurement. *Instruments and Control Systems*, **41**, 71–76, 1968.

Czysz, P. and Dixon, W.P. Quantitative heat transfer measurement using thermographic phosphors. *SPIE Journal*, 77–79, 1969.

Dever, M., Bugos, A., Dyer, F., Cates, M., Tobin, K., Beshears, D. and Capps, G.

Measurement of the surface of textiles during microwave drying using a thermographic phosphor. *Journal of Microwave Power and Electromagnetic Energy*, **25**, 230–235, 1990.

Donovan, J.F., Morris, M.J., Pal, A., Benne, M.E. and Crites, R.C. Data analysis techniques for pressure and temperature sensitive paints. AIAA Paper 93–0176, 1993.

Edge, A.C., Laufer, G. and Krauss, R.H. Surface temperature-field imaging with laser-induced thermographic phosphorescence. *Appl. Optics*, **39**(4), 546–553, 2000.

Ervin, J., Murawski, C., Macarthur, C., Chyu, M. and Bizzak, D. Temperature-measurement of a curved surface using thermographic phosphors. *Experimental Thermal and Fluid Science*, **11**(4), 387–394, 1995.

Fairchild, C.O. and Peters, M.F. Characteristics of pyrometric cones. *Journal of the American Ceramic Society*, **9**, 701–743, 1926.

Farina, D.J., Hacker, J.M., Moffat, R.J. and Eaton, J.K. Illuminant invariant calibration of thermochromic liquid crystals. *Experimental Thermal and Fluid Science*, **9**:(1), 1–12, 1994.

Feist, J.P. and Heyes, A.L. The characterization of $Y_2O_2S:Sm$ powder as a thermographic phosphor for high temperature applications. *Measurement Science & Technology*, **11**, 942–947, 2000.

Fergason, J.L. Liquid crystals. *Scientific American*, **211**, 76–86, 1964.

Fonger, W.H. *J. Chem. Phys.*, **52**, 6364, 1970.

Fonger, W.H. and Struck, C.W. Eu^{+3} ^5D resonance quenching to the charge-transfer states in Y_2O_2 S, La_2O_2S and LaOCL. *Journal of Chemical Physics*, **52**, 6364–6372, 1970.

Fonk, D.A. and Vukovish, M. Deformation behaviour of pyrometric cones and the testing of self-supporting cones. *Bulletin of the American Ceramic Society*, **53**, 156–158, 1974.

Gallery, J., Gouterman, M., Callis, J., Khalil, G., McLachlan, B. and Bell, J. Luminescent thermometry for aerodynamic measurements. *Review of Scientific Instruments*, **65**, 712–720, 1994.

Hamner, M., Kelble, C.A., Owens, L.R., and Popernack, T.G. Application of temperature sensitive paint technology to boundary layer analysis. AIAA Paper 97-5536, 1997.

Hay, J.L. and Hollingsworth, D.K. A comparison of trichromic systems for use in the calibration of polymer dispersed thermochromic liquid crystals. *Experimental Thermal and Fluid Science*, **12**, 1–12, 1996.

Hoffs, A., Bolcs, A. and Harasgama, S.P. Transient heat transfer experiments in a linear cascade via an insertion mechanism using the liquid crystal technique. ASME Paper 95-GT-8, 1995.

Hubner, J.P., Caroll, B.F., Schanze, K.S., Ji, H.F. and Holden, M.S. Temperature and pressure sensitive paint measurements in short duration hypersonic flows. AIAA Paper 99-0388, 1999.

Ireland, P.T. and Jones, T.V. The response time of a surface thermometer employing encapsulated thermochromic liquid crystals. *Journal of Physics E.*, **20**, 1195–1199, 1987.

Ireland, P.T. and Jones, T.V. Liquid crystal measurement of heat transfer and surface shear stress. *Meas. Sci. Technol.*, **11**, 969–986, 2000.

Kusama, H. *Jpn. Appl. Phys.*, **15**, 2349, 1976.

Lee, S.J. and Yoon, J.H. Temperature field measurement of heated ventilation flow in a vehicle interior. *International Journal of Vehicle Design*, **19**, 228–243, 1998.

Neumann R.D. Aerothermodynamic Instrumentation, AGARD Report No. 761, Special Course on Aerothermodynamics of Hypersonic Vehicles, pp. 4(1–40), 1989.

Noel, B.W., Borella, H.M., Lewis, W., Turley, W.D., Beshears, D.L., Capps, G.J., Cates, G.J., Cates, M.R., Muhs, J.D. and Tobin, K.W. Evaluating thermographic phosphors in an operating turbine engine. *Journal of Engineering for Gas Turbines and Power*, **113**, 242–245, 1991.

Noel, B.W., Turley, W.D., Lewis, W., Tobin, K.W. and Beshears, D.L. Phosphor thermometry on turbine engine blades and vanes. In Schooley, J.F. (Editor), *Temperature. Its Measurement and Control in Science and Industry*, Vol. 6, pp. 1249–1254. American Institute of Physics, 1992.

Ou, S., Rivir, R., Meininger, M., Soechting, F. and Tabbita, M. Transient liquid crystal measurement of leading edge film cooling effectiveness and heat transfer with high free stream turbulence. ASME Paper 2000-GT-0245, 2000.

Roberts G.T. and East R.A. Liquid crystal thermography for heat transfer measurement in hypersonic flows: a review. *Journal of Spacecraft and Rockets*, **33**(6), 761–768, 1996.

Sholes, R.R. and Small, J.G. Fluorescent decay thermometer with biological applications. *Rev. Sci. Instrumftl.*, **51**, 882–886, 1980.

Simonich, J.C. and Moffat R.J. Liquid crystal visualisation of surface heat transfer on a concavely curved turbulent boundary layer. *Journal of Engineering for Gas Turbines and Power*, **106**, 619–627, 1984.

Simons, A.J., McClean, I.P. and Stevens, R. Phosphors for remote thermograph sensing in lower temperature ranges. *Electronics Letters*, **32**(3), 253–254, 1996.

Smith, L. Practical notes on pyrometry. 1995. http://art.sdsu.edu/ceramicsweb/articles/pyrometry/pyrometry.html

Tanda, G., Stasiek, J. and Collins, M.W. An experimental study by liquid crystals of forced convection heat transfer from a flat plate with vortex generators. *C510/106/95*, 141–145, 1995.

Tobin, K.W., Allison S.W., Cates M.R., Capss, G.J., Beshears, D.L., Cyr, M. and Boel, B.W. High-temperature phosphor thermometry of rotating turbine blades. *AIAA Journal*, **28**(8), 1485–1490, 1990.

Wang, Z., Ireland, P.T., Jones, T.V. and Davenport, R. A colour image processing system for transient liquid crystal heat transfer experiments. ASME Paper 94-GT-290, 1994.

Wang, Z., Ireland, P.T., Kohler, S.T. and Chew, J.W. Heat transfer measurements to a gas turbine cooling passage with inclined ribs. *Journal of Turbomachinery*, **120**, 63–69, 1998.

Woodmansee, W.E. Cholesteric liquid crystals and their application to thermal nondestructive testing. *Materials Evaluation*, 564–572, 1966.

Standards

ASTM. Standard method for pyrometric cone equivalent (PCE) of refractory materials. Part 17. 1984 Annual Book of ASTM Standards. ANSI/ASTM C24-97.

BS 1041: Part 7: 1988 British Standard Temperature Measurement Part 7. Guide to selection and use of temperature/time indicators.

Web sites

At the time of going to press the world wide web contained useful information relating to this chapter at the following sites.

digitalfire.com/magic/cones.htm
hocmem.com/HOC/Cones.htm
http://cyclops-mac.larc.nasa.gov/AFCwww/testTek/
testTek.html#anchor61000166
http://members.wbs.net/homepages/y/d/x/ydx/index.html
http://www.arnold.af.mil/aedc/systems/73-366.htm
http://www.bjwe.com/
http://www.bjwe.com/tempil.htm
http://www.eng.ox.ac.uk/lc/noframes/research/intro.html
http://www.eng.ox.ac.uk/lc/noframes/worldlinks.html
http://www.hallcrest.com/chcm.htm
http://www.mv.com/ipusers/paperthermometer/
http://www.nv.doe.gov/business/capabilities/SciPhoto/ThermoPhosphors.htm
http://www.ortonceramic.com/
http://www.porphyrin.com/company/catalog.html
http://www.tempil.com/
www.clayartcenter.com/orton/correct_useof.html
www.testo.co.uk

Nomenclature

a_{CTS} = charge transfer state rate
a_j = probability rate
A = factor related to a_j
$b(\lambda)$ = filter transmissivity
B = blue
$E(\lambda)$ = lighting spectral distribution
E = energy
$g(\lambda)$ = filter transmissivity
G = green
H = hue
I = intensity
k = Boltzmann's constant
$r(\lambda)$ = filter transmissivity
$R(\lambda)$ = surface reflectance
R = red
S = saturation
T = temperature
λ = wavelength.

9

Infrared thermometry

Infrared thermometry is the most widely used form of non-invasive temperature measurement. The aims of this chapter are to introduce the physical phenomena exploited and the wide range of infrared thermometer and thermal imagers available.

9.1 Introduction

Heat transfer can occur by means of three fundamental mechanisms: conduction, convection and radiation. The last form can be stated more fully as the transfer of heat energy by means of electromagnetic radiation, which is also known as thermal radiation. The emission of energy in the form of electromagnetic radiation can be exploited to undertake a measurement of temperature. A typical infrared measurement system might comprise the source, the medium though which the radiant energy is transmitted, an optical system to gather the electromagnetic radiation, a transducer to convert the radiation into a signal related to temperature and amplification and interface circuitry to control, display and record the measurement.

As infrared radiation can pass across a vacuum or through a fluid medium it is not necessary for the temperature measurement to be made with the sensor in direct contact with the medium of interest. Temperature can therefore be measured in a non-invasive manner with the thermometer located some distance away from the target. Applications for infrared thermometry are extensive from remote objects such as stars, to moving surfaces and harsh environments such as furnaces. Infrared thermometers can be used for a wide range of temperatures from approximately 50 K to 6000 K. The measurement principle used is based on a fundamental thermodynamic relationship and can provide a means of measuring high temperatures where other devices cannot function or would cause too much disturbance to the medium of interest. Infrared thermometers are used to define the part of the ITS-90 above the freezing point of silver, 961.78°C. There is a wide range of infrared thermometers and the uncertainty associated with them is highly dependent on an adequate accounting of the effects of the measurement environment. It is therefore important to have some understanding of the fundamental concepts of thermal radiation and these are described in Section 9.2. Infrared thermometers can be broadly classified into five categories: spectral band thermometers, total radiation thermometers, ratio thermometers, fibre-optic thermometers and thermal imagers and these are introduced in Sections 9.3 to 9.8. Measurement of temperature using infrared thermometers is subject to a number of sources of error. These are introduced with specific reference to spectral band thermometers in Section 9.4.2 but the principles are applicable to the majority of infrared-based thermometers. Similarly, the typical use and operation procedure for an infrared thermometer is introduced in Section 9.4.3. The necessity and procedures for calibration are described in Section 9.9. Considerations for the selection of a particular kind of infrared thermometer are provided in Section 9.10.

9.2 Fundamentals of thermal radiation

Radiation can be defined as energy that spreads out as it travels. Electromagnetic radiation consists of interacting self-sustaining electric and magnetic fields that propagate through vacuum at a speed of 299 792 458 m/s (Cohen and Taylor, 1999). It does not necessarily need a medium for transmission and can travel through vacuum or through a body of fluid. It will, however, travel at a lower speed through a fluid and can become scattered. The known electromagnetic spectrum comprises electromagnetic waves with wavelengths as illustrated in Figures 9.1 and 9.2. The classifications are for convenience and the band between 0.1 μm and 100 μm is known as thermal

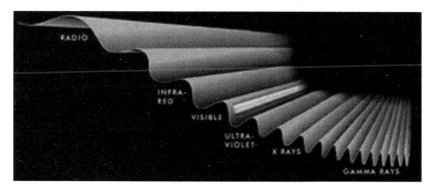

Figure 9.1 Schematic illustrating the electromagnetic spectrum. Courtesy of IPAC (the Infrared Processing and Analysis Centre) http://www.ipac.caltech.edu/Outreach/Edu/infrared.html

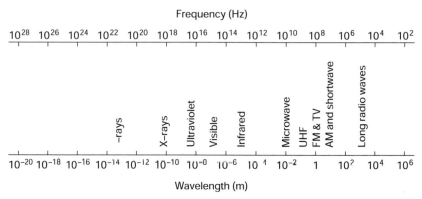

Figure 9.2 The electromagnetic spectrum

radiation and that between $0.78\,\mu m$ and $1000\,\mu m$ as infrared radiation. Infrared is an invisible portion of the electromagnetic spectrum. All substances, regardless of temperature, will emit some energy in the form of electromagnetic radiation due to their temperature.

As the temperature rises the quantity of heat transferred by means of thermal radiation increases and it is a common experience that as the temperature increases the brighter an object becomes. As examples the perceived colour of objects which radiate within the visible portion of the electromagnetic spectrum at different temperatures are listed in Table 9.1. This information can be used to estimate crudely the temperature of an object and is still used to this day by blacksmiths and engineers as an approximate check that a metal has reached the desired temperature to within $\pm 50°C$. It should be noted, though, that just because our eyes cannot detect radiation at

Table 9.1 Temperature versus perceived colour.
(After Nicholas and White, 1994)

Temperature (°C)	Perceived colour
500	Red just visible
700	Dull red
900	Cerise
1000	Bright cerise
1100	Dull orange red
1250	Bright orange yellow
1500	White
1800	Dazzling white

temperatures below 500°C, it does not mean that objects are not emitting thermal radiation. Any object with a temperature above 0 K will emit thermal radiation.

9.2.1 Planck's law

Any object will emit energy due to its temperature. The term 'thermal radiation' is used to describe the energy that is emitted in the form of electromagnetic radiation at the surface of a body that has been thermally excited. Thermal radiation is emitted in all directions and if it strikes another body, part may be reflected, part may be absorbed and part may be transmitted through it. Absorbed radiation will appear as heat within the body. As it is an electromagnetic wave, thermal radiation can be transferred between two bodies, without the need for a medium of transport between them. For example, if two bodies are at different temperatures and are separated by a vacuum between them, then heat will still be transferred from the hotter object to the colder by means of thermal radiation. Even in the case of thermal equilibrium, the temperature of both bodies the same, an energy exchange will occur although the net exchange will be zero.

The thermal radiation emitted by a surface is not equally distributed over all wavelengths. Similarly, the radiation incident, reflected or absorbed by a surface may also be wavelength dependent. The wavelength dependency of any radiative quantity or surface property is referred to as spectral dependency. The adjective monochromatic is used to refer to a radiative quantity at a single wavelength.

The term 'emissive power' is used for the thermal radiation leaving a surface per unit surface area of the surface. This is qualified depending on whether it is summed over all wavelengths or whether it is emitted at a particular wavelength. Total hemispherical emissive power or just total emissive power, normally denoted by the symbol E, defines the emitted energy summed over all directions and all wavelengths. The total emissive power is dependent on the temperature of the emitting body, the type of material and the nature of the surface features such as its roughness. The spectral emissive power, E_λ, defines the emissive power emitted at a particular wavelength

When dealing with thermal radiation from and to real surfaces it is useful to use the concept of an ideal surface as a comparison baseline. This ideal surface is given the name 'blackbody'. A blackbody surface has the following properties:

- A blackbody absorbs all incident radiation regardless of wavelength and direction.
- For a given wavelength and temperature, no surface can emit more energy than a blackbody.
- Radiation emitted by a blackbody is independent of direction.

The spectral emissive power for a blackbody can be determined using Planck's law, (equation 9.1) (see Planck (1959)):

$$E_{\lambda,b} = \frac{C_1}{\lambda^5[\exp(C_2/\lambda T) - 1]} \tag{9.1}$$

where: $E_{\lambda,b}$ = spectral emissive power for a blackbody (W/m^3)
C_1 = the first radiation constant = $3.7417749 \times 10^{-16}$ W.m^2 (Cohen and Taylor, 1999),
λ = wavelength (m)
C_2 = the second radiation constant = 0.01438769 m.K (Cohen and Taylor, 1999)
T = absolute temperature (K).

This equation is plotted for a range of wavelengths and temperatures in Figure 9.3. This figure illustrates how the energy radiated for a given temperature is a function of the wavelength at which the energy is radiated. In the figure the wavelength has been plotted in μm and emissive power

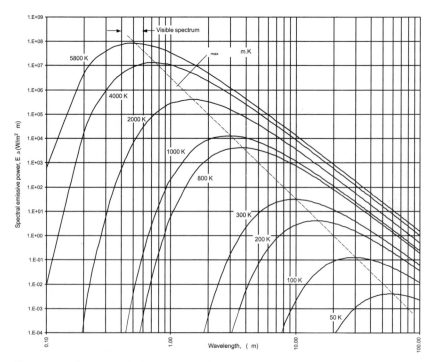

Figure 9.3 Spectral blackbody emissive power

calculated in $W/m^2/\mu m$ as wavelengths are commonly referred to using the unit μm. The first and second radiation constants in equation (9.1) become $C_1 = 3.7417749 \times 10^8$ $W \cdot \mu m^4/m^2$ and $C_2 = 14387.69$ $\mu m \cdot K$ accordingly.

9.2.2 Wien's displacement law

Examination of Figure 9.3 reveals that blackbody spectral emission has a maximum. The corresponding wavelength at which this maximum occurs is related to the absolute temperature. The relationship between wavelength and temperature can be determined by differentiating equation (9.1) with respect to wavelength and setting the result equal to zero. The result is given by equation (9.2) and is known as Wien's displacement law:

$$\lambda_{max} T - 0.028978 \text{ m} \cdot K \tag{9.2}$$

9.2.3 Wien's law

In many applications in infrared thermometry the exponential term in equation (9.1) is much greater than unity and the −1 term can be neglected. This approximation, known as Wien's law, is stated in equation (9.3). Wien's law gives a reasonable approximation to equation (9.1) for wavelengths in the range $\lambda < 5\lambda_{max}$:

$$E_{\lambda, b} = \frac{C_1}{\lambda^5 \exp(C_2/\lambda T)} \tag{9.3}$$

9.2.4 Stefan–Boltzmann's law

Integrating equation (9.1) over all wavelengths gives the total emissive power for a blackbody:

$$E_b = \int_0^\infty \frac{C_1}{\lambda^5 [\exp(C_2/\lambda T) - 1]} \, d\lambda \tag{9.4}$$

The result of this integration is

$$E_b = \sigma T^4 \tag{9.5}$$

where σ is the Stefan–Boltzmann constant and has the numerical value 5.67051×10^{-8} $Wm^{-2}K^{-4}$ (Cohen and Taylor, 1999).

Equation (9.5) is known as Stefan–Boltzmann's law. This equation enables a calculation to be made to determine the amount of radiation emitted in all directions over all wavelengths simply from knowledge of the temperature. It is, however, only valid for an ideal blackbody surface and herein lies many of the problems and errors associated with infrared thermometer-based temperature measurements. Very few surfaces even approach this ideal and frequently the surface properties relevant to the calculation of thermal radiation are not well known or not constant. The next section serves to introduce the modelling of non-ideal surfaces.

9.2.5 Grey bodies and grey body radiation

The magnitude of radiation from a real surface is a function of both the temperature of the surface and the surface properties. The surface property limiting the quantity of radiation is called the emissivity, ε. Emissivity is defined on a scale from 0 to 1 and is the ratio of the electromagnetic flux that is emitted from a surface to the flux that would be emitted from an ideal blackbody at the same temperature. Emissivity is generally wavelength dependent. The spectral emissive power of a real surface at a given temperature differs from that of a blackbody at the same temperature in terms of the magnitude of the emissive power and the spectral distribution. A non-blackbody surface has an absorptivity of less than unity and its value may be dependent on the wavelength of the incident radiation. Real surfaces may also exhibit non-diffuse behaviour with properties varying according to direction.

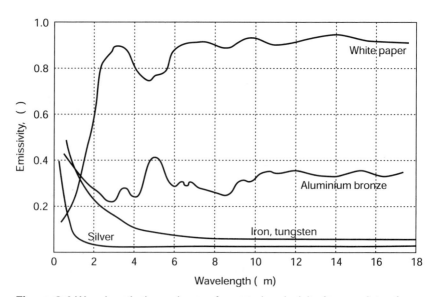

Figure 9.4 Wavelength dependency of spectral emissivity for a variety of materials. (Reproduced from Fraden, 1996)

A rigorous approach to the analysis of grey bodies is to define the monochromatic directional properties and then determine the total and hemispherical properties by integration over the spectrum and all directions. This is not always necessary and an analysis where directional dependencies are assumed negligible and the material is assumed to be diffuse is often adequate.

The hemispherical spectral emissivity is defined as the ratio of the hemispherical emissive power of a grey body to that of a blackbody at the same temperature and wavelength, (equation (9.6)). The term 'hemispherical' is used to denote the radiation emitted in all directions from a surface, as defined by a hemisphere:

$$\varepsilon_{\lambda,T} = \frac{E_\lambda\ (\lambda,T)}{E_{\lambda,b}\ (\lambda,T)} \tag{9.6}$$

where: $\varepsilon_{\lambda,T}$ = spectral emissivity
E_λ = spectral emissive power for a grey body (W/m^3).

The hemispherical total emissivity is defined by the ratio of the total emissive power of a grey body to that of a blackbody at the same temperature,

$$\varepsilon(T) = \frac{E(T)}{E_b(T)} \tag{9.7}$$

where: ε = total emissivity
E = emissive power for a grey body (W/m^2).

The emissivity of a material is a function of its dielectric constant and subsequently its refractive index. As noted, emissivity is generally wavelength dependent and this is illustrated by example in Figure 9.4 for a number of materials. Tables 9.2 and 9.3 list typical total emissivities for a number of different materials. This kind of information is essential in the use of most infrared thermometers, as the devices normally need adjustment depending on the emissivity of the target. Many infrared thermometer manufacturers are able to supply information on emissivity for a given application.

The emissive power of a grey body at a given temperature can be readily calculated by a modification to the Stefan–Boltzmann law:

$$E = \varepsilon\sigma T^4 \tag{9.8}$$

Because a surface can emit, absorb, reflect and transmit emissive power the analysis and measurement of temperature by means of infrared thermometry can be a complex undertaking. This is particularly the case when dealing with surfaces where the emissivity is unknown or a complex

Table 9.2 Typical total emissivities for a variety of metals. (After Brewster, 1992)

Material	300 K	500 K	800 K	1600 K
Aluminium (smooth, polished)	0.04	0.05	0.08	0.19
Aluminium (smooth, oxidized)	0.11	0.12	0.18	–
Aluminium (rough, oxidized)	0.2	0.3	–	–
Aluminium (anodized)	0.9	0.7	0.6	0.3
Brass (highly polished)		0.03	'	
Brass (polished)	0.1	0.1		
Brass (oxidized)	0.6			
Chromium (polished)	0.08	0.17	0.26	0.4
Copper (polished)	0.04	0.05	0.18	0.17
Copper (oxidized)	0.87	0.83	0.77	
Gold (highly polished)	0.02		0.035	
Iron (polished)	0.06	0.08	0.1	0.2
Iron (oxidized)	0.6	0.7	0.8	
Stainless steel (polished)	0.1	0.2		
Stainless steel (aged)	0.5 to 0.8			
Mild steel (polished)	0.1		0.3	
Mild steel (oxidized)	0.8	0.8		
Lead (polished)	0.05	0.08		
Lead (oxidized)	0.6	0.6		
Mercury (clean)	0.1			
Magnesium (polished)	0.07	0.13	0.18	0.24
Nickel (polished)	0.05	0.07		
Nickel (oxidized)	0.4	0.5		
Platinum (polished)	0.05		0.1	
Platinum (oxidized)	0.07		0.1	
Silver (polished)	0.01	0.02	0.03	
Silver (oxidized)	0.02		0.04	
Tin (polished)	0.05			
Tungsten filament	0.032	0.053	0.088	0.35 (3500 K)
Zinc (polished)	0.02	0.03		
Zinc (oxidized)				
Zinc (galvanized)	0.02 to 0.03	0.1		

function of wavelength and temperature or when the target is in the vicinity of other objects at different temperatures. In this case an assessment needs to be made of the radiation exchange between the various objects involved. It is possible for a target surface at a temperature T_{target} to reflect radiation from another object thereby significantly distorting the temperature measurement given by an infrared thermometer. An assessment of the radiation exchange between different surfaces can be undertaken. This can require determination of the view factors for the objects concerned and the reader is

Table 9.3 Typical total emissivities for a variety of non-metals. (After Brewster, 1992)

Material	300 K	500 K	800 K	1600 K
Aluminum oxide		0.7	0.6	0.4
Asbestos	0.95			
Asphalt	0.93			
Brick (alumina refractory)			0.4	0.33
Brick (fireclay)	0.9		0.8	0.8
Brick (kaolin insulating)			0.7	0.53
Brick (magnesite refractory)	0.9			0.4
Brick (red, rough)	0.9			
Brick (silica)	0.9		0.8	0.8
Concrete (rough)	0.94			
Glass	0.95	0.9	0.7	
Graphite	0.7			0.8
Ice (smooth)	0.97 (273 K)			
Ice (rough)	0.99 (273 K)			
Limestone	0.9	0.8		
Marble (white)	0.95			
Mica	0.75			
Paint (aluminized)	0.3 to 0.6			
Paint (most others, including white)	0.9			
Paper	0.9 to 0.98			
Porcelain (glazed)	0.92			
Pyrex	0.82	0.8	0.72	0.6
Quartz	0.9		0.6	
Rubber (hard)	0.95			
Rubber (soft, grey, rough)	0.86			
Sand (silica)	0.9			
Silicon carbide		0.9		0.8
Skin	0.95			
Snow	0.8 to 0.9			
Soil	0.93 to 0.96			
Rocks	0.88 to 0.95			
Teflon	0.85	0.92		
Vegetation	0.92 to 0.96			
Water (>0.1 mm thick)	0.96			
Wood	0.8 to 0.9			

referred to a standard heat transfer text for an explanation of this technique such as Bejan (1993), Incropera and Dewitt (1996), Chapman (1987) or Kreith and Bohn (1996). The text by Howell (1982) provides a comprehensive listing of view factors for a wide variety of geometries. An alternative to this is to eliminate the sources of error in the temperature-measuring environment, (see Section 9.4.2).

9.3 Detector classification

Infrared thermometers can be classified in a number of groups based on the principle of operation:

- spectral band thermometers
- total radiation thermometers
- ratio thermometers
- multiwaveband thermometers
- special-purpose thermometers and methods
- thermal imagers.

Spectral band thermometers are the most common form of infrared temperature measurement. These devices measure radiant energy across a narrow waveband. Examples include devices which use silicon or germanium detectors. These fall into the category known as short-wavelength thermometers, with a waveband approximately between 0.5 and 2 μm. In this spectral region, short wavelengths have the advantage that the rate of change of radiant energy with temperature is high (up to 2–3%/°C) and they are therefore highly sensitive. Optical windows can be used to limit the wavelengths reaching the detector within the thermometer. This technique is useful for observing the temperatures of semitransparent materials. For example, most glasses are transparent to radiation at wavelengths up to 2 μm but become opaque and therefore visible to the detector at wavelengths greater than 4.5 μm. In a similar manner many plastics are transparent over wide bands of the spectrum but have high opacity and low reflectivity at known spectral bands. Thermometers with special filters and detectors sensitive in the appropriate waveband can be selected for these applications.

Total radiation thermometers measure radiant energy over a broad waveband (several micrometres). As a broadband of wavelengths will contain a large amount of energy at any temperature these are normally used for general purposes or low-temperature application temperature measurements where the energy emitted is relatively low.

Ratio thermometers measure the radiance at two wavelengths and determine the ratio. If it is assumed that the absorption and emissivity are constant over the wavelength range, then the ratio is a function of temperature only and this can be determined. The advantage of ratio thermometers, which are also known as dual-wavelength or two-colour thermometers, is that it is not necessary to know the emissivity of the target provided the assumption that its value is constant over the range of wavelengths used. Ratio thermometers are, however, less sensitive than spectral band thermometers.

Multiwaveband thermometers allow measurement of surface temperature when the emissivity is not constant at different wavelengths. Special-purpose infrared thermometers and methods include auxiliary reflector methods, hot source methods, polaradiometer methods, reflectance methods, temperature-invariant methods and fibre-optic thermometers. These techniques have been mostly developed in an attempt to provide a measurement of temperature that is emissivity independent.

Thermal imagers can be used to measure the temperature distribution over a target area. An image of the object can be formed by either using an array of detectors or by directing a portion of the image onto a single detector using an optical system in a fashion similar to the raster pattern used in televisions. Some thermal imagers are equipped with a liquid crystal display allowing a direct visual picture of the temperature distribution. The data can be stored and subsequently transferred to a computer for analysis.

9.4 Spectral band thermometers

The majority of infrared temperature measurement is performed using spectral band thermometers. These devices measure the radiance over a relatively narrow band of wavelengths normally within the range of 0.5–25 μm. They are also known as single-waveband, narrow-waveband or monochromatic thermometers. The choice of wavelength selected depends for a given application on the temperature range, the environment and the properties of the target surface. Figure 9.5 illustrates the principal components making up a spectral band thermometer and a commercial example is shown in Figure 9.6. The principle of operation is to collect radiation from the surface, filter it to select the wavelengths of interest, and then measure the radiation with a detector and process the information. The two apertures illustrated in Figure 9.5 are used to define the target area (also known as the spot size or field of view) and the acceptance angle for the thermometer. The purpose of the lens is to focus an image of the target area onto the target-defining aperture.

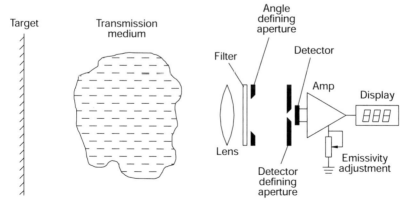

Figure 9.5 Schematic showing the principal components of a spectral band thermometer

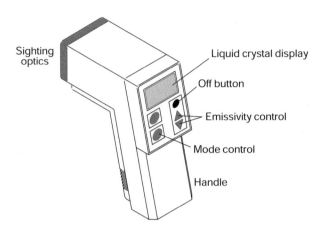

Figure 9.6 An example of a handheld spectral band thermometer. Model shown courtesy of Land Infrared

If the lens is not properly focused then the boundary of the target area will not be well defined.

Equation (9.1) can be integrated between specific wavelengths to identify the proportion of radiation emitted between those wavelengths. Alternatively the concept of an effective wavelength representative of the bandwidth of wavelengths can be used (Kostkowski and Lee, 1962; Gardner, 1980). This avoids the need to integrate equation (9.1) and the output from the thermometer can be represented by

$$S(T) = GE_{\lambda,T} \qquad (9.9)$$

where: $S(T)$ = the output of the thermometer at a given temperature
G = a factor incorporating geometrical and responsivity parameters.

The sensitivity of a spectral band thermometer can be determined by differentiating equation (9.9) with respect to T. Using Wien's approximation for $E_{\lambda,T}$, (equation (9.3) this gives

$$\frac{\Delta E_\lambda}{E_\lambda} = \frac{C_2 \Delta T}{\lambda T^2} \qquad (9.10)$$

Examination of equation (9.10) indicates that the sensitivity, $\Delta E_\lambda / \Delta T$, increases as the wavelength decreases. In addition it also reveals that equal errors in measuring the spectral emission produce larger temperature errors at longer wavelengths. In other words the uncertainty of spectral waveband thermometers increases with increasing wavelength. When considering uncertainty and sensitivity therefore, the shortest possible wavelength is normally chosen. In some applications, however, errors due to uncertainties of target properties, such as emissivity, dominate and the selection of the waveband must be based on different criteria.

9.4.1 Detectors

An essential component of an infrared thermometer is the transducer used to convert the absorbed electromagnetic radiation into a signal that can be measured. These devices are called detectors or thermal detectors and function by converting the absorbed radiation into heat energy causing the detector temperature to rise accompanied by the output. Detectors for thermal radiation can be classified into two broad categories: quantum and thermal.

Quantum, photon or photoelectric detectors measure the direct excitation of electrons to conduction states by incident photons. That is, when a photon strikes the surface of a conductor it can result in the generation of a free electron. Types of quantum detector include photoemissive, photoconductive and photovoltaic. Photon detectors respond to individual photons by releasing or displacing electrical charge carriers by the photoelectric effect (vacuum photocells, photomultipliers), photoconductive effect, photovoltaic effect or the photoelectromagnetic effect.

The operating ranges for some infrared detectors are illustrated in Figure 9.7 and Table 9.4. Photon detectors have much higher spectral detectivities

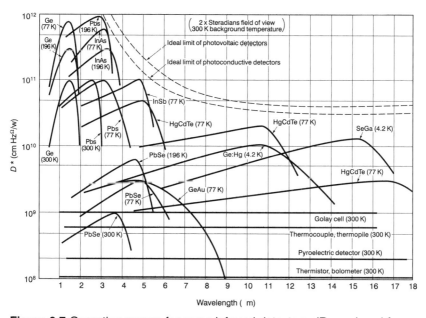

Figure 9.7 Operating ranges for some infrared detectors. (Reproduced from Fraden, 1999)

Table 9.4 Operating ranges and characteristics of some infrared detectors. (After Ricolfi and Barber, 1990; Michalski et al., 1991; Hackforth, 1960)

Detector	Type of thermometer	Waveband (μm)	Characteristic
Photomultiplier	Primary standard		DC or photon-counting operation at room temperature. High cost.
Al_2O_3		0.2–5	
CaF_2		0.13–9.5	
Ge	Industrial spectral band and ratio	1–12	Photoconductor or photovoltaic. DC or AC operation at room temperature. Good linearity and stability. Minimum target temperature 200°C. Low cost. (Michalski et al., 1991)
HgCdTe	Special spectral band		Photoconductor. AC operation, cooled. Possibility of tailoring the peak wavelength within a wide range. Minimum target temperature −50°C. High cost.
InAs	Special spectral band	1–3.8	Photovoltaic. AC operation, cooled. Minimum target temperature 0°C, High cost. http://www.egginc.com
InSb	Special spectral band	1–5.5	Photovoltaic. AC operation, cooled. Minimum target temperature 0°C, High cost. http://www.vigo.com
KBr		0.2–32	
KCl		0.21–25	

Material	Application	Range (µm)	Characteristics
PbS	Industrial spectral band and ratio	2.5	Photoconductor. AC operation at room temperature or cooled. High-temperature coefficient of responsivity. Minimum target temperature 100°C. Low, medium or high cost depending on temperature range.
PbSe	Industrial spectral band and ratio	2–5	Photoconductor. AC operation at room temperature or cooled. Good temperature coefficient of responsivity. Minimum target temperature 50°C. Low, medium or high cost depending on temperature range. http://www.rmtltd.ru
Si	Primary and secondary standards. Industrial spectral band and ratio	0.3–1.08	Photoconductor or photovoltaic. DC or AC operation at room temperature. Best among infrared detectors for detectivity, linearity and stability. Minimum target temperature 400°C. Low cost.
SiO_2		0.2–4	
TiO_2		0.4–5.2	
ZnS		3–12	http://www.mortoncvd.com
Pyroelectric	Industrial spectral band and total radiation	Filtered or all	Thermal detector. AC operation at room temperature. Minimum target temperature 0°C. Low–medium cost
Thin-film thermopile	Industrial spectral band and total radiation	Filtered or all	Thermal detector. AC operation at room temperature. Minimum target temperature 0°C. Low–medium cost
Thermistor	Industrial spectral band and total radiation	Filtered or all	Thermal detector. AC operation at room temperature. Minimum target temperature 0°C. Low–medium cost

than thermal detectors and a faster response. Quantum detectors are not perfect, however, because some of the incident photons are not absorbed or are lost and some excited electrons return to the ground state. For measurements of objects emitting photons in the range of 2 eV or more, quantum detectors at room temperature are generally used. For smaller energies, longer wavelengths and narrower bandwidths are necessary. Quantum detectors with narrow bandwidths tend to have relatively high intrinsic noise at room temperature in comparison to the photoconductive signal. The noise level is temperature dependent and can therefore be reduced by cooling. Cryogenic methods of cooling are used with lead sulphide (PbS), indium arsenide (InAs), germanium (Ge), lead selenide (PbSe) and mercury–cadmium–telluride (HgCdTe) detectors.

Thermal detectors convert the absorbed electromagnetic radiation into heat energy causing the detector temperature to rise. This can be sensed by its effects on certain physical properties, such as: electrical resistance used by bolometers; thermoelectric emf used by thermocouple and thermopile detectors; electrical polarization used by pyroelectric detectors. The principal application of thermal detectors is for measurement of low temperatures where there is limited radiant flux and the peak of the Planck curve is well into the infrared. Thermal detectors offer wide spectral response, by detecting the emitted radiation across the whole spectrum, at the expense of sensitivity and response speed. For higher temperatures, devices with a narrower spectral bandwidth are more suitable.

Bolometers are thermal detectors in which the incident thermal radiation produces a change in temperature of a resistance temperature device, which may be an RTD or a thermistor. Bolometers can, however, be comparatively slow with time constants of 10–100 ms. An alternative to the use of a resistance temperature device in a thermal detector is to use a thermopile. This consists of a number of series-connected thermocouples arranged so that there is a temperature difference between each pair of thermocouple junctions. In an infrared thermal detector the thermopile is arranged so that half of the junctions are maintained at a constant temperature by being in contact with a component with relatively large thermal inertia. The radiant energy heats the other junctions generating a thermoelectric emf. A sectional view of a thermopile sensor is shown in Figure 9.8. The output from a thermopile is a function of the temperature difference between the hot and the cold junctions. The sensor consists of a base where the cold junctions are placed with a relatively large thermal mass and a membrane where the hot junctions are located. The number of junctions used varies from a few tens to several hundred. Thermopile sensors can be manufactured using conventional thermoelements or semiconductors (Schieferdecker et al., 1995). The Seebeck coefficients for crystalline and polycrystalline silicon, for example, are high and the resistivity combined with the use of microelectronics fabrication techniques can be used to produce a highly sensitive device.

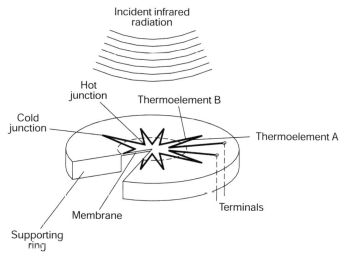

Figure 9.8 Thermopile sensor. A differential temperature between the hot and cold junctions is generated by using a geometry which conducts heat from the hot and cold junction at different rates or by using materials with different thermal conductivity

Pyroelectric detectors are manufactured using crystal wafers such as triglycerine sulphate or lithium tantalate which produce surface electric charges when heated. The electrical signal corresponds to the removal of charge by conducting electrons deposited on the crystal. A change in the temperature of the crystal due to the absorption of radiation in a certain time period produces a change in the polarization charge. The detector produces an electrical signal proportional to the rate of change of charge and therefore cannot be used to measure a continuous heat flux. However, addition of a mechanical chopper system allows steady-state flexibility with the chopper being used to interrupt the radiation from a target at a fixed frequency. Types include blackened and mirror choppers. Pyroelectric detectors have a wide spectral response similar to bolometers and thermopiles but have a faster response time. The uncertainty of these devices is low, of the order of 0.2% over the temperature range.

As listed in Table 9.4, there are infrared detectors that are sensitive across the range of the thermal radiation spectrum. The sensitivity of a detector decreases with the area that receives photons. In addition, the sensitivity of photon detectors varies with wavelength. Depending on the temperature range to be observed the detector and the spectral characteristics of the optics can be chosen so that their sensitivity matches this. The required spectral range of a detector dictates the type of material to be used. Quantum detectors are more spectrally selective, sensitive and faster than thermal detectors.

The principal factors that should be considered in the selection of a detector are listed below (after Ricolfi, 1990):

1 *Spectral responsivity.* This factor gives a measure of the relative performance with wavelength of the different types of detector. It is displayed in terms of the detectivity for a number of detectors in Figure 9.7.

2 *Detectivity.* The temperature resolution of an infrared thermometer can be represented by the detectivity. This is a figure of merit quantity and is defined in equation (9.11). Detectivity traditionally has the units $cm \cdot Hz^{0.5} W^{-1}$ for historical reasons:

$$D_\lambda = \frac{\sqrt{A \Delta f}}{NEP} \tag{9.11}$$

where: D_λ = detectivity ($cm \cdot Hz^{0.5} W^{-1}$)
         A = the effective sensitive area of the detector (cm^2)
         Δf = frequency bandwidth of the post-detector electronics (Hz)
         NEP = noise equivalent power, a figure of merit value describing the level of rms incident radiation required to produce a signal-to-noise ratio of unity.

3 *Stability of response.* This factor determines the reproduceability of measurements. Short-term instabilities are generally due to the dependence of the responsivity of the detector on ambient temperature. Silicon detectors, for example, are known to have a relatively low temperature coefficient with a 1°C change in the ambient temperature causing less than 0.1°C variation in the indicated target temperature. Lead sulphide detectors, however, can indicate a 1°C change in the target temperature for a 1°C change in the ambient temperature of the detector. Compensation for this can be achieved by means of controlling the temperature of the detector, correction of the output or use of a reference source in the thermometer. Long-term stability involves drift of the indicated temperature with time. This is highly undesirable. Silicon detectors give good long-term stability and have been demonstrated to drift less than 0.2°C in a year (Brown, 1986).

4 *Linearity.* A linear relationship between temperature and detector output can be desirable as it reduces the complexity of computations for calibration, interpolation and extrapolation. With modern processing, however, it is not essential. Silicon provides the best linearity as do germanium and indium antimonide although to a lesser extent.

5 *Speed of response.* The response time of detectors varies from a few nonoseconds for silicon detectors to a few milliseconds for thin-film thermopiles and thermistor bolometers. If the temperature of the application is varying rapidly then the choice for detectors is reduced.

6 *Operating mode.* This refers to whether the radiation signal received by the detector is continuous (DC) or deliberately varied or chopped (AC). Some detectors such as silicon and germanium can be operated in both modes. Pyroelectric detectors can only sense temperature variations and must therefore be operated in the AC mode using, for example, a rotating disc with a number of holes in front of the detector to vary the signal received with time.

7 *Operating temperature.* The detectivity of most photon detectors decreases with operating temperature. These devices can be cooled in order to improve the temperature resolution. An exception is silicon, which has a positive temperature coefficient at wavelengths of more than approximately 0.7 μm. As a result, these detectors are usually operated near to room temperature.

8 *Cost.* This is usually a function of the facilities available on an infrared thermometer and the type of detector used. The necessity to cool the detector can increase the cost substantially.

9.4.2 Sources of error

Measurements of temperature using spectral band thermometers and most other types of infrared thermometers are subject to a large number of potential sources of error. This is a consequence of the complexity of the radiation heat transfer process where the radiation is transmitted across a medium between the target and the sensor with the associated risk that some of this energy may be absorbed or redirected and the need to know surface, environment and transfer medium properties. Errors associated with infrared thermometers can be categorized into three main groupings:

• Characterization of the radiation process: surface emissivity, reflections, fluorescence
• Transmission path errors: absorption, scattering, size of object effects and vignetting
• Signal processing errors.

In undertaking a measurement using a spectral thermometer it is necessary to know the surface emissivity. As described in Section 9.2.5, emissivity can vary with wavelength and surface finish. Figure 9.4 illustrates this for a variety of materials. In practice this need not cause too much of a problem. Most materials can be readily identified, along with the condition of the surface. Considerations include whether the surface is polished, rough or oxidized. Tabulated information is normally available from the manufacturer of the spectral band thermometer for the emissivity of many materials over the relevant waveband. This allows an estimate to be made for the emissivity to within ±0.05 (Nicholas and White, 1994). Alternatively, information for the emissivity, such as that given in Tables 9.2 and 9.3, can be sourced from reference texts such as Touloukian *et al.* (1970, 1972a,b).

The error in the measurement of temperature caused by uncertainty in the assumed value for emissivity can be modelled by equations (9.12) and (9.13) (Corwin and Rodenburgh, 1994):

$$\delta\varepsilon = \frac{E_{12}\,(T) - E_{12}[(\delta T + 1)T]}{E_{12}[(\delta T + 1)T]} \tag{9.12}$$

where: $\delta\varepsilon = (\varepsilon_m - \varepsilon)/\varepsilon$
   ε_m = assumed emissivity
   ε = true emissivity
   E_{12} = integral of Planck's relation over the waveband λ_1 to λ_2 (W/m^2)
   $\delta T = (T_m - T)/T$
   T_m = assumed temperature (K)
   T = true temperature (K).

$$\delta\varepsilon = \frac{\exp\left\{(C_2/C_3)/[(\delta T + 1)(\lambda_c/\lambda_m)]\right\} - 1}{\exp[(C_2/C_3)/(\lambda_c/\lambda_m)] - 1} - 1 \tag{9.13}$$

where: λ_c = centre wavelength of bandwidth (μm)
   λ_m = wavelength at which spectral radiance is an optimum (μm)
   C_3 = 2898 μm.K.

Equation (9.13) has been plotted in Figure 9.9 for a range of emissivity uncertainties illustrating the variation in temperature error against the ratio λ_c/λ_m. For a narrow spectral band thermometer these results show that the temperature error becomes much less than the emissivity error when λ_c is less than λ_m. The errors as a result of uncertainty in emissivity for a range of common bandwidths for a target at selected temperatures are plotted in Figure 9.10. These graphs, determined using equation (9.12), illustrate that the effect uncertainty in emissivity on temperature error is significantly less at either lower wavelength or at lower temperature. This analysis does not take into account other errors inherent to some infrared thermometers such as detector noise and atmospheric absorption.

The data in Figure 9.10 suggests that there is an optimum value to use for the emissivity in order to minimize the error in temperature. This can be determined from

$$\varepsilon_o = \sqrt{\varepsilon_u \, \varepsilon_1} \tag{9.14}$$

where: ε_o = optimum value to use for emissivity to minimize the error in temperature
   ε_u = upper estimate of emissivity
   ε_1 = lower estimate of emissivity.

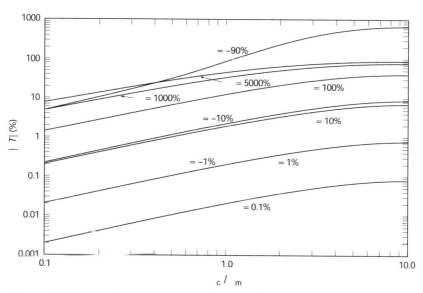

Figure 9.9 Temperature error versus the ratio of central wavelength to peak wavelength for a range of emissivity uncertainties. (After Corwin and Rodenburgh, 1994)

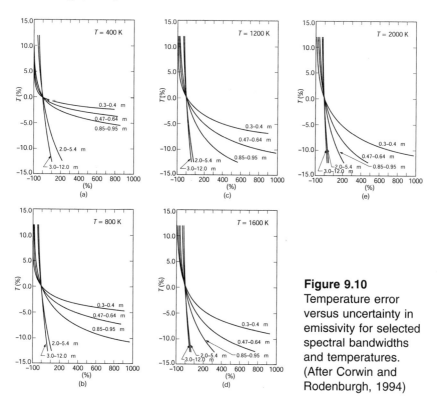

Figure 9.10 Temperature error versus uncertainty in emissivity for selected spectral bandwidths and temperatures. (After Corwin and Rodenburgh, 1994)

The maximum temperature error is given by

$$\delta T = \pm \frac{\lambda_c T_m}{2C_2} \ln \frac{\varepsilon_u}{\varepsilon_l} \qquad (9.15)$$

Values for the optimum emissivity based on a lower and an upper estimate to minimize the temperature error for a range of emissivities are listed in Table 9.5.

When undertaking a measurement of thermal radiation it is not possible to discriminate directly between radiation that has originated from the target surface of interest and any radiation from another source that is incident on the thermometer. The example shown in Figure 9.11 should serve to illustrate this. In the figure the surface of interest has an emissivity below 1 and is at a lower temperature than an object in the vicinity. As the emissivity is below unity, from Kirchhoff's law, which for thermal equilibrium in an isothermal enclosure states that the absorptivity is equal to the emissivity, not all the radiation emitted from the hot object and incident on the target surface will be absorbed and some will be reflected. Some of this reflected radiation will be incident on the infrared thermometer. The incident radiation at the thermometer will therefore comprise radiation emitting from the target surface and radiation reflected by the target surface.

Any radiation emitted by sources other than the target is referred to as background radiation. Background radiation will not, however, cause significant problems unless its intensity is high. If the brightness of the background sources across the detected waveband is comparable to or greater than that of the target then the extra contribution to the measured radiation

Table 9.5 Optimum emissivity and corresponding temperature error for selected upper and lower estimates of emissivity. (After Corwin and Rodenburgh, 1994)

ε_l to ε_u	λ_c (μm)	T_m (K)	ε_o	δT (%)
0.4–0.7	0.65	1800	0.53	±2.3
0.4–0.7	0.35	1800	0.53	±1.2
0.75–0.95	0.7	1200	0.84	±0.7
0.75–0.95	0.7	2100	0.84	±1.2
0.02–0.05	0.7	1200	0.032	±2.7
0.02–0.05	0.7	2100	0.032	±4.7
0.02–0.1	0.7	1200	0.045	±4.7
0.1–0.96	0.7	1200	0.31	±6.6
0.1–0.96	0.7	1200	0.14	±11.3

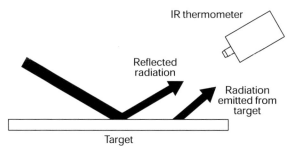

Figure 9.11 Background radiation reflected from the surface of interest onto the infrared thermometer distorting the desired temperature measurement reading

may be enough to increase the indicated temperature. As background radiation cannot be seen it is not immediately obvious. As a result an assessment of possible sources of background thermal radiation should be made when undertaking a temperature measurement using an infrared thermometer. This can be achieved initially by simply estimating the approximate temperatures of the target and its surroundings. For example, if the target is within a room and the target is heated well above room temperature, then the radiation emitted by objects at room temperature will be barely noticeable in comparison to the radiation emitted by the target. If the room is lit by tungsten filament bulbs, however, the radiation emitted by these could contribute significantly to that emitted by the target. Other relatively hot objects near to the target can also cause problems.

A number of general principles can be applied to minimize the effects of background radiation.

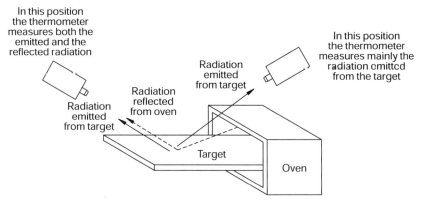

Figure 9.12 Repositioning of an infrared thermometer to limit the effects of reflected radiation

1 Try to ensure that the target completely fills the field of view. If this is the case then no background radiation can be detected unless the surface has reflective or transmissive properties.
2 Select an infrared thermometer with a spectral bandwidth corresponding to a spectral region where the target has a high emissivity. Generally instruments should be selected so that the target has no transmission and as little reflection as possible. This in part explains why there are so many options for infrared thermometers.
3 If possible locate the infrared thermometer to minimize or eliminate reflection effects.

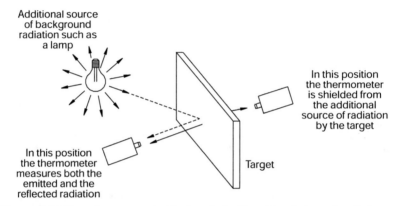

Figure 9.13 Use of a screen to block out significant background radiation

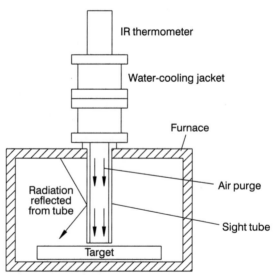

Figure 9.14 Use of a cooled screen to block out significant background radiation in a furnace application

This last principle is illustrated by the examples in Figures 9.12 to 9.14. Figure 9.12 illustrates an example where repositioning the thermometer will avoid the problem of reflected radiation from a hot source other than the target. In Figure 9.13 a screen is used around the target to limit the reflected radiation. Figure 9.14 also illustrates the use of a screen to totally block out reflected radiation from the surfaces of a hot oven. For this kind of application it may be necessary to cool the screen to avoid high temperatures at the interface between the thermometer and the screen.

Example 9.1

A furnace is illustrated in Figure 9.14. In the absence of the cooled screen shown in the figure, radiation from the furnace walls will be incident on the target. Some of this will be absorbed and some will be reflected depending on the emissivity of the target material. If the temperature of the furnace walls, $T_{furnace} = 1000°C$, the temperature of the target, $T_{target} = 700°C$, the total emissivity of the target is $\varepsilon = 0.75$ and there is no transmittance in the spectral region, estimate the difference between the indicated and the actual temperature. In addition, determine the difference between the indicated and the actual temperature of the target if the temperature of the furnace walls is 400°C, 500°C, 600°C, 700°C, 800°C, or 900°C.

Solution

The total radiation detected by the infrared thermometer is given by $E_{total} = E_{target} + E_{background}$. As the target has a total emissivity of 0.75, $E_{target} = \varepsilon E_b$. If the oven is much larger than the target, it can be assumed that it approximates to blackbody characteristics, absorbing and emitting 100% of the radiation. Since the target is surrounded by the oven walls, it is not possible to position the sensor to avoid the effects of reflected radiation. If the transmittance is zero then the reflectivity is given by $1 - \varepsilon = 0.25$ and $E_{background} = E_{reflected} = 0.25\sigma T_{walls}^4$. Therefore

$$E_{total} = \sigma T_{indicated}^4 = 0.75\sigma T_{target}^4 + 0.25\sigma T_{walls}^4$$

$$T_{indicated} = (0.75(973.15)^4 + 0.25(1273.15)^4)^{0.25} = 1073.8\,K$$

Table 9.6 shows values for the error as a function of the furnace temperature in the absence of a screen.

The measurement of radiation emitted from a surface requires that the radiation reaches the detector. The presence of a medium between the target and the detector leads to a number of complicating factors, namely absorption, scattering and fluorescence. Most gases, including air, are not completely transparent and absorb some radiation or are even opaque and absorb all radiation at certain wavelengths. This property can be quantified in terms of

Table 9.6 Error as a function of the furnace temperature in the absence of a screen.

$T_{background}$ (°C)	$T_{indicated} - T_{target}$ (°C)
1000	100.6
900	61.5
800	27.9
700	0
600	−22.1
500	−38.9
400	50.7

the transmissivity, the proportion of radiation that is transmitted across a certain distance of fluid. Data is presented in Figure 9.15 for the transmissivity over a 300 m path of air. Examination of the data shows that there are windows, for example near 0.65 μm, 0.9 μm, 1.05 μm, 1.35 μm, 1.6 μm, 2.2 μm, 4 μm and 10 μm where the transmittance is very high and the target can be 'seen' by the detector. It would therefore be appropriate to select a detector that is sensitive for a window of wavelengths where the transmittance is high for a given application.

Most of the absorption in air is due to water vapour and CO_2. This can be a particular problem when monitoring temperatures near flames where water and CO_2 concentrations are high. The CO_2 and water vapour will absorb some

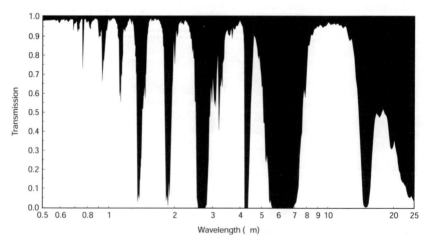

Figure 9.15 Transmissivity for air over a 300 m distance. (Reproduced from Nicholas and White, 1994) The shaded areas indicate that the air is opaque, unshaded areas indicate that the air is transparent

of the radiation emitted and may cause the value of temperature indicated to be low. Conversely, some flames emit radiation and cause an elevation in the indicated temperature measurement.

It is also necessary to consider any other medium that may be between the target and the detector. It is sometimes necessary to separate the detector from the medium in order to protect equipment. Examples include dealing with pressurized or corrosive gases. In these cases it may be possible to use a window through which the infrared thermometer is focused on the target. The material used for the window must transmit most of the thermal radiation in the spectral band of the thermometer if an effective temperature measurement is to be made. Figure 9.16 shows the transmittance for a variety of window materials commonly used in infrared thermometry. It is possible to compensate a reading for low transmittance through a window provided the temperature of the window is low relative to the target. This can be achieved by incorporating a factor for the window transmittance into an effective emissivity for the target surface,

$$\varepsilon_{eff} = \tau_{window}\varepsilon_{target} \tag{9.16}$$

where: ε_{eff} = effective emissivity

          τ_{window} = transmittance of the window

          ε_{target} = total emissivity of the target.

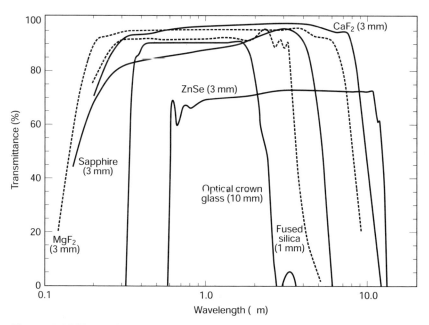

Figure 9.16 Transmittance for a variety of window materials. Reproduced from Ballico (1998)

For example, if the transmittance of the window material is 0.9 for the waveband of the thermometer and the target total emissivity is 0.8, then an effective emissivity of $0.9 \times 0.8 = 0.72$ could be used.

Losses through windows arise from both transmission and reflection. Reflection losses tend to dominate in the passband of the window and therefore need to be carefully considered.

Suspended material such as dust and smoke will generally act as a grey absorber. Where dust and smoke are present use of a ratio thermometer may overcome this problem (see Section 9.6). An alternative is to use a purge tube, blowing, say, nitrogen, to keep the transmission environment between the thermometer and target clear of any solid particles.

Infrared thermometers do not provide a point measurement of temperature but instead collect radiation from a conical region in front of the optical system as illustrated in Figure 9.17 and as a result will produce an averaged temperature reading for the zone concerned. The size of the target, or spot size, is defined by the angle and detector defining apertures, and is known as the field of view. Manufacturers typically supply information with infrared thermometers defining the field of view characteristics for each device (Figure 9.17). Ideally the target zone will be well defined with a sharp distinct boundary as shown in Figure 9.18 so that no radiation from outside the boundary distorts the temperature reading. In practice a number of effects including flare, poor focus, optical aberrations and alignment contribute to blur the boundaries of the target zone.

Obstruction of the field of view is known as vignetting. As the output from an infrared thermometer is a function of the radiation reaching the thermometer, anything that restricts the amount of radiation reaching the

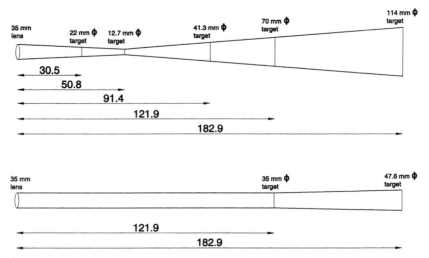

Figure 9.17 Spot size for a wide- and narrow-angle infrared thermometer

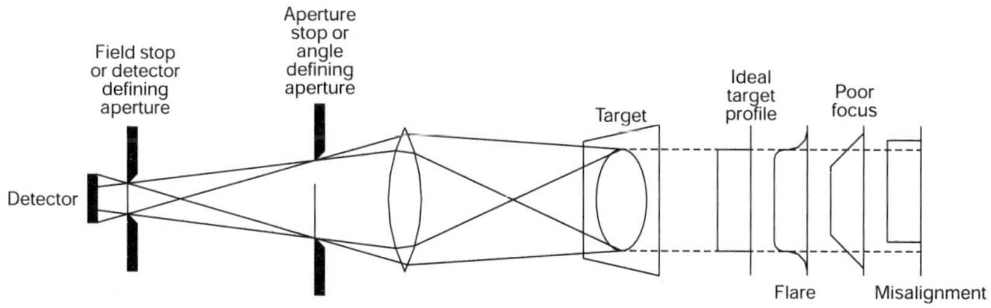

Figure 9.18 Target area definition

thermometer will cause, in the case of a hotter target, a reduction in the indicated temperature.

In order to compensate for the actual target emissivity, most spectral band thermometers include an emissivity adjustment somewhere on the device. This is sometimes in the form of a dial or a digital panel display with press up and down buttons. In the case of a dial the likely precision for the emissivity setting is about ±0.02 and for the digital alternative about ±0.005. This uncertainty is in addition to any uncertainty for the surface emissivity characteristics. Some fixed-installation thermometers will have the emissivity set by the manufacturer to specification prior to delivery.

The following general principles are suggested in the selection of the waveband for a spectral band thermometer:

1 For general-purpose thermometers, the effect of emissivity increases with an increase in wavelength and therefore short wavelengths are more suitable.
2 For metals, where the emissivity decreases as the wavelength increases, the error in temperature measurement will reduce with a shorter wavelength spectral band.
3 Most glasses and some ceramics are transparent at short wavelengths and therefore will not be 'seen' by a short wavelength spectral band thermometer; use a longer wavelength thermometer.
4 The emissivities of some plastics exhibit peaks at certain wavelengths; operation within these wavebands is essential.
5 Certain gases exhibit minima in atmospheric absorption and operation within these wavebands is essential.

9.4.3 Use and operation

The following considerations can form the basis of a procedure in the use of an infrared thermometer:

● Identify the surface material and condition so that an estimate of the emissivity can be made. The emissivity setting of the thermometer should be adjusted to this value.

- Systematically inspect the hemisphere above the target surface for sources of additional radiation such as sunlight, flames, heaters and incandescent lamps that could corrupt the desired radiation measurement. If possible shield the target from any additional sources.
- Avoid dusty environments. Dust can dirty the optics and reduce the indicated reading as well as acting as a grey body absorber in the medium between the target and the thermometer.
- Ensure, if possible, that there are no windows in the field of view of the thermometer.
- Check that the field of view is completely filled.
- Avoid exposing the infrared thermometer to too high an ambient temperature. The manufacturer will have supplied guidelines for the limit on its maximum operating temperature. In addition to considering the maximum temperature, because lenses are precision optical devices, they should not be exposed to too rapid temperature transients as this can produce thermal shocks and cause a lens to distort or shatter.
- Care must be taken when exposing an infrared thermometer to bright sources. Some radiation sources, such as the sun, contain a lot of ultraviolet radiation and exposure to this can cause permanent damage to the thermometer by, for example, changing the detector characteristics. In addition, it is quite possible to cause unintentional damage to the human eye when sighting very bright objects.
- Keep a log of both the temperature measurement and the use of the thermometer. For the thermometer, minimum and maximum temperatures should be recorded and information about the application along with dates, times and any observations. This information serves as a useful record to check the function of the device and identify any malfunctions.
- Check that the thermometer is providing a sensible reading against a blackbody emitter periodically.

9.4.4 The disappearing-filament thermometer

The disappearing-filament thermometer was one of the earliest forms of spectral band thermometers and the principal components for this kind of thermometer are illustrated in Figure 9.19. Other names for this form of device include optical pyrometer and monochromatic brightness radiation thermometer. The disappearing-filament optical pyrometer is similar to a refractory telescope, the difference being that an electrically heated tungsten filament is placed in the focal plane of the objective lens and a red filter is located between the lamp and the eyepiece (Fairchild and Hoover, 1923). The pyrometer is sighted on the target and the image is formed in the same plane as the lamp filament. The magnified image of the lamp filament is superimposed on the target. By adjusting the current through the filament, its luminance or brightness can be matched to that of the target. The red filter ensures that the image is nearly monochromatic so no colour difference is

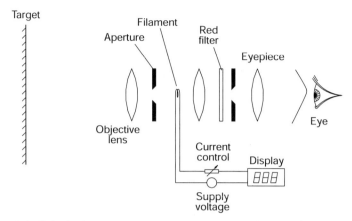

Figure 9.19 Principal components of a disappearing-filament thermometer

observed and the image appears to disappear against the target. By viewing a blackbody at known temperatures a calibration of the device can be obtained. The uncertainty of disappearing-filament thermometers is comparatively low for temperatures over 700°C. The uncertainty of commercial devices can be ±1°C at 775°C and ±5°C at 1225°C. For temperatures above 1300–1400°C, a grey absorbing filter may be placed between the lamp and the objective lens to reduce the radiance to a level at which the eye is comfortable and the range can be extended to about 4000°C. The error in temperature measurement can be calculated from (Doebelin, 1990)

$$\frac{dT_t}{T_t} = -\frac{\lambda_f \, T_t}{C_2} \frac{d\varepsilon_{\lambda f}}{\varepsilon_{\lambda f}} \tag{9.17}$$

where: T_t = target temperature (°C)
λ_f = effective wavelength of the filter (μm or m)
$\varepsilon_{\lambda f}$ = emissivity of the target at wavelength λ_f.

For example, if the uncertainty in the emissivity is 10% at the filter's wavelength, say 0.65 μm, then the resulting error in temperature for a target at 1000°C is 0.58% or ±7°C.

9.5 Total radiation thermometers

Total radiation thermometers aim to measure radiation emitted from a target over a broad band of wavelengths, typically several micrometres. In principle a total radiation thermometer need only consist of a collector and detector as illustrated in Figure 9.20. The basis of the temperature measurement is the

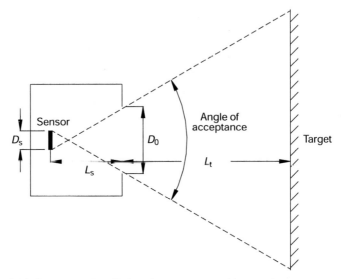

Figure 9.20 Simple total radiation thermometer without a lens

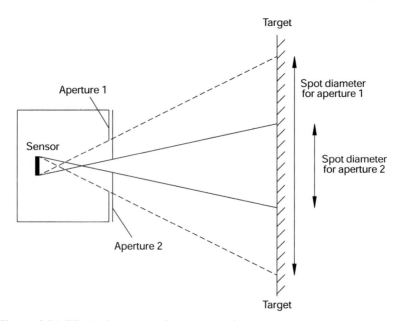

Figure 9.21 Effect of aperture size on spot size

Stefan–Boltzmann law, (equations (9.5) and (9.8). The radiation that reaches the detector will be a function of the aperture size and there is a direct trade-off between sensitivity and spot size. As the aperture size is reduced, the spot size reduces (Figure 9.21) and so does the amount of radiation reaching the detector, thereby reducing the sensitivity of the device. Distance from the target does not, however, affect the sensitivity but as the distance increases, so does the spot size (Figure 9.22). In order to overcome the problem of reduced sensitivity with reduced spot size typical radiation thermometers utilize an optical system using either lenses or mirrors to focus radiation onto a detector (Figure 9.23). For a single-lens system the distance of the detector behind the lens depends on the focal length of the lens and the distance between the lens and the target. The lens position can often be adjusted allowing the focus to

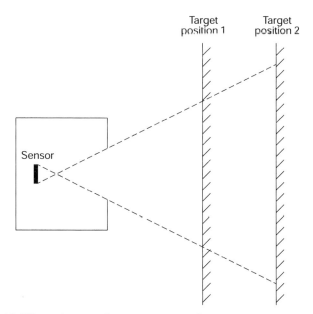

Figure 9.22 Effect of target distance on spot size

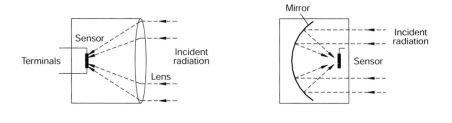

Figure 9.23 Use of a lens or mirror to focus radiation on a detector

be altered for different target distances, although the manufacturer's limitations should be followed. The relationship between the spot size, lens position and focal length (see Figure 9.24) can be determined using simple lens equations,

$$\frac{1}{L_1} + \frac{1}{L_2} = \frac{1}{f} \qquad (9.18)$$

$$\frac{D_t}{D_d} = \frac{L_2}{L_1} \qquad (9.19)$$

where: L_1 = distance between the lens and the detector (m)
L_2 = distance between the lens and the target (m)
f = focal length (m)
D_t = spot diameter (m)
D_d = diameter of the detector (m).

The transmittance of lenses and windows serves to cut off radiation above or below a certain wavelength, leaving a broad band of wavelengths that do reach the detector. As the detector does not see all the wavelengths, these devices are therefore also known as broadband thermometers. Total radiation thermometers have a wide range of applications and can be used for temperatures between about 200°C and 1800°C. With particular care they can produce low uncertainty results. Quinn and Martin (1996) report the use of a

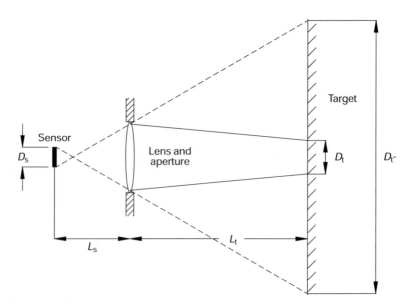

Figure 9.24 Relationship between lens location and spot size

specially designed total radiation thermometer to measure thermodynamic temperature with an uncertainty of 0.3 mK in the room temperature range. The detectors used tend to be thermopiles, thermistors, bolometers or pyroelectric sensors. The speed of response depends on the system but typical values are from 0.1 s for high-temperature applications to 2 s for low-temperature ones.

With total radiation thermometers it is normally necessary to take account of surface emissivity when determining the temperature. From equation (9.8), the error in temperature can be related to an uncertainty in total emissivity by

$$\frac{\Delta T}{T} \approx \frac{1}{4} \frac{\Delta \varepsilon}{\varepsilon} \tag{9.20}$$

where: ΔT = temperature error (K)
   T = temperature (K)
   $\Delta \varepsilon$ = tolerance on the total emissivity
   ε = actual total emissivity.

From equation (9.20) a 10% overestimate in emissivity, for example, would produce a 2.5% underestimate in temperature.

An innovative form of total radiation thermometer is illustrated in Figure 9.25. Here a gold-plated hemisphere is placed on the surface of interest and serves to form a blackbody collecting nearly all the radiation emitting from the surface. A small window in the hemisphere allows radiation to be exchanged between the detector and the cavity. For this device it is not

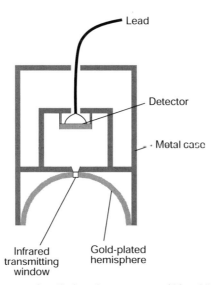

Lead

Detector

Metal case

Infrared transmitting window

Gold-plated hemisphere

Figure 9.25 Gold cup total radiation thermometer. (After Nicholas and White, 1994)

necessary to correct for the surface emissivity as the hemisphere enclosure forms near blackbody conditions. Errors can, however, result due to thermal disturbance. The cavity serves to insulate the surface and therefore the surface temperature rises above the undisturbed value. This is an example where an infrared thermometer cannot be classified as a non-invasive device.

9.6 Ratio/dual-wavelength/two-colour thermometers

A ratio pyrometer, also known as a dual-wavelength or two-colour pyrometer or thermometer, measures the emitted radiation around two fixed wavelengths. The ratio of the quantities of emitted radiation is given by

$$R = \frac{\varepsilon(\lambda_1)\lambda_2^5 \, (e^{C_2/\lambda_2 T} - 1)}{\varepsilon(\lambda_2)\lambda_1^5 \, (e^{C_2/\lambda_1 T} - 1)} \tag{9.21}$$

If the target temperature is low, then the term $e^{C_2/\lambda T}$ will be much greater than 1 so,

$$R = \frac{\varepsilon(\lambda_1)\lambda_2^5 e^{C_2/\lambda_2 T}}{\varepsilon(\lambda_2)\lambda_1^5 e^{C_2/\lambda_1 T}} = \frac{\varepsilon(\lambda_1)\lambda_2^5}{\varepsilon(\lambda_2)\lambda_1^5} e^{(C_2/\lambda_2 T - C_2/\lambda_1 T)} \tag{9.22}$$

If $\varepsilon(\lambda_1)/\varepsilon(\lambda_2) \approx 1$, then the measurement becomes independent of the surface emissivity, and

$$R \left(\frac{\lambda_1}{\lambda_2}\right)^5 = e^{C_2\left(\frac{\lambda_1 - \lambda_2}{\lambda_1 \lambda_2 T}\right)} \tag{9.23}$$

$$\ln\left[R \left(\frac{\lambda_1}{\lambda_2}\right)^5\right] = C_2 \left(\frac{\lambda_1 - \lambda_2}{\lambda_1 \lambda_2 T}\right) \tag{9.24}$$

Hence,

$$T = \frac{C_2 \, (\lambda_1 - \lambda_2)/ \, \lambda_1 \lambda_2}{\ln\left[R \left(\frac{\lambda_1}{\lambda_2}\right)^5\right]} \tag{9.25}$$

It should be noted that although emissivity uncertainty is eliminated or reduced, ratio thermometers are less sensitive than spectral band thermometers.

Example 9.2

A ratio pyrometer uses filters to obtain radiation around two discrete frequencies, 2.2 μm and 2.6 μm. If the ratio of the radiation readings is 0.9, determine the temperature of the target.

Solution

From equation (9.25),

$$T = \left| \frac{14387.69(2.2 - 2.6)/(2.2 \times 2.6)}{\ln(0.9(2.2/2.6)^5)} \right| = 1069.6\,\text{K}$$

Example 9.3

Determine the temperature indicated by a ratio thermometer in the previous example if the ratio of emissivities is 0.8.

Solution

In this case then the ratio of emissivities must be included and the temperature can be determined from:

$$T = \frac{C_2\,(\lambda_1 - \lambda_2)/\lambda_1\lambda_2}{\ln\left[R\left(\dfrac{\varepsilon(\lambda_2)}{\varepsilon(\lambda_1)} \right)\left(\dfrac{\lambda_1}{\lambda_2} \right)^5 \right]}$$

So,

$$T = \left| \frac{14387.69(2.2 - 2.6)/(2.2 \times 2.6)}{\ln(0.9 \times 0.8(2.2/2.6)^5)} \right| = 864.5\,\text{K}$$

9.7 Fibre-optic thermometers

For monitoring surface temperatures or gas temperatures such as in combustion processes it is possible to use an optical fibre to channel thermal radiation into a narrow wavelength band from the location of concern to a measurement sensor using an optical fibre (Dils, 1983). Typical fibre-optic sensors include those based upon optical reflection, scattering, interference, absorption, fluorescence and thermally generated radiation. One commercial system consists of a cavity built onto the end of an optical fibre (Figure 9.26). The blackbody emitter is integrated onto the tip of a single-crystal sapphire optical cavity joined to an optical fibre. The cavity closely approximates a blackbody and the optical fibre transmits the radiant energy to a photodiode or photomultiplier. This measures the intensity of the radiation emitted at a particular wavelength and converts the signal using the laws of radiant emission. These devices can measure temperatures from above 100°C to approximately 4000°C. Different methods for various applications are available such as phosphor-tipped fibre-optic temperature

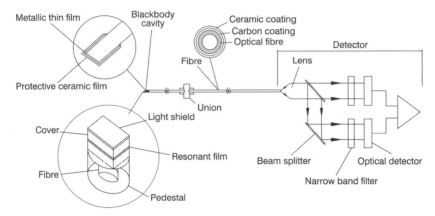

Figure 9.26 Fibre-optic temperature probe for a thin film blackbody sensor. (After Childs *et al.*, 2000)

sensors for the measurement of blackbody radiation and interferometric sensors for the measurement of phase differences between transmitted and received laser light (Saaski and Hartl, 1992). The uncertainty of these devices is dependent on the type of sensor used. For a sapphire rod device at 1000°C, McGee (1988) reports an uncertainty of 1°C, but the uncertainty is limited to that of the temperature standard. For high-temperature probes, (100–1600°C) Ewan (1998) describes the design of two types: a water-cooled low-temperature fibre and an all ceramic-construction probe. A comprehensive introduction to this technology is given by Grattan and Zhang (1995).

9.8 Thermal imaging

Thermography or thermal imaging involves determining the spatial distribution of thermal energy emitted from the surface of an object. This information can be manipulated to provide qualitative and quantitative data of the distribution of temperature on a surface. In their usual form they comprise an optical system, a detector, processing electronics and a display. A typical handheld thermal imager is shown in Figure 9.27 and images resulting from the use of such devices in Figures 9.28 and 9.29. Thermal imagers do not require any form of illumination in order to operate and this makes them highly attractive to military and surveillance users. However, these devices tend to optimized to produce an image rather than quantitative information on the distribution of temperatures.

It is possible to make use of a single detector with some form of scanner to transmit the radiation signal from specific regions of the optical system to enable a two-dimensional image of the temperature distribution to be built up,

Figure 9.27 The FLIR Systems PM 675 thermal imager. Photograph courtesy of FLIR Systems Ltd

Figure 9.28 Thermal distribution for the outer surface of a glazed building. Photograph courtesy of FLIR Systems Ltd

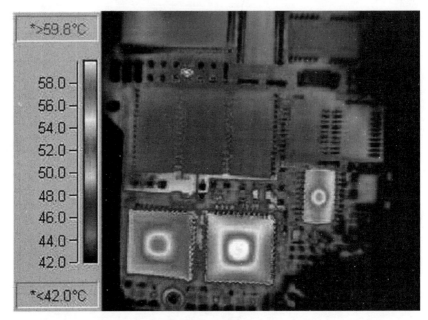

Figure 9.29 Thermal distribution for a printed circuit board. Photograph courtesy of FLIR Systems Ltd

(Figure 9.30). This principle can be extended by the use of a linear detector array as illustrated in Figure 9.31. An alternative is to use an array of detectors and this form is known as a staring array (Figure 9.32). It is usually necessary to cool single and linear array detectors because of the level of incident energy. This can be achieved in a number of ways ranging from introducing liquid nitrogen into a dewar incorporated into the device to the use of a miniature Sterling engine.

The optimum waveband for a thermal imager, as for most other infrared thermometers, is dictated by the wavelength distribution of the emitted radiation, the transmission characteristics of the atmospheric environment between the imager and the target and by the characteristics of the available detector technology. The optical windows for air shown in Figure 9.15 between 3 and 5 μm, with a notch at 4.2 μm due to CO_2 absorption, and between 7.5 and 14 μm make these a common choice for the selection of detectors. The band 3–5 μm is commonly referred to as medium-wave infrared (MWIR) and that between 7.5 and 14 μm as long-wave infrared (LWIR). The emissivity of most naturally occurring objects and organic paints is high in the long-wave infrared but is lower and more variable in the medium-wave infrared. Metallic surfaces tend to have lower emissivity in either band. As such, the use of a thermal image to provide quantitative information for the temperature distribution, particularly of a surface

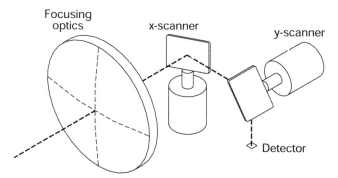

Figure 9.30 Two-dimensional scanning for a small detector array or single-element detector

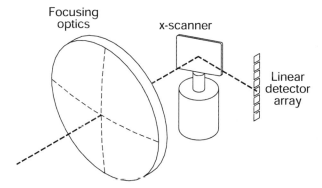

Figure 9.31 One-dimensional scanning for a linear detector array

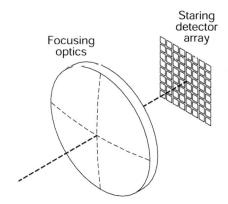

Figure 9.32 A staring array for a thermal imager without scanning

Table 9.7 Some specifications for a selection of commercially available thermal imagers

Manufacturer	Resolution	Detector	Description
Mitsubishi IR M700	801 × 512 pixels	PtSi	Uses a Stirling engine to provide continuous cooling and therefore continuous operation. Size 128 × 250 × 131 mm. Mass 5 kg
Mitsubishi IR G600	512 × 512 pixels	PtSi	Suited to mounting on gimbals and turrets. Size 185 × 210 × 210 mm. Mass 6.8 kg
FLIR Systems ThermaCAM PM 695	320 × 240 pixels	Bolometer, Spectral response 7.5–13 μm	This is an uncooled handheld or tripod mounted thermal imaging camera. Size 220 mm × 133 mm × 140 mm. Mass 2.3 kg. Temperature range −40 to +2000°C
Compix PC2000	244 × 193 pixels	PbSe 3–5 μm	PC-based thermal imaging system. Thermoelectrically cooled detector. 17 to 150°C, 5–240°C, 17 to 1000°C. Size 140 × 430 × 110 mm. Mass 2 kg
Indigo Systems Merlin-Mid	320 × 256 pixels	InSb 1–5.5 μm	The camera may be operated by a button panel, or via a PC-based graphical inter-face application. Mass <2.8 kg. Size 140 × 127 × 249 mm. Cool-down start time less than 10 minutes at 30°C. Temperature range 0–350°C or 300–2000°C
Indigo Systems Merlin-Long	320 × 256 pixels	QWIP (Quantum well infrared photodetector) 8–9.2 μm spectral response	A long-wave infrared camera system. Mass < 2.8 kg. Size 140 × 127 × 249 mm. Temperature range 0–2000°C
Quest Integrated Inc TAM Model 200	12.5 μm	2.4 to 5.5 μm	A microthermography system for identifying temperature distribution on a microscopic scale

comprising different materials, has be carefully managed. Without correction for local emissivity values the thermal imager will assume a default value and apply this to the whole image.

Thermal imagers remain expensive devices although their cost is falling. Prices range from over £71 500 ($100 000 year 2000 prices) for high-performance military imagers to a few thousand pounds for uncooled imagers. The price reflects the performance, ruggedness and image-processing capability. Some compact imagers can be readily hand held whilst other systems are designed to be mounted on a platform with the imager weighing as much as 100 kg. The uncertainty associated with the temperature measurement is specific to the device but typical figures are ±2 K or ±2% full-scale output (Runciman, 1999). Table 9.7 provides some information about commercial imagers available at the time of press. Most of the devices now commercially available are particularly easy to use. They must simply be aimed at the target of interest and the image captured by pressing a button on the camera. Alternatively, some devices allow the image to be continuously streamed to memory on board the camera or via a PC link.

9.9 Calibration

As with any thermometer it is necessary to establish the relationship between indicated and thermodynamic temperatures and this is achieved in the calibration process. Infrared thermometers can be calibrated by a

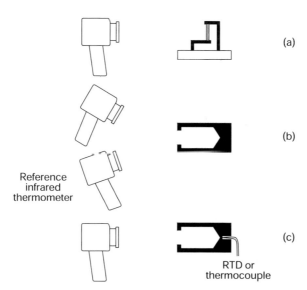

Figure 9.33 Calibration techniques for infrared thermometers. (a) Using a tungsten strip lamp. (b) Comparison. (c) Using a PRT or thermocouple

Table 9.8 Factors affecting the selection of an infraed thermometer. (After Ricolfi and Barber, 1990)

Factor	Suggestion
Target material, condition and stability	If the target is stable and the emissivity is above 0.5 use a single-waveband short-wavelength thermometer.
	If the target material has a low or variable emissivity consider the use of a ratio thermometer or a single-waveband short-wavelength thermometer.
Temperature span	If a wide temperature span is needed then it is necessary to compromise on the short-wavelength requirement. The shortest wavelength infrared thermometer that meets the minimum temperature requirement should be selected.
Target size, range and position stability	If the target is small or if it is not in a constant position then a rectangular field of view system or a ratio thermometer might be appropriate.
Distance range from thermometer	If the target distance is large or highly variable, ensure that a thermometer is chosen that is free from atmospheric absorption effects and that the field of view is filled at all distances.
Rate of temperature change	If the temperature of the target can vary rapidly ensure that the speed of response of the thermometer system is fast enough to provide the required measurement.
Surroundings colder than target	If the surroundings are much colder than the target, use the shortest wavelength thermometer that meets the temperature span requirement.
Surroundings same temperature as target	If the target is in thermal equilibrium with its surroundings, any thermometer can be used with the emissivity set at 1. If the temperature is simply near to that of the surroundings, care needs to be taken to set the emissivity at the correct value for the target.
Surroundings hotter than target	If the surroundings are hotter than the target, use the compensation method (two-sensor system) or if the target is moving shield the measurement point.

Atmospheric absorbtion and emission	If the optical path between the target and the thermometer contains a significant quantity of dust particles or droplets then the radiant energy may be absorbed or may excite the particles and cause emission. If the obscuration is intermittent a peak-picking signal processing can be used. If the obscuration is permanent then a ratio thermometer may be appropriate.
Target periodically obscured	If the line of sight between the thermometer and the target is periodically obscured (e.g. by moving components or machinery) or if the target's intermittent (e.g. components on a conveyer) then adopt a signal-processing method to synchronize the temperature measurement with the target position or line of sight.
Electrical fields	If there are strong electrical fields in the vicinity of the thermometer, consider the use of a fibre-optic cable alternative or if inappropriate shield the thermometer against the electrical field.
Size restraints on thermometer	If there is not sufficient room to position an infrared thermometer then use of fibre-optic cable system could be considered or a miniature infrared thermometer.
Ambient temperature range	If the ambient temperature of the thermometer exceeds safe limits for the equipment then an air- or water-cooled jacket can be specified around the thermometer.
Cleanliness of environment	If the environment is dirty then the optical path and external optics can become partially obscured. A lens-purging system can be used to keep external optics clean.
Optical window	An optical window is often necessary in order to protect an infrared thermometer from the application (e.g. pressurized vessels, vacuum applications, corrosive gases, etc.). Glass or silica windows should be used for short-wavelength thermometers. At low temperatures where long-wavelength thermometers are necessary professional advice should be sought.
Output signal	A number of options are possible for the output signal: analogue, digital, panel meter display, etc. An output compatible with the needs of the application should be specified.

number of methods as listed below and illustrated schematically in Figure 9.33:

- Using a tungsten lamp.
- By comparison with a transfer standard infrared thermometer.
- Against a thermocouple or PRT.

Tungsten strip lamps comprise a ribbon of tungsten approximately 5 mm wide and 50 mm long supported in a pyrex or silica bulb. The bulb is either evacuated or filled with an inert gas to prevent oxidation or other contamination of the filament. Evacuated bulbs are suitable for operation between 700°C and 1700°C, whilst inert gas filled bulbs can be used from 1500°C to 2300°C. Tungsten lamps are calibrated in terms of the current necessary in order to achieve a specified radiance temperature at a particular wavelength, often 655 nm. They can be used in the calibration of an infrared thermometer by focusing the thermometer on the filament at a known current. The blocks illustrated in Figures 9.33(b) and (c) are known as blackbody enclosures. These have specially formed interiors comprising pyramidal large-scale roughness which serves the purpose of scattering radiation. The blackbody enclosures can be heated up using an external electric coil heater element or within a furnace. As the enclosures act as blackbodies they are perfect emitters and absorbers. They can be used to compare the reading of a pre-calibrated infrared thermometer and that of the device to be calibrated. Alternatively, a calibrated thermocouple or PRT located in close thermal contact with the blackbody can be used to define the calibration temperature.

The output of an infrared thermometer will need to be related to the temperature. For narrow spectral band thermometers the effective wavelength can be approximated by

$$\lambda_e = A + \frac{B}{T} \tag{9.26}$$

where: λ_e = the effective wavelength (μm)
T = temperature (K)
A and B are constants.

This can be substituted into Wien's law giving a simple relationship between voltage output and temperature:

$$V_T = C \exp\left(\frac{-C_2}{AT + B}\right) \tag{9.27}$$

where: V_T = voltage (V)
C_2 = the second radiation constant
A, B and C are constants.

Equation (9.27) is suitable for many thermometers including some transfer standard thermometers. For spectral thermometers with bandwidths of less than about 50 nm equation (9.27) will fit the response to within a few tenths of a degree. For wide-band thermometers equation (9.28) is more suitable:

$$\log(V_T) = A + \frac{B}{T} + \frac{C}{T^2} + \frac{D}{T^3} \tag{9.28}$$

where A, B, C and D are constants.

9.10 Selection

The principal criteria to be considered in the selection of an individual infrared temperature measurement system include the temperature range, atmospheric conditions, spectral sensitivity range, optical signal strength, desired signal level, maximum acceptable noise, cooling constraints, the spectral passband, field of view, the resolution, speed of response, stability, the reference standard, geometry and cost. Guidance for a selection according to a variety of considerations is given in Table 9.8.

References

Books and papers

Ballico, M. Radiation thermometry. Chapter 4, in Bentley, R.E. (Editor), *Handbook of Temperature Measurement,* Vol. 1, Springer, 1998.

Bejan, A. *Heat Transfer.* Wiley, 1993.

Brewster, M.Q. *Thermal Radiative Transfer and Properties*, Wiley, 1992.

Brown, M.E. *Proc. Symp. on Major Problems on Present Day Radiation Thermometry*, IMEKO TC12, Moscow, pp. 51–61, 1986.

Chapman, A.J. *Fundamentals of Heat Transfer.* Macmillan, 1987.

Childs, P.R.N., Greenwood, J.R. and Long, C.A. Review of temperature measurement. *Review of Scientific Instruments*, **71**, 2959–2978, 2000.

Cohen, E.R. and Taylor, B.N. The fundamental physical constants. *Physics Today*, BG5-BG9, 1999.

Corwin, R.R. and Rodenburgh, A. Temperature error in radiation thermometry caused by emissivity and reflectance measurement error. *Applied Optics*, **33**(10), 1950–1957, 1994.

Dils, R.R. High temperature optical fiber thermometer. *J. Applied Phys.*, **54**, 1198–1201, 1983.

Doebelin, E.O. *Measurement Systems*, 4th edition. McGraw-Hill, 1990.

Ewan, B.C.R. A study of two optical fibre probe designs for use in high-temperature combustion gases. *Measurement Science and Technology*, **9**, 1330–1335, 1998.

Fairchild, C.O. and Hoover, W.H. Disappearance of the filament and diffraction effects in improved forms of an optical pyrometer. *Journal of the Optical Society of America*, **7**, 543–579, 1923.

Fraden, J. *Handbook of Modern Sensors*, 2nd edition Springer. 1996.

Fraden, J. Infrared thermometers. Section 32.6 in Webster, J.G. (Editor), *The Measurement, Instrumentation and Sensors Handbook*. CRC Press, 1999.

Gardner, J.L. *Applied Optics*, **19**, 3088–3091, 1980.

Grattan, K.T.V. and Zhang, Z. *Fibre Optic Fluorescence Thermometry*. Chapman and Hall, 1995.

Hackforth, H.L. *Infrared Radiation*. McGraw-Hill, 1960.

Howell, J.R. *A Catalog of Radiation Configuration Factors*. McGraw-Hill, 1982.

Incropera, F.P. and DeWitt, D.P. *Fundamentals of Heat and Mass Transfer*, 4th edition. Wiley, 1996.

Kostkowski, H.J. and Lee, R.D. Theory and methods of optical pyrometry. In Herzfield, C.M. (Editor), *Temperature. Its Measurement and Control in Science and Industry*, Vol. 3(1), pp. 449–481. Reinhold, 1962.

Kreith, F. and Bohn, M.S. *Principles of Heat Transfer*, 5th edition. Harper and Row, 1996.

McGee, T.D. *Principles and Methods of Temperature Measurement*. Wiley, 1988.

Michalski, L., Eckersdorf, K. and McGhee, J. *Temperature Measurement*. Wiley, 1991.

Nicholas, J.V. and White, D.R. *Traceable Temperatures: An Introduction to Temperature Measurement and Calibration*. Wiley, 1994.

Planck, M. *The Theory of Heat Radiation*. Dover, 1959.

Quinn, T.J. and Martin, J.E. Total radiation measurements of thermodynamic temperature. *Metrologia*, **33**, 375–381, 1996.

Ricolfi, T. and Barber, R. Radiation thermometers, In Gopel, W., Hesses J. and Zemel, J.N. (Editors), *Sensors. A comprehensive survey*, Vol. 4, *Thermal Sensors*, pp. 163–223. VCH, 1990.

Runciman, H.M. Thermal imaging. Section 35.1, in Webster, J.G. (Editor), *The Measurement, Instrumentation and Sensors Handbook*. CRC Press, 1999.

Saaski, E.W. and Hartl, J.C. Thin-film Fabry–Perot temperature sensors. In: Schooley, J.F. (Editor), *Temperature. Its Measurement and Control in Science and Industry*, Vol. 6(2), pp. 731–734. American Institute of Physics, 1992.

Schieferdecker, J., Quad, R., Holzenkampfer, E. and Schulze, M. Infrared thermopile sensors with high sensitivity and very low temperature coefficient. *Sensors and Actuators A*, **46–47**, 442–427, 1995.

Touloukian, Y.S. and DeWitt, D.P. *Thermophysical Properties of Matter*, Vol. 7. *Thermal Radiative Properties. Metallic elements and alloys*. IFI/Plenum, 1970.

Touloukian, Y.S. and DeWitt, D.P. *Thermophysical Properties of Matter*, Vol. 8. *Thermal Radiative Properties. Nonmetallic solids*. IFI/Plenum, 1972a.

Touloukian, Y.S., DeWitt, D.P. and Hernicz, R.S. *Thermophysical Properties of Matter*, Vol. 9. *Thermal Radiative Properties. Coatings*. IFI/Plenum, 1972b.

Yates, H.W. and Taylor, J.H. *NRL Report 5453*, US Naval Research Laboratory, Washington DC, 1960.

Standards

BS 1041: Part 5: 1989. British Standard. Temperature measurement. Part 5. Guide to selection and use of radiation pyrometers

Japanese Industrial Standard JIS C 1612–1988. General rules for expression of the performance of radiation thermometers.

ASTM E 1256–95 Standard test methods for radiation thermometers (single waveband type).

Web sites

At the time of going to press the world wide web contained useful information relating to this chapter at the following sites.

http://e2t.com/prd_07.htm
http://mbcontrol.com/prt_fra.htm
http://temperatures.com/rts.html
http://www.copas.com.au/testo/temp.htm
http://www.dera.gov.uk/html/case/fire/
http://www.dwyer-inst.com/temperature/t-toc-process.html
http://www.eosael.com/overview.html
http://www.flir.com/
http://www.infrared.com/history.html
http://www.ipac.caltech.edu/Outreach/Edu/infrared.html
http://www.Ircon.com
http://www.landinst.com/
http://www.mikroninst.com
http://www.mitsubishi-imaging.com/products/thermal_imagers.html
http://www.ncdc.noaa.gov/pub/software/lowtran/
http://www.newportelect.com/Databook/g8-9.htm
http://www.optronics.co.uk/thermal.htm
http://www.prwalker.com/prwalker/temperat.htm#Terms
http://www.quantumlogic.com/portable.html
http://www.raytek.com/
http://www.sofradir.fr/pages/products/fiches/pg_old_products.htm
http://www.wintron.com/Infrared/guideIR.htm
http://www3.mediagalaxy.co.jp/smm/tech/b_2_5e.html
http://www-lmt.phast.umass.edu/pub/012/node1.html
http://www-vsbm.plh.af.mil/soft/modtran.html
http:www.luxtron.com/prod

Nomenclature

A	=	area (m^2 or cm^2)
C_1	=	first radiation constant = $3.7417749 \times 10^{-16}$ W·m^2 (Cohen and Taylor, 1999)
C_2	=	second radiation constant = 0.01438769 m·K (Cohen and Taylor, 1999)
D_t	=	spot diameter (m)
D_d	=	diameter of the detector (m)
D_λ	=	detectivity (cm·Hz$^{0.5}$W^{-1})
E	=	emissive power for a grey body (W/m^2)
E_b	=	total emissive power for a blackbody (W/m^2)
E_λ	=	spectral emissive power for a grey body (W/m^3)

$E_{\lambda,b}$	=	spectral emissive power for a blackbody (W/m^3)
f	=	focal length (m)
G	=	factor incorporating geometrical and responsivity parameters
IR	=	infrared
L	=	distance (m)
LWIR	=	long-wave infrared
MWIR	=	medium-wave infrared
NEP	=	noise equivalent power
R	=	ratio
$S(T)$	=	output of the thermometer at a given temperature
T	=	temperature (K)
T_t	=	target temperature (°C)
T_m	=	assumed temperature (K)
V_T	=	voltage (V)
ε	=	total emissivity
ε_{eff}	=	effective emissivity
ε_l	=	lower estimate of emissivity
ε_m	=	assumed emissivity
ε_o	=	optimum value to use for emissivity to minimize the error in temperature
ε_u	=	upper estimate of emissivity
$\varepsilon_{\lambda f}$	=	emissivity of the target at wavelength λ_f
$\varepsilon_{\lambda,T}$	=	spectral emissivity
λ	=	wavelength (μm or m)
λ_c	=	centre wavelength of bandwidth (μm or m)
λ_e	=	effective wavelength (μm)
λ_f	=	effective wavelength of the filter (μm or m)
λ_m	=	wavelength at which spectral radiance is an optimum (μm or m)
σ	=	the Stefan–Boltzmann constant = 5.67051×10^{-8} Wm^{-2}K^{-4} (Cohen and Taylor, 1999)
τ_{window}	=	transmittance of the window
Δf	=	frequency bandwidth of the post-detector electronics (Hz)
ΔT	=	temperature error (K)
$\Delta\varepsilon$	=	tolerance on the total emissivity

10

Other non-invasive temperature measurement techniques

A number of non-invasive measurement techniques have been developed in addition to the infrared methods described in Chapter 9. These non-invasive techniques are particularly useful for monitoring gas or plasma temperatures. The aims of this chapter are to describe the techniques and their typical applications.

10.1 Introduction

Measurement of temperature in very hot mediums such as flames or plasmas represents a particular challenge. Most invasive types of instruments cannot survive for very long if at all at these temperatures, or would require such elaborate use of cooling jackets to make the measurement impracticable. A variety of techniques have, however, been developed based on observing the variation of refractive index, absorption or emission spectroscopy, scattering, fluorescence and acoustics. Several of these methods have been made possible by the introduction of the laser, improved optics and high-speed data processing. They tend to require specialist skills to assemble and integrate the equipment and are, as a result, highly expensive in terms of both initial outlay and running costs. Nevertheless, the possibility of non-invasive, high-speed data in a gas or plasma may make the investment worth while.

10.2 Refractive index methods

In compressible gas flows the density varies sufficiently to give measurable results in the variation of the refractive index and from measurements of this the temperature can be inferred. For many practical systems the density of the gas can be determined using the ideal gas equation, $p = \rho RT$. If the pressure of the system can be considered constant then the temperature of the gas can be determined by

$$T = \frac{n_{i,0} - 1}{n_i - 1} \frac{p}{p_0} T_0 \qquad (10.1)$$

where: n_i = refractive index at a particular location
n_{10} = refractive index at a reference location
p = pressure (Pa)
p_0 = pressure at a reference condition (Pa)
T_0 = temperature at a reference condition (K).

A number of techniques are available that exploit the variation of refractive index with temperature including the schlieren, shadowgraph and inter-ferometric methods. The typical application for these methods is the determination of combustion process temperatures. Each of these techniques involves monitoring the variation of refractive index with position in a transparent medium in a test section through which light passes. Each method, however, involves the measurement of a different quantity. Interferometers measure the differences in the optical path lengths between two light beams. The schlieren method involves the determination of the first derivative of the index of refraction and the shadowgraph method the second derivative.

The schlieren system is designed to measure the angular deflection of a light beam that has passed through a test section where the index of refraction is varying. A typical system based on optical lenses is illustrated in Figure 10.1. It is also possible to use focusing mirrors in place of the lenses. A light source is located at the focus of the lens L_1 and provides a parallel beam passing through the test section. The deviation of the light due to a possible variation in the index of refraction in the test section is illustrated by the dotted line. The light is collected by the second lens L_2 at the focus of which a knife edge is located. The image can be projected

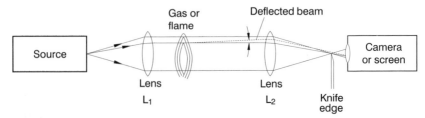

Figure 10.1 A typical schlieren system based on optical lenses

onto a screen or a camera, located after the knife edge, can be focused on the centre of the test section (see Goldstein, 1970; Gaydon and Wolfhard, 1979).

By assuming for simplicity that the index of refraction varies only in the y direction, then the angular deflection of a light beam is given by

$$\alpha = \frac{1}{n_{i,air}} \int \frac{\partial n_i}{\partial y} dz \qquad (10.2)$$

where $n_{i,air}$ is the refractive index of ambient air. Knowing the angle of deflection allows the refractive index to be evaluated and hence the temperature.

The temperature range for schlieren measurements is approximately from 0°C to 2000°C, with a sensitivity of the order of 0.1°C and an accuracy of 10% of the range. Tomographic or temperature mapping utilizing schlieren methods has been reported by Schwarz (1996).

In a shadowgraph system (Figure 10.2) the second derivative of the index of refraction is measured (equation (10.3)). A shadowgraph system measures the displacement of the disturbed light beam rather than the angular deflection as in a schlieren system. The displacement, however, tends to be small and difficult to measure and as a result the contrast is used.

$$\frac{\partial^2 \rho}{\partial y^2} = \frac{\rho_0}{n_{i,0} - 1} \frac{\partial^2 n_i}{\partial y^2} \qquad (10.3)$$

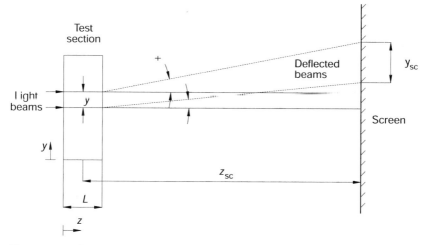

Figure 10.2 Optical path for a shadowgraph system

If the pressure can be assumed constant and the ideal gas law is valid for the system concerned, then

$$\frac{\partial^2 n_i}{\partial y^2} = C \left[-\frac{\rho}{T} \frac{\partial^2 T}{\partial y^2} + \frac{2\rho}{T^2} \left(\frac{\partial T}{\partial y} \right)^2 \right] \tag{10.4}$$

where: C = the Gladstone–Dale constant
y = coordinate (m).

The contrast between light and dark patterns is given by

$$\frac{\Delta I}{I_i} = -\frac{z_{sc}}{n_{i,air}} \int C \left[-\frac{\rho}{T} \frac{\partial^2 T}{\partial y^2} + \frac{2\rho}{T^2} \left(\frac{\partial T}{\partial y} \right)^2 \right] dz \tag{10.5}$$

where: I_i = the initial illumination on the screen in the shadowgraph system
z_{sc} = distance between the screen and test section (m).

In interferometry the differences in the optical path lengths of two beams are measured. A schematic of a Mach–Zehnder interferometer is shown in Figure 10.3. If the two light beams originating from a monochromatic source pass through uniform media then the recombined beam should be uniformly bright. If, however, the temperature in the test section is elevated above that of the ambient air then the optical path lines will no longer be the same and the recombined light beam will exhibit interference fringes with bright and dark patterns.

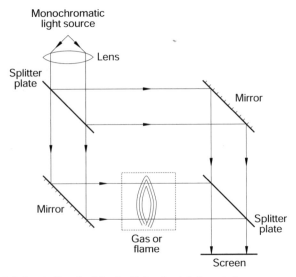

Figure 10.3 Schematic of a Mach–Zehnder inteferometer

If the thermal field is two-dimensional and the variation of the index of refraction occurs only perpendicular to the light beam then the fringe shift can be expressed by (Goldstein *et al.*, 1998)

$$\varepsilon = \frac{L(n_i - n_{i,0})}{\lambda_0} \tag{10.6}$$

where: L = length of the test section (m)
λ_o = wavelength of the monochromatic light source (m)
n_i = refractive index field to be determined
$n_{i,0}$ = reference refractive index in reference beam 2.

If the density of the gas can be determined from the ideal gas law and the pressure in the test section is maintained constant and the index of refraction is only a function of density, then the temperature distribution can be evaluated from (Goldstein *et al.*, 1998)

$$T(x,y) = \left(\frac{\lambda_0 \mathfrak{R}}{pCLM} \, \varepsilon + \frac{1}{T_0} \right)^{-1} \tag{10.7}$$

where: $\mathfrak{R}$ = the universal gas constant
C = the Gladstone–Dale constant
L = length (m)
M = molecular weight
p = absolute pressure of the test section (Pa).

Holographic interferometers utilize the same principle as an interferometer with the exception that the fringe pattern between the object beam and the reference beam are shown on a hologram plate.

10.3 Absorption and emission spectroscopy

Absorption and emission spectroscopy are useful techniques for mapping of the temperature distribution in flames and gases at high temperatures. Atoms emit electromagnetic radiation if an electron in an excited state makes a transition to a lower energy state and the band of wavelengths emitted from a particular species or substance is known as the emission spectrum.

Emission spectroscopy involves measurements of this emission spectrum and can be achieved utilizing an atom cell, light-detection system, mono-chrometer and a photomultiplier detection system (Metcalfe, 1987). Conversely, atoms with electrons in their ground state can absorb electromagnetic radiation at specific wavelengths and the corresponding wavelengths are known as the absorption spectrum. Absorption spectroscopy relies on measurements of the wavelength dependence of the absorption of a pump source such as a tuneable laser due to one or more molecular transitions. In order to evaluate the temperature it is necessary to fit the observed spectrum to a theoretical model, which normally involves prior knowledge of molecular parameters such as oscillator strength and pressure-broadened linewidths. The temperature can then be calculated from the ratio of the heights of two spectral lines utilizing the Boltzmann distribution. The typical uncertainty for these techniques is of the order of 15% of the absolute temperature. Examples of the use of these methods in flames and gases are given by Hall and Bonczyk (1990) and Uchiyama *et al.* (1985).

10.4 Line reversal

The line reversal method involves viewing a sample of reference gas at a known temperature through a test section containing a sample of gas at an unknown temperature (Figure 10.4) and comparing the spectral line. If the temperature of the gas is less than the temperature of the brightness continuum then the line will appear in absorption, that is, dark against the background. If the temperature of the test section is higher than the comparison brightness temperature, the spectral line will appear in emission or bright against the background (Carlson, 1962; Gaydon and Wolfhard, 1979).

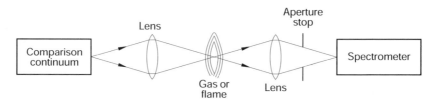

Figure 10.4 Line reversal technique for gas temperature measurement

The temperature of the gas can be determined by adjusting the temperature of the brightness continuum until reversal of brightness occurs. This method can be used to measure the static temperature of a gas in the temperature range from approximately 1000 K to 2800 K with an uncertainty of approximately ±10 K to ±15 K. Applications have included combustion chambers, flames, rocket exhausts and shock waves.

10.5 Spontaneous Rayleigh and Raman scattering

An alternative technique for monitoring temperature in gases is the observation of spontaneous Rayleigh and Raman scattering. Scattering in this context is the absorption and re-emission of electromagnetic radiation by atoms and molecules.

Rayleigh scattering is the elastic scattering of light by molecules or very small particles less than about 0.3 μm in size. The intensity of Rayleigh scattering is proportional to total number of particles, N, and the irradiance I_L:

$$I_R = CI_L N \sum_i x_i \sigma_i \tag{10.8}$$

where: C = a calibration constant for the optical system
   x_i = the mole fraction
   σ_i = the effective Rayleigh scattering cross-section of each species.

By using the ideal gas law,

$$I_R - CI_L \frac{pV}{RT} \sigma \tag{10.9}$$

where

$$\sigma = \sum_i x_i \sigma_i \tag{10.10}$$

If the Rayleigh cross-section is kept constant, the temperature of the probed volume can be found. The calibration constant, C, can be determined by measuring a reference temperature under known conditions and the measured temperature related to this, assuming that the probe volumes for the reference and test conditions are equal, by

$$T = \frac{I_L}{I_{L,ref}} \frac{p}{p_{ref}} \frac{\sigma}{\sigma_{ref}} \frac{I_R}{I_{ref}} T_{ref} \tag{10.11}$$

Rayleigh spectra can be obtained using continuous wave and pulsed lasers to excite the flow. The principal components for scattering based measurements consist of an optical system and spectrometer to observe the gas sample and a pulsed or continuous wave laser to excite the particles in the gas. A typical system is illustrated in Figure 10.5. In Rayleigh scattering the collected signal will typically be a factor of 10^9 smaller than the pump signal from the laser, making it susceptible to corruption from other processes such as Mie scattering, optical effects and background radiation. Also, in order to analyse the spectra it is usually necessary to know the individual concentrations of the species in the flow. As a result this technique must be undertaken very carefully. The range for Rayleigh scattering is from approximately 293 K to 9000 K (e.g. see Farmer and Haddad, 1988). Applications have included plasmas (Murphy and Farmer, 1992; Bentley, 1996), combustor flames (Barat et al., 1991), sooting flames (Hoffman et al., 1996), and supersonic flows (Miles and Lempert, 1990).

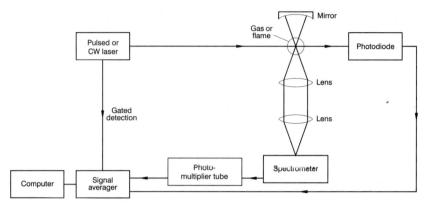

Figure 10.5 Schematic of the usual experimental set-up for the measurement of temperature by scattering methods (Iinuma *et al.*, 1987)

When a molecule is promoted by incident radiation from the ground state to a higher unstable vibrational state it can either return to the original state, which is classed as Rayleigh scattering as discussed above, or to a different vibrational state which is classed as Raman scattering. Raman scattering gives rise to Stokes lines on the observed spectra (see Kittel, 1986). Alternatively, if a molecule is in an excited state it can be promoted to a higher unstable state and then subsequently return to the ground state. This process is also classed as Raman scattering and gives rise to an anti-Stokes line on the observed spectrum. Raman scattering involves the inelastic scattering of light from molecules. There are two basic methods for determining the temperature by Raman scattering: the Stokes–Raman method and the Stokes to anti-Stokes ratio method. The Stokes–Raman method is based on measurements of the density of the non-reactive species assuming uniform pressure and ideal gas conditions. The Stokes to anti-Stokes ratio method involves measurement of the scattering strengths of the Stokes to anti-Stokes signals of the same spectral line. The temperature can then be calculated utilizing the Boltzmann occupation factors for the lines in question (Edwards, 1997). This process is generally only suitable for high combustion temperatures due to the relative weakness of the anti-Stokes signal. The uncertainties in temperature measurement utilizing Raman spectroscopy are discussed by Laplant *et al.* (1996). Raman scattering spectra can be observed by an optical system and spectrometer and using a pulsed or continuous wave laser to excite the flow. The setup is similar to that for Rayleigh scattering as illustrated in Figure 10.5. The range and accuracy for Raman scattering are approximately 20 to 2230°C and 7% respectively. Applications have included: reactive flows (Dibble *et al.*, 1990); flames (Burlbaw and Armstrong, 1983); atmospheric temperature observation (Vaughan *et al.*, 1993).

10.6 Coherent anti-Stokes–Raman scattering (CARS)

The CARS (coherent anti-Stokes–Raman scattering) technique permits the non-invasive monitoring of local temperatures in gases, flames and plasmas. CARS involves irradiating a gas or flame with two collinear laser beams, a pump beam at frequency ω_1 and a probe beam at ω_2 and determining the temperature from the population distribution. The temperature range of this method is from 20°C to 2000°C and the uncertainty of the resulting temperature is approximately 5%.

In a typical system a Nd:YAG laser (532 nm) is split into three beams, two of which are focused on a small volume (<0.5 mm cube) within the target. The third beam is used to pump a broadband dye-laser, referred to as the Stokes laser, at a frequency ω_2 with a wavelength of about 606 nm. This beam is also focused onto the target volume as illustrated in Figure 10.6. The pump and probe beam frequencies are selected so that $\omega_1 - \omega_2$ is equal to the vibrational frequency of a Raman active transition of the irradiated molecules so that a new source of light is generated within the medium as indicated in Figures 10.6 and 10.7. The signal generated is produced in the form of a

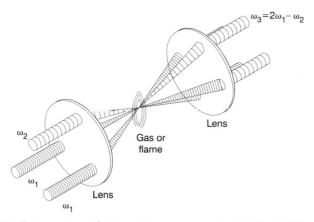

Figure 10.6 Coherent anti-Stokes–Raman scattering (CARS). (After Eckbreth, 1988)

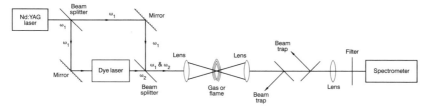

Figure 10.7 Coherent anti-Stokes–Raman scattering (CARS). (After Iinuma *et al.*, 1987)

coherent, laser-like beam that can be separated physically and spectrally from sources of interference. The emitted light contains information about the population distribution within the rotational energy levels of the target molecule and its intensity is proportional to the density. The population distribution is proportional to temperature. As the signal appears on the high-frequency side of the pump beam, i.e. an anti-Stokes spectrum, and because it is observable only if the molecular vibrations are Raman active, this mechanism is called coherent anti-Stokes–Raman scattering (Attal-Tretout *et al.*, 1990). As temperature is related to the rotational state of molecules, the anti-Stokes lines increase in intensity with increasing temperature. CARS is a non-linear Raman technique so the scattered signal is not linearly related to the input laser intensity. Because the interaction is non-linear, high laser powers are necessary and this can be attained using pulsed lasers, typically 200 mJ in 10 ns; 20 MW. More recent applications of CARS have made use of an XeCl excimer laser in place of the Nd:YAG laser as an XeCl excimer laser can provide increased flexibility in the laser repetition rate which can be moderated to match a periodicity in an application, such as engine speed. Precise alignment and focusing of the three target beams and the detector are necessary. This usually limits the effective range of the technique to about 0.5 m from the last focusing lens.

Either narrow- or broadband CARS is used depending on the nature of the flow. Narrow-band CARS requires the use of a narrowband laser and fine wavelength tuning must be undertaken in order to match an excitation wavelength to obtain sufficient output signal. Broadband CARS utilize a broadband laser and does not need to be tuned to a specific excitation peak.

The merits associated with CARS are the capability to measure temperature non-invasively and at high temperatures, up to several thousand degrees Celsius. The disadvantages of the technique are its complexity, requiring specialist skills to set the system up and use it, and its high price, up to several hundred thousand pounds at (year 2000 prices). CARS is generally suitable to highly luminous or particulate laden systems. Applications of CARS have included flames (Farrow *et al.*, 1982; Porter and Greenhalgh, 1985), internal combustion engine in-cylinder flows (Alessandretti and Violino, 1983), combustion and plasma diagnostics (Anderson *et al.*, 1986), jet engine exhausts (Eckbreth *et al.*, 1984), low-pressure unsteady flows (Herlin *et al.*, 1991) and supersonic combustion (Antcliff *et al.*, 1991).

10.7 Degenerative four wave mixing (DFWM)

The degenerative four wave Mixing (DFWM) method is similar to CARS but uses three input beams. The difference is that all three input beams and hence the output signal as well have the same frequency. The advantages of DFWM over CARS are that phase matching conditions are satisfied, the process is Doppler free, beam aberrations are lower and signal levels are greater. The use of DFWM for measuring flame temperatures is reported by Herring *et al.* (1996).

10.8 Laser-induced fluorescence

When an atom or molecule has been excited it will tend to return to its ground state by decreasing its energy level and one of the means of achieving this is by fluorescence. Fluorescence is the emission of light that occurs between energy states of the same electronic spin states. Fluorescence can be induced by a number of methods. Laser-induced fluorescence (LIF) is the optical emission from molecules that have been excited to higher energy levels by the absorption of laser radiation. Lasers tend to be the preferred means of inducing fluorescence due to their ability to reach high temporal, spatial and spectral resolutions. It is used to measure concentration and local temperature in flames by exciting molecules and atoms in specific species, for example NO, SiO, OH, N_2 and O_2.

Two different strategies are available: one using two laser beams at different excitation wavelengths (the two-line method) and the other a single laser beam. The two-line method uses a pair of excitation wavelengths in order to produce two fluorescence signals corresponding to two distinct lower states of the same species. If the two signals have the same upper state then any difference in quenching is avoided. This method can be applied to molecules, e.g. OH, O_2 and NO or to atoms such as In, Th, Sn or Pb which may occur naturally in the medium or can be seeded into it. Uncertainties of 5% can be achieved but the requirement for two lasers and two ICCD cameras makes the technique expensive. The measurement of flame temperatures using this technique is reported by Dec and Keller (1986), and for the temperature of an argon jet by Kido *et al.* (1998). The use of a single laser beam requires a constant or known mole fraction of the species to excite (Seitzman *et al.*, 1985, 1993). This can be achieved by using pre-mixed gases of known concentrations or by seeding the flow with a specific species.

For species with concentrations below about 100 parts per million the density is not high enough to produce a sufficiently strong scattered signal and temperature measurement attempts based on the Raman process are inappropriate. In these situations it may be possible to use laser-induced fluorescence. The temperature range of LIF is approximately from 200 K to 3000 K. LIF has been applied extensively to combustion measurements in flames (Chan and Daily, 1980), in cylinder flows (Andresen *et al.*, 1990) and diesel sprays (Megahed, 1993). The uncertainty associated with LIF is approximately ±5% at 2000 K.

10.9 Acoustic thermography

Acoustic thermography is based on the variation of the speed of sound with temperature in both fluids and solids. The speed of sound in an ideal gas is given by

$$c = \sqrt{\gamma R T} \qquad (10.12)$$

where: γ = the isentropic index
  R = the characteristic gas constant (J/kg.K)
  T = temperature (K).

The speed of sound can be measured by using a pulse echo or a pulse transit time technique (Ballico, 1998). A transducer on one side of the test section of interest is used to generate the pulse, which travels across the test section, to be monitored by a microphone on the opposite side at a known separation distance. For a homogenous gas the time taken for the pulse transit is $1/\sqrt{T}$ along the signal path. If the temperature within the test section is not constant then a number of additional detectors can be installed to determine the weighted average for different paths within the test section.

In liquids the speed of sound is related to the bulk modulus by $(K/\rho)^{\frac{1}{2}}$. The velocity of sound in different liquids is tabulated in Lynnworth and Carnevale (1972). In a solid, the speed of sound is related to the Young's modulus for the material by $(E/\rho)^{\frac{1}{2}}$. This technique can be used to monitor temperature in rapid thermal processing where an electric pulse across a transducer generates an acoustic wave guided by a quartz pin (Lee et al., 1994). This results in the generation of Lamb waves (a type of ultrasonic wave propagation in which the wave is guided between two parallel surfaces of the test object), which propagate across the medium. Temperatures can be measured from 20°C to 1000°C (with a proposed use up to approximately 1800°C) (Auld, 1990).

Acoustic thermometry can be relatively inexpensive to apply in comparison to most other non-invasive techniques. The technique is versatile with a temperature range capability for acoustic thermography from a few kelvin to over 5000°C. Acoustic thermometry has been used to detect changes in ocean temperature by receiving low-frequency sounds (below 100 Hertz) transmitted across an ocean basin (Forbes, 1994). The measurement of internal temperatures in steel and aluminium billets is reported by Wadley (1986). The speed of sound in a medium is a function of the local temperature so any variation in a medium will distort the reading. An uncertainty of ±5°C is reported by Auld (1990) for a solid-based application. The method is reviewed by Colclough (1992), Green (1986) and Moore (1984).

References

Books and papers

Alessandretti, G.C. and Violino, P. Thermometry by CARS in an automobile engine. *J. Phys. D: Appl. Phys*, **16**, 1583–1594, 1983.

Anderson, T.J., Dobbs, G.M. and Eckbreth, A.C. Mobile CARS instrument for combustion and plasma diagnostics. *Applied Optics*, **25**, 4076–4085, 1986.

Andresen, P., Meijer, G., Schlüter, H., Voges, H., Koch, A., Hentschel, W., Oppermann, W. and Rothe, E. Fluorescence imaging inside an internal combustion engine using tunable excimer lasers. *Applied Optics*, **29**(16), 2392–2404, 1990.

Antcliff, R.R., Smith, M.W., Jarrett, O., Northam, G.B., Cutler, A.D. and Taylor, D.J. A hardened CARS system utilised for temperature measurements in a supersonic combustor. 9th Aerospace Sciences Meeting, Reno, Nevada, AIAA-91–0457, 1991.

Attal-Tretout, B., Bouchardy, P., Magre, P., Pealat, M. and Taran, J.P. CARS in combustion: prospects and problems. *Applied Physics B*, **51**, 17–24, 1990.

Auld, B.A. *Acoustic Fields and Waves in Solids*, 2nd edition. Wiley, 1990.

Ballico, M.J. Unconventional thermometry. Chapter 5 in Bentley, R.E. (Editor), *Handbook of Temperature Measurement*, Vol. 1. Springer, 1998.

Barat, R.B., Longwell, J.P., Sarfim, A.F., Smith, S.P. and Bar-Ziv, E. Laser Rayleigh scattering for flame thermometry in a toroidal jet stirred combustor. *Applied Optics*, **30**, 3003–3010, 1991.

Bentley, R.E. Integrated Thomson–Rayleigh scattering as a means of measuring temperatures to 15000 K: in the PLASCON rig. DAP Confidential Report DAP-C00158, CSIRO Telecommunications and Industrial Physics, 1996.

Burlbaw, E.J. and Armstrong, R.L. Rotational Raman interferometric measurement of flame temperatures. *Applied Optics*, **22**, 2860–2866, 1983.

Carlson, D.J. Static temperature measurements in hot gas particle flows. In Herzfield, C.M. (Editor), *Temperature. Its Measurement and Control in Science and Industry*. Vol. 3, Part 2, pp. 535–550. Rheinhold, 1962.

Chan, C. and Daily, J.W. Measurement of temperature in flames using laser induced fluorescence spectroscopy of OH, *Applied Optics*, **19**(12), 1963–1968, 1980.

Colclough, A.R. Primary acoustic thermometry: Principles and current trends. In Schooley J.F. (Editor), *Temperature. Its Measurement and Control in Science and Industry*, Vol. 6(2), pp. 65–75. American Institute of Physics, 1992.

Dec, J.E. and Keller, J.O. High speed thermometry using two-line atomic fluorescence. 21st International Symposium on Combustion, The Combustion Institute, pp. 1737–1745, 1986.

Dibble, R.W., Starner, S.H., Masri, A.R. and Barlow, R.S. An improved method of data reduction for laser Raman–Rayleigh and fluorescence scattering form multispecies. *Appl. Phys. B*, **51**, 39–43, 1990.

Eckbreth, A.C. *Laser Diagnostics for Combustion Temperature and Species*. Abacus Press, 1988.

Eckbreth, A.C., Dobbs, G.M., Stufflebeam, J.H. and Tellex, P.A. CARS temperature and species measurements in augmented jet engine exhausts. *Applied Optics*, **23**, 1328–1339, 1984.

Edwards, G.J. Review of the status, traceability and industrial application of gas temperature measurement techniques. NPL Report CBTM S1, 1997.

Farmer, A.J.D. and Haddad, G.N. Rayleigh scattering measurements in a free burning argon arc. *J. Phys. D, Appl. Phys.*, **21**, 426–431, 1988.

Farrow, R.L., Mattern, P.L. and Rahn, L.A. Comparison between CARS and corrected thermocouple temperature measurements in a diffusion flame. *Applied Optics*, **21**, 3119–3125, 1982.

Forbes, A., Acoustic monitoring of global ocean climate. *Sea Technology*, **35**, No.5, 65–67, 1994.

Gaydon, A.G. and Wolfhard, H.G. *Flames. Their structure, radiation and temperature.* Chapman and Hall, 1979.

Goldstein, R.J. Optical measurement of temperature. In Eckert, E.R.G. and Goldstein, R.J. (Editors), *Measurement Techniques in Heat Transfer.* AGARD 130, pp. 177–228, 1970.

Goldstein, R.J., Chen, P.H. and Chiang, H.D. Measurement of temperature and heat transfer. Chapter 16 in Rohsenow, W.M, Hartnett, J.P., and Cho, Y.I. (Editors). *Handbook of heat transfer*, 3rd edition. McGraw-Hill, 1998.

Green, S.F. Acoustic temperature and velocity measurement in combustion gases. *Proceedings of the International Heat Transfer Conference*, Vol. 2, pp. 555–560, 1986.

Hall, R.J. and Bonczyk, P.A. Sooting flame thermometry using emission/absorption tomography. *Applied Optics*, **29**, 4590–4598, 1990.

Herlin, N., Pealat, M., Lefebvre, M., Alnot, P. and Perrin, J. Rotational energy transfer on a hot surface in a low pressure flow studied by CARS. *Surface Science*, **258**, 381–388, 1991.

Herring, G.C., Roberts, W.L., Brown, M.S. and DeBarber, P.A. Temperature measurement by degenerate four wave mixing with strong absorption of the excitation beams. *Applied Optics*, **35**, 6544–6547, 1996.

Hoffman, D., Munch, K.U. and Leipertz, A. Two dimensional temperature determination in sooting flames by filtered Rayleigh scattering. *Optics Letters*, **21**, 525–527, 1996.

Iinuma, K., Asanuma, T., Ohsawa, T. and Doi, J. (Editors). *Laser Diagnostics and Modelling of Combustion.* Springer, 1987.

Kido, A., Kubota, S., Ogawa, H. and Miyamoto, N. Simultaneous measurements of concentration and temperature distributions in unsteady gas jets by an iodine LIF method. SAE Paper 980146, 1998.

Kittel, C. *Introduction to Solid State Physics*, 6th edition. Wiley, 1986.

Laplant, F., Laurence, G. and Ben-Amotz, D. Theoretical and experimental uncertainty in temperature measurement of materials by Raman spectroscopy. *Applied Spectroscopy*, **50**, 1034–1038, 1996.

Lee, Y.J., Khuriyakub, B.T. and Saraswat, K.C. Temperature measurement in rapid thermal processing using acoustic techniques. *Review of Scientific Instruments*, **65**(4), 974–976, 1994.

Lynnworth, L.C. and Carnevale, E.H. Ultrasonic thermometry using pulse techniques. In Plumb H.H. (Editor), *Temperature. Its Measurement and Control in Science and Industry*, Vol. 4(1), pp. 715–732. American Institute of Physics, 1972.

Megahed, M. Estimation of the potential of a fluorescence thermometer for Diesel spray studies. *Applied Optics*, **32**(25), 4790–4796, 1993.

Metcalfe, E. *Atomic Absorption and Emission Spectroscopy.* Wiley, 1987.

Miles, R. and Lempert, W. Two dimensional measurement of density, velocity and temperature in turbulent high speed air flows by UV Rayleigh scattering. *Appl. Phys B*, **51**, 1–7, 1990.

Moore, G. Acoustic thermometry – a sound way to measure temperature. *Electronics and Power*, 675–677, 1984.

Murphy, A.B. and Farmer, A.J.D. Temperature measurement in thermal plasmas by Rayleigh scattering. *J. Phys. D: Appl. Phys.*, **25**, 634–643, 1992.

Porter F.M. and Greenhalgh D.A. Applications of the laser optical technique CARS to heat transfer and combustion. UK Atomic Energy Authority, Harwell, AERE-R 11824, 1985.

Schwarz, A. Multi-tomographic flame analysis with a schlieren apparatus. *Meas. Sci. Technol.*, **7**, 406–413, 1996.

Seitzman, J.M. and Hanson, R.K. Planar fluorescence imaging in gases. In Taylor A.M.K.P. (Editor), *Instrumentation for Flows with Combustion*, pp. 405–466. Academic Press, 1993.

Seitzman, J.M., Kychakoff, G. and Hanson, R.K. Instantaneous temperature field measurements using planar laser-induced fluorescence. *Optics Letters*, **10**, 439–441, 1985.

Uchiyama, H., Nakajima, M. and Yuta, S., Measurement of flame temperature distribution by IR emission computed tomography. *Applied Optics*, **24**, 4111–4116, 1985.

Vaughan, G., Wareing, D.P., Pepler, S.J., Thomas, L. and Mitev, V. Atmospheric temperature measurements made by rotational Raman scattering. *Applied Optics*, **32**, 2758–2764, 1993.

Wadley, H.N.G. An ultrasonic method for measuring internal temperatures in steel and aluminium. *Proc. Aluminum Assoc. on Sensors*, Atlanta, USA, 1986.

Websites

At the time of going to press the world wide web contained useful information relating to this chapter at the following sites.

http://www.lerc.nasa.gov/Other_Groups/OptInstr/schl.html
http://www.microphotonics.com/pyrit.html
http://www.rit.edu/~andpph/text-schlieren.html
http://www.scimedia.com/chem-ed/spec/spectros.htm
http://www.netaccess.on.ca/~dbc/cic_hamilton/spect.html
http://www.netaccess.on.ca/~dbc/cic_hamilton/atomic.html
http://www.netaccess.on.ca/~dbc/cic_hamilton/raman.html
http://www.kosi.com/tutorial/main.html
http://www.anu.edu.au/Physics/honours/physics/honours/aldir/html/
    aldir-cars.html
http://mstb.larc.nasa.gov/tech/CARS.html
http://mstb.larc.nasa.gov/tech/DFWM.html
http://www.netaccess.on.ca/~dbc/cic_hamilton/fpcsepc.html
http://www.asnt.org/publications/materialseval/solution/jan98solutions/
    jan98solfig1–4.htm
http://www.onr.navy.mil/onrasia/oceans/ali12.html
http://www.nist.gov/cstl/div836/Yearly/04.html
http://www.meto.govt.uk/sec5/CWINDE/RASS.html

Nomenclature

C	=	the Gladstone–Dale constant ($= (n-1)/\rho$)
E	=	Young's modulus (N/m^2)
I_i	=	initial illumination
I_L	=	irradiance
K	=	bulk modulus (N/m^2)
L	=	length of the test section (m)
M	=	molecular weight
n	–	refractive index
n_i	=	refractive index at a particular location
$n_{i,air}$	=	the refractive index of ambient air
n_{i0}	=	refractive index at a reference location
N	=	number of particles
p	=	pressure (Pa)
p_0	=	pressure at a reference condition (Pa)
R	=	characteristic gas constant (J/kg·K)
$\mathfrak{R}$	=	the universal gas constant (J/mol·K)
T	=	temperature (K)
T_0	=	temperature at a reference condition (K)
V	=	volume (m^3)
x_i	=	the mole fraction
y	=	location (m)
z	=	distance from test section datum (m)
z_{sc}	=	distance between the screen and test section (m)
α	=	angular deflection
ϵ	=	fringe shift
γ	=	the isentropic index
λ	=	wavelength (m)
ρ	=	density (kg/m^3)
σ_i	=	the effective Rayleigh scattering cross-section of each species
ω	=	frequency

11

Technique selection

As evident from the descriptions within this text there is a wide variety of techniques for the measurement of temperature. Furthermore, within each category there is a large number of devices from which the selection must be made. In the absence of experience this choice can seem formidable. The aim of this chapter is to provide guidelines for the initial selection of a technique for temperature measurement. Using this information the choices within an individual category can then be explored.

11.1 Introduction

The undertaking of any task in science and industry is usually subject to a specification. If this does not formally exist then it should be explored and defined. Aspects that should be considered in the selection of a temperature measurement system, as illustrated in Figure 11.1, broadly include:

- size and shape
- nature of contact
- temperature range
- uncertainty
- response
- protection
- disturbance
- output
- commercial availability
- user constraints
- multiple locations
- cost.

In defining the specification for the temperature measurement requirement for a given application bounds should be developed for each of these categories so that a given solution can be assessed as to whether it fulfils the specification. If there are deficiencies between a proposed solution and a specification, then the impact of these should be assessed and, if unsatisfactory, another solution developed.

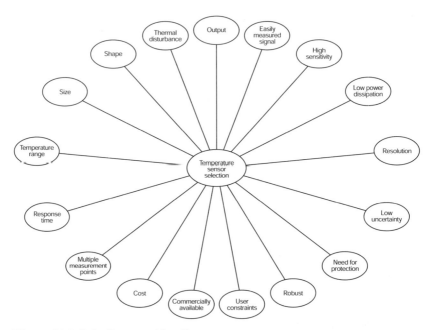

Figure 11.1 Selection considerations

11.1.1 Size and shape

The size and shape of the components of a temperature measurement system, including the transducer, any connection leads and any processing electronics and display can be a significant consideration. It is important to size not only the transducer but also the sensing system. The size of the transducer and its thermal properties will determine the thermal distortion of the temperature field (see Chapter 2, Sections 2.2.2 and 2.2.3) and the transient response of the sensor (see Chapter 2, Section 2.2.4). Many transducers, such as PRTs and thermocouples, are available in the form of a cylindrical probe that incorporates a protective cover, with diameters typically ranging from 0.25 mm to over 15 mm (see Chapter 5, Section 5.4.2 and Chapter 6, Section 6.2.3). In addition to considering whether the diameter of the sensor and its length can be accommodated, thought should also be given to the path of any connection leads, their size and whether it is necessary to protect these from the environment through which they pass. The weight and physical dimensions of any sensing electronics and display can be a factor. Is a plug-in board for a computer desirable, or is a standalone display wanted? If a display is wanted how big does it need to be and can it be located in the available space? A rack-mounted system may be appropriate for some applications. Many of these considerations are, of course, common sense, but the best of professionals can end up with an inconvenient system due to an oversight of such considerations. This is especially true when considering cable runs between the transducer and sensing electronics. For example, during the development of a gas turbine engine, multiple thermocouples may be installed on a turbine blade. The lead-out routes for sensors is often the most challenging aspect. In such an application the leads must have a sufficiently small diameter so that they can be installed in shallow slots and fed through small holes in order to minimize thermal disturbance.

11.1.2 Nature of contact

The various temperature measurement techniques have, for convenience, been categorized according to the level of physical contact between the sensor and the medium of interest. Traditionally, invasive methods have been the first choice for the majority of applications because of economic considerations and the ability of many contact sensors to give a point measurement. The reduction in cost of some of the non-invasive measurement techniques, particularly infrared methods (see Chapter 9), combined with additional merits of often insignificant thermal disturbance, has made the decision based on economy and performance more competitive. An example is the measurement of food temperature in shop refrigeration display units. Originally a liquid-in-glass thermometer might have been used, although legislation in many countries has forced the replacement of these by RTDs (see Chapter 6), which can more readily be sterilized to remove harmful bacteria and are more environmentally friendly. An infrared thermometer in such applications could lessen the risk of bacterial contamination even further and could therefore justify the likely five times first-cost price. In some remote or high-temperature applications, non-invasive techniques are the only option and disadvantages of complexity, cost and possibly increased uncertainty have to be accepted.

11.1.3 Temperature range

Generally a sensor should be selected that can survive the temperatures to which it is exposed. The data given in Figure 11.2 illustrates the temperature range capabilities of various techniques. It should be noted, however, that the majority of the spans presented cannot be achieved with a single sensor. The thermocouple range, for example, shows capability up to 3300°C. A type T thermocouple is useful in the range −262°C to 400°C whilst a type K is useful for the range −250°C to about 1300°C. Figure 11.2 therefore only serves to indicate capability and as an initial guideline. In the case of thermocouples use of Table 5.2 and Figure 5.27 could be made to help to determine which type should be selected.

It is possible and sometimes necessary to use a transducer that ultimately degrades above a certain temperature. An example is the measurement of combustion chamber temperatures. One possibility is to use a MIMS thermocouple (see Chapter 5, Section 5.4.2), which can survive in the environment at extended temperatures for only a limited time before the

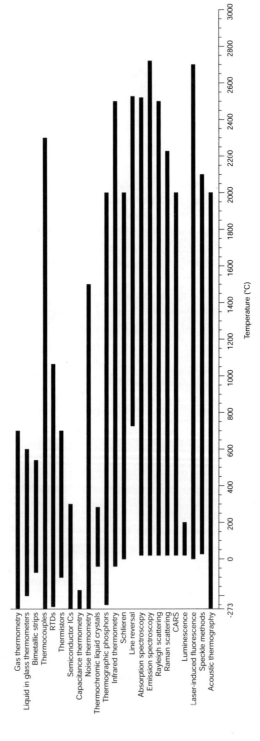

Figure 11.2 Normal temperature range capability of various measurement techniques

sheath or insulation degrades. This period of function before degradation may be long enough to provide a meaningful measurement. Once the transducer has ceased to function it can be replaced.

11.1.4 Uncertainty

The level of uncertainty to which a temperature must be determined should be explored before a sensor is selected. It should be ascertained whether an absolute value of temperature is desired or a temperature difference. If a temperature difference is required, i.e. a comparative measurement, then it should be noted that for some sensors the gradient of the output temperature characteristic remains more stable with time than does the origin. If a low uncertainty is required, then the sensor will need to be calibrated on a regular basis, or replaced by a calibrated substitute (see Chapter 2, Section 2.3). Figure 11.3 provides an indication of the uncertainties of various techniques. The bars show two bands, the best practically achievable, and a more general practical assumed uncertainty.

11.1.5 Response

It is quite possible to install a temperature sensor in an application and not even realize that the temperature is in fact fluctuating significantly because

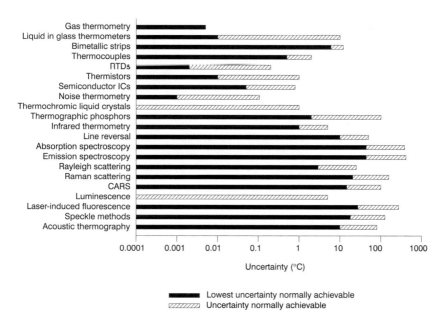

Figure 11.3 Uncertainty capabilities of various temperature measurement techniques. Inner bar represents the lowest uncertainty normally attainable. The outer bar represents normal limits on uncertainty

the transient response of the sensor is so slow that it merely indicates a weighted average temperature. An assessment of the likely temperature fluctuations in an application should be made along with what kind of information is actually required from the measurement. If a weighted average is all that is required then a sensor with a thermal mass low enough to pick this up would be quite appropriate. If, however, an indication of the transient fluctuation is necessary then a sensor with a time constant compatible with the fluctuations should be selected. The lower the product of mass and specific heat capacity, the faster the sensor. If a sensor has a time constant lower than the fluctuation in the temperature of the application there will still be a phase lag between the sensor output and the temperature of the application as well as an amplitude difference between the indicated and the actual temperatures (see Chapter 2, Example 2.8). The smaller the time constant, the lower the phase lag and amplitude difference. Time constants depend on the sensor physical properties, local heat transfer and any signal processing times. For a specific sensor the time constant should be determined taking the boundary conditions of the actual application into consideration.

11.1.6 Protection

For some applications it may be necessary to isolate the sensor from the medium of interest using a protective cover or enclosure. A typical example is the use of a thermowell that protrudes into a fluid (see Chapter 2, Section 2.2, Chapter 4, Section 4.1, Chapter 5, Section 5.4.3 and Chapter 6, Section 6.2.3). The probe can be installed directly into the well and attain a temperature related to the fluid temperature. Some form of protection for sensors must generally be considered when dealing with corrosive environments such as reducing or oxidizing atmospheres, or extreme temperatures and pressures. Any form of protective coating will have implications on thermal response, thermal disturbance and cost of the system.

11.1.7 Disturbance

The insertion of a temperature probe into an application will distort the temperature distribution as the thermal properties of the sensor are unlikely to match those of the application and the presence of the sensor may distort the local thermal boundary conditions (see Chapter 2, Section 2.2). A common maxim is that a temperature sensor measures its own temperature. The magnitude of the thermal distortion will depend on the properties of the sensor and its effect on the boundary conditions. Generally, as small a sensor as possible is desirable, preferably with thermal properties similar to those of the application. In addition, care should be taken to minimize the effects of any lead wires on the local system boundary conditions.

11.1.8 Output

The output of each type of transducer is different depending on the type of device. Liquid-in-glass thermometers (see Chapter 4) provide a visual indication of temperature, whilst an RTD gives an output in terms of resistance (see Chapter 6). Figure 11.4 illustrates the typical outputs for four types of temperature measurement device.

The form of output desired will depend on a number of factors:

1 Is the output to be used for control? If so, an electrical transducer, such as a PRT or thermistor (see Chapter 6), or mechanical transducer, such as a bimetallic thermometer (see Chapter 3), may be suitable.
2 Is the output to be recorded? If yes, then a sensor with an electrical output is likely to be most suitable.
3 Must the temperature be displayed at some distance from the point of measurement? If so, then a sensor with a high output, not sensitive to transmission errors, will be most sensible; e.g. an RTD or semiconductor-based sensor (see Chapter 6).

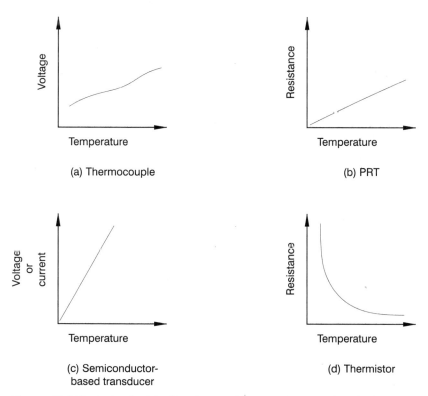

(a) Thermocouple

(b) PRT

(c) Semiconductor-based transducer

(d) Thermistor

Figure 11.4 Typical output for four types of temperature-sensing device

4 Is high sensitivity, output change/degree temperature change desirable? If yes then an RTD or semiconductor-based device may be suitable (see Chapter 6).

11.1.9 Commercial availability

It is good general practice to use temperature-measuring sensors that are widely commercially available. Such sensors can be readily obtained and replaced as necessary. In particular some sensors are available as stock items. Examples include type J, K, N, R and T thermocouples (see Chapter 5, Section 5.3), standardized thermistors (see Chapter 6, Section 6.5) and a number of semiconductor-based sensors (see Chapter 6, Section 6.6). These can be ordered and supplied within a day in many cases. Some of the non-invasive techniques (see Chapter 10) are specialist undertakings, requiring sophisticated equipment and skills that can take months to build up and develop. Even some of the infrared thermometry devices (see Chapter 9), despite their increased use, can require order times of several months. The selection of a technique must therefore be undertaken with a knowledge of the availability of the devices and the requisite knowledge to install and use them.

11.1.10 User constraints

The level of expertise required to utilize the various measurement techniques described is highly variable. Even highly sophisticated techniques can in some cases be configured in a turn-key system where the user need only switch on the device to monitor the temperatures of interest. Nevertheless the number of considerations described in this chapter give an indication of the possible complexity of temperature measurement. Due thought must therefore be given to user constraints, local expertise, speed of system use, customary usage and traditions, and resistance to rough handling, among others.

11.1.11 Multiple locations

It is often desirable to be able to monitor the temperature at a number of locations on, say, a surface. In such cases a number of the smaller invasive devices such as thermocouples (see Chapter 5), thermistors or PRTs (see Chapter 6) may be suitable. An alternative is to consider a non-invasive or semi-invasive technique (see Chapters 8–10). One example is the use of an infrared thermometer with a traverse system, repositioning the field of view on each area of interest. Another possibility is to coat the surface with thermochromic liquid crystals and use of a video camera to record the spatially distributed temperatures (see Chapter 8, Section 8.2.1).

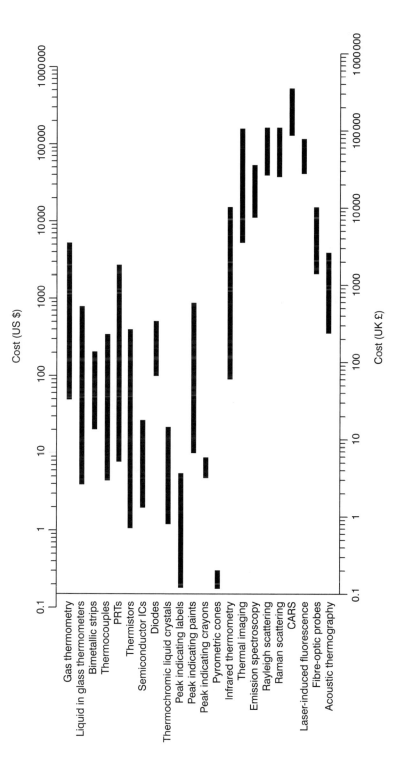

Figure 11.5 Comparison of purchase costs for a variety of temperature-measurement techniques

11.1.12 Cost

Cost is the ever-present driver of many if not all decisions in science and industry. The aim of business is production of profitable money. When considering a temperature measurement, the value of the data should be assessed in order to determine what maximum budget should be allocated to the task. The cost of the various temperature-measurement techniques considered here varies significantly, as does the cost of particular sensors within an individual category. When determining the cost of a temperature measurement, the entire system, the sensor, any connectors and sensing electronics, should be assessed along with the costs of calibration, servicing and use. Figure 11.5 provides an indication of the cost of a variety of techniques. In this figure it should be noted that in some case the cost of the sensor alone is considered, as in the case of a thermocouple, whilst in others the entire measurement system is considered as for infrared thermometers.

11.2 Applications

The range of applications of temperature measurement is extensive. One method of classifying these is according to the medium of interest, i.e. solid, liquid or gaseous and these applications are considered in Sections 11.2.1 to 11.2.3.

11.2.1 Solid temperature measurement

In order to measure the temperature at some location in a solid body it is normally necessary to embed a sensor within the material (see Chapter 2, Section 2.2.2). Common choices include thermocouples and RTDs (see Chapter 5, Section 5.4.3 and Chapter 6 respectively). Thermocouples tend to be the most versatile for this purpose because of their small size and flexibility of the cables. Care must be taken to ensure that the sensor assumes a temperature near to that of the undisturbed medium. This entails ensuring good thermal contact between the sensor and the local material and minimization of unwanted thermal conduction along any lead wires and protective sheaths. A relatively simple rule of thumb is that the Biot number can be used to assess whether thermal gradients are likely to be a significant concern when measuring the temperature of a solid. The Biot number is defined by

$$Bi = \frac{hL}{k} \tag{11.1}$$

where: h = heat transfer coefficient (W/m^2·K)
           L = smallest dimension of the solid, e.g. the thickness (m)
           k = thermal conductivity of the solid (W/m·K).

If the Biot number is less than 0.2, then no significant thermal gradients will be expected in the solid and measurement of the temperature anywhere in the solid will give similar results regardless of the size and configuration of the sensor. If, however, the Biot number is greater than 0.2, then thermal gradients will occur in the solid and the size and location of the point of temperature measurement requires further consideration as the temperature within the solid will vary from one location to another. In addition, the insertion of the temperature probe is likely to distort the temperature field. In order to determine a sensible size for a sensor under these conditions, use can be made of Fourier's conduction law,

$$q = -k \frac{\partial T}{\partial n} \ (\text{W/m}^2) \tag{11.2}$$

where: q = heat flux (W/m^2)
       k = thermal conductivity of the solid (W/m·K)
      $\partial T/\partial n$ = temperature gradient at the surface (K/m).

The length of any of the dimensions of the sensing element should not be greater than the distance between two points of the solid that are at a different temperature by more than the acceptable measurement error.

The measurement of surface temperature can be undertaken by a variety of means, depending on the requirements of the measurement. For approximate indications, bimetallic thermometers with flat bases, spring clips or permanent magnets can be used (see Chapter 3). Thermocouples and RTDs (see Chapters 5 and 6) provide significantly lower uncertainty, especially if installed with flat or suitably shaped pads. Unfortunately, such configurations will alter the temperature of the surface (see Chapter 2, Section 2.2). The best solution for low uncertainty applications is normally to embed an RTD, thermistor or thermocouple in a shallow recess in the surface. An alternative is to consider the use of a semi-invasive or non-invasive temperature measurement technique such as liquid crystals (see Chapter 8, Section 8.2.1) or infrared thermometry (see Chapter 9). Liquid crystals or other thermal paints provide a minimal, or in any case uniform, thermal disturbance to the surface and permit either a visual indication or one that can be captured and interpreted using a CCD camera and suitable software. Infrared techniques using a spectral band thermometer (see Chapter 9, Section 9.4) or even a thermal imaging system (see Chapter 9, Section 9.8) can provide an excellent indication of surface temperature. The assessment of surface emissivity is usually a concern with infrared thermometry and costs can be significant especially in the case of thermal imaging systems.

The measurement of the temperature of moving surfaces presents a particular challenge, common to processing industries and science. For a linear motion, where either the surface is moving and the sensor stationary or when the surface is stationary and the sensor is moving, infrared thermometers can provide an elegant solution with the optical system focused on the region of interest (see Chapter 9). An alternative is to coat the surface with, say, liquid crystals and use a video camera to record the image as the surface moves relative to the camera (see Chapter 8, Section 8.2.1). If the surface is rotating, any of the above techniques could be used. It also possible to use invasive instrumentation in conjunction with a slip ring or telemetry unit. A slip ring unit consists of a series of rings, which rotate with the rotating machinery. Electrical signals from the instrumentation are transmitted to the stationary frame by means of brushes. A telemetry unit converts the electrical signal on board the moving component to a radio wave or infrared signal that is then transmitted to a stationary receiver.

11.2.2 Liquid temperature measurement

Measurement of a liquid temperature usually involves the use of an immersed sensor (see Chapter 2, Section 2.2.3). Almost any invasive sensor could be used provided it has adequate protection from the liquid. If a simple visual indication alone is required then liquid-in-glass thermometers are excellent (see Chapter 4). If a larger visual indication is required then a bimetallic thermometer with a dial indicator might be suitable (see Chapter 3). Bimetallic thermometers also have the advantage that they can be used as control devices. Both liquid-in-glass and bimetallic thermometers have the benefit that they are self-contained without the need for a power supply. The cost of thermocouples (see Chapter 5) and RTD (see Chapter 6) temperature measurement systems, including the sensor, lead wires and display is relatively low and can be comparable to the cost of a liquid-in-glass thermometer. These devices have the advantage that the signal is available for recording, processing and control purposes. They do, however, require a power supply for either or both the sensing electronics and energisation.

It is also possible to use some of the non-invasive techniques such as infrared thermometry to measure liquid surface temperatures. Care must be taken to assess the emissivity, transmissivity and reflectivity of the surfaces. The emissivity of water, for example, varies according to the frequency and magnitude of ripples on the surface. To map temperature variations in liquid bodies, acoustic methods can be used (see Chapter 10, Section 10.9).

11.2.3 Gas temperature measurement

In a similar fashion to liquids, the measurement of a gas temperature can involve immersion of a probe. The heat transfer coefficient for flow over a sensor at similar velocities will be significantly lower for gases in comparison to liquids and care must be taken to ensure that unwanted effects of conduction along the sensor leads and support do not distort the temperature measurement (see Chapter 2, Section 2.2.3).

The choice for gas temperature measurement includes liquid-in-glass thermometers, if a simple visual indication is required (see Chapter 4). If a sensor with lower thermal disturbance or data recording and control capability is needed then the options include RTDs, thermistors, transistor-based sensors (see Chapter 6) and thermocouples (see Chapter 5). Thermocouple and resistance lead wires can be made from very small diameter cable to minimize conduction and disturbance effects.

A variety of non-invasive techniques can be used for gas temperature measurement including absorption and emission spectroscopy, line reversal, spontaneous Rayleigh and Raman scattering, coherent anti-Stokes–Raman scattering, degenerative four-wave mixing, laser-induced fluorescence, speckle methods and acoustic thermography (see Chapter 10). These methods are generally associated with the measurement of relatively high temperatures, for example combustion applications, where the energetic molecular motions produce observable density differences or electromagnetic effects.

11.3 Selection overview

Considerations in the selection of a method for temperature measurement and the associated equipment to suit a particular application include: temperature range, likely maximum temperature, heating rate, response, desired uncertainty, stability, sensitivity, ruggedness, service life, safety, environment and contact methods. The selection of an appropriate technique requires an appreciation of a wide range of different technologies, what is possible and what is available. The descriptions and references given in Chapters 3–10 are intended to serve this purpose. The specific requirements of an application can limit the choice of suitable instrumentation. Some applications, for example, preclude the use of invasive instrumentation. Sometimes a full-field temperature map may be required; alternatively, point temperature measurements may be acceptable. Very low uncertainty measurements may or may not be worth the investment in equipment. The information given in Table 11.1, which is based on a wide range of common selection criteria, can be used to assist in the initial choice of an appropriate technique. Table 11.2 provides an overview of typical applications for each type of temperature measuring technique.

In addition to this text a substantial number of sources of information on temperature are available. For general review articles on the subject the reader is referred to Liptak (1995), Webster (1999) and Childs et al. (2000). For cryogenic applications, temperatures below 0°C, the reviews by Rubin et al. (1970, 1982, 1997) provide a thorough overview. For very high temperature measurements the reader is referred to Farmer (1998). A number of informative textbooks are available including Bentley (1998) and McGee (1988) giving general overviews, Quinn (1990) concentrating on the fundamental aspects of temperature, Kerlin and Shepard (1982) giving an

Table 11.1 Guide to the performance of various types of temperature measuring techniques and sensors. After Childs *et al.* (2000)

Method	Minimum temperature	Maximum temperature	Response	Transient capability	Sensitivity	Uncertainty	High signal	Stability/ repeatability	Low thermal disturbance	Commercially available	Relative cost
Gas thermometer	about −269°C	700°C	slow	×	–	A standard	✓	✓	×	×	v high
Liquid-in-glass thermometer	−200°C	600°C	slow	✓	1°C	±0.02 – ±10°C (ind) ±0.01°C (lab)	✓	✓	×	✓	v low
Bimetallic thermometer	−73°C	540°C	mid	✓	–	±1°C	✓	✓	✓	✓	low
Thermocouple	−270°C	2300°C	very fast	✓	±10 µV/°C	±0.5°C – ±2°C	×	✓	✓	✓	v low
Suction pyrometer	−200°C	1900°C+	very fast	✓	–	±5°C of reading	✓	✓	×	✓	mid high
Electrical resistance device	−260°C	1064°C	fast	✓	0.1 Ω/°C	the standard above 13.81K	✓	✓	✓	✓	mid low
Thermistors	−100°C	700°C	fast	✓	10 mV/K	±0.01 – ±0.05°C	✓	✓	✓	✓	mid low
Semiconductor devices	−272°C	300°C	very fast	✓	±1%	±0.1°C	✓	✓	×	✓	low
Fibre-optic probes	−200°C	2000°C	fast	✓	10 mV/°C	0.5°C	✓	✓	✓	✓	mid high
Capacitance	−272°C	−170°C	fast	✓	good	poor	✓	×	✓	✓	mid
Noise	−273°C	1500°C	fast	✓	good	good	×	✓	✓	×	high
Chemical sampling	5°C	2100°C	slow	×	–	±25 K	×	✓	×	✓	mid
Thermochromic liquid crystals	−40°C	283°C	mid	✓	±0.1°C	±1°C	–	✓	✓	✓	low mid
Thermographic phosphors	−250°C	2000°C	very fast	✓	~0.05°C	0.1% –5%	✓	✓	✓	✓	high

Method										
Heat-sensitive paints	300°C	1300°C	slow	×	–	±5°C	✓	✓	✓	✓	mid
Infrared thermometer	−40°C	2000°C	very fast	✓	~0.1°C	±2°C	✓	✓	✓	✓	v high
Two colour	150°C	2500°C	very fast	✓	1°C/mV	±1%(±10°C)	✓	✓	✓	×	v high
Line scanner	100°C	1300°C	very fast	✓	–	±2°C	✓	✓	✓	✓	v high
Schlieren	0°C	2000°C	fast	✓	n/a	n/a	visual	✓	✓	✓	mid
Shadowgraph	0°C	2000°C	fast	✓	n/a	n/a	visual	✓	✓	✓	mid
Interferometry	0°C	2000°C	fast	✓	n/a	n/a	✓	✓	✓	✓	high
Line reversal	727°C	2527°C	very fast	×	line of sight av.	±10–15 K	✓	✓	✓	✓	low
Absorption spectroscopy	20°C	2500°C	very fast	×	line of sight av.	15%	✓	✓	✓	✓	low
Emission spectroscopy	20°C	2700°C	very fast	✓	line of sight av.	15%	✓	✓	✓	✓	low
Rayleigh scattering	20°C	2500°C	very fast	×	0.1 mm³ in 100°C	1%	✓	✓	✓	×	v high
Raman scattering	20°C	2227°C	very fast	×	0.1 mm³ in 100°C	7%	✓	✓	✓	×	v high
CARS	20°C	2000°C	fast	✓	1 mm³ in 50°C	5%	✓	@atm	✓	✓	v high
Degenerative four wave mixing	270°C	2600°C	very fast	✓	1 mm³ in 50°C	10%	✓	@atm	✓	×	v v high
Luminescence	20°C	200°C	fast	✓	1.5 nm in 200°C	±5°C	✓	✓	✓	×	high
Laser-induced fluorescence	0°C	2700°C	very fast	×	–	10%	✓	✓	✓	×	v high
Speckle methods	27°C	2100°C	very fast	×	–	6%	✓	✓	✓	×	v high
Acoustic thermography	−269°C	2000°C	very fast	✓	–	4%	✓	✓	✓	×	high

Table 11.2 An overview of typical applications

Medium	Technique	Typical applications
Solid	Liquid-in-glass thermometers	Body temperatures
	Bimetallic thermometers	Surface and body temperatures
	Thermocouples	Surface and body temperatures
	RTDs	Surface and body temperatures
	Thermistors	Surface and body temperatures
	Semiconductor-based sensors	Body temperatures
	Capacitance	Sample temperatures
	Noise thermometry	Sample temperatures
	Quartz thermometers	Surface and body temperatures
	Paramagnetic thermometry	Sample temperatures
	Nuclear magnetic resonance thermometry	Sample temperatures
	Liquid crystals	Surface temperatures
	Thermographic phosphors	Surface temperatures
	Heat-sensitive paints	Surface temperatures
	Infrared thermometers	Surface temperatures
	Acoustic thermography	Thermal processing
Liquid	Liquid-in-glass thermometers	Typically installed in a thermowell
	Bimetallic thermometers	Typically installed in a thermowell
	Thermocouples	Often integrated into a total temperature probe or installed in a thermowell
	RTDs	Often integrated into a total temperature probe or installed in a thermowell
	Thermistors	Often integrated into a total temperature probe or installed in a thermowell

	Method	Application
	Semiconductor-based sensors	Often integrated into a total temperature probe or installed in a thermowell
	Acoustic thermography	Liquid body, oceans
Gas	Liquid-in-glass thermometers	Typically installed in a thermowell
	Bimetallic thermometers	Typically installed in a thermowell
	Thermocouples	Often integrated into a total temperature probe. Combustion chambers
	RTDs	Often integrated into a total temperature probe or installed in a thermowell
	Thermistors	Often integrated into a total temperature probe or installed in a thermowell
	Semiconductor-based sensors	Often integrated into a total temperature probe or installed in a thermowell
	Suction pyrometer	Hot gases, flames
	Chemical sampling	Flames
	Schlieren	Hot gases, flames
	Shadowgraph	Hot gases, flames
	Interferometry	Hot gases, flames
	Line reversal	Flames, hot gases
	Absorption spectroscopy	Flames
	Emission spectroscopy	Flames
	Rayleigh scattering	Plasmas, combustion processes, sooting flames, supersonic flows
	Raman scattering	Reactive flows, flames, atmospheric temperature measurement
	CARS	Flames, combustion chambers, combustion and plasma diagnostics, exhausts, unsteady flows
	Degenerative four-wave mixing	Flames
	Gas thermometer	Gas sample
	Acoustic thermography	Gaseous medium
	Laser-induced fluorescence	Flames, IC engine cylinder measurements
	Speckle methods	Flames
	Luminescence	Flames

accessible introduction to temperature measurement and Nicholas and White (1994) introducing the subject of traceability in measurement. For specialist articles on temperature measurement the series entitled *Temperature. Its Measurement and Control in Science and Industry* (Herzfeld, 1962; Plumb, 1972; Schooley, 1982, 1992), is an excellent starting point for research.

References

Bentley, R.E. (Editor). *Handbook of Temperature Measurement*, Vols. 1–3. Springer, 1998.

Childs, P.R.N., Greenwood, J.R. and Long, C.A. Review of temperature measurement. *Review of Scientific Instruments*, **71**, 2959–2978, 2000.

Farmer, A.J. Plasma temperature measurement. In Bentley, R.E. (Editor), *Handbook of Temperature Measurement*, Vol. 1. Springer, 1998.

Herzfeld, C.H. (Editor). *Temperature. Its Measurement and Control in Science and Industry*, Vol. 3. Reinhold, 1962.

Kerlin, T.W. and Shepard, R.L. *Industrial Temperature Measurement*. ISA, 1982.

Liptak, B.G. (Editor). Temperature measurement. Section 4 in Liptak, B.G. (Editor), *Instrument Engineer's Handbook: Process measurement and analysis*. Chilton Book Co., 1995.

McGee, T.D. *Principles and Methods of Temperature Measurement*. Wiley, 1988.

Nicholas J.V. and White D.R. Traceable temperatures. *An Introduction to Temperature Measurement and Calibration*. Wiley, 1994.

Plumb, H.H. (Editor). *Temperature. Its Measurement and Control in Science and Industry*. Vol. 4. Instrument Society of America, 1972.

Quinn, T.J. *Temperature*, 2nd edition. Academic Press, 1990.

Rubin, L.G. Cryogenic thermometry: a review of recent progress. *Cryogenics*, **10**, 14–20, 1970.

Rubin, L.G. Cryogenic thermometry: a review of progress since 1982. *Cryogenics*, **37**, 341–356, 1997.

Rubin, L.G., Brandt, B.L. and Sample, H.H. Cryogenic thermometry: a review of recent progress II. *Cryogenics*, **22**, 491–503, 1982.

Schooley, J.F. (Editor). *Temperature. Its Measurement and Control in Science and Industry*, Vol. 5. American Institute of Physics, 1982.

Schooley, J.F. (Editor). *Temperature. Its Measurement and Control in Science and Industry*, Vol. 6. American Institute of Physics, 1992.

Webster, J.G. (Editor). Section 32, Temperature. In: *The Measurement Instrumentation and Sensors Handbook*. CRC Press, 1999.

Web sites

At the time of going to press the world wide web contained useful information relating to this chapter at the following sites.

http://www.advancedtelemetrics.com/mpx.htm
http://www.fabricast.com/
http://www.focaltech.ns.ca/product-esr.html

http://www.idmelectronics.co.uk/
http://www.litton-ps.com/SlipRings/slipring.html
http://www.michsci.com/slip.htm
http://www.tcal.com/eck_slip.htm

Nomenclature

Bi = Biot number
h = heat transfer coefficient (W/m^2·K)
k = thermal conductivity (W/m·K)
L = characteristic length (m)
n = coordinate (m)
q = heat flux (W/m^2)
T = temperature (K)

12

Heat flux measurement

The aims of this chapter are to review the various different methods available for the measurement of heat flux. These include methods based on temperature difference, heat balance, energy supply and a mass transfer analogy. The structure and content of this chapter follows an article published by Childs *et al.* (1999). Permission to reproduce aspects of this by the Institution of Mechanical Engineers is gratefully acknowledged.

12.1 Introduction

Heat flux can be defined as the energy in transit due to a temperature difference per unit cross-sectional area normal to the direction of the flux. There are no devices that can measure energy and hence heat or heat flux directly. Instead the effects of energy in transit must be monitored and the heat flux inferred. The physical methods currently available for this are principally based on the monitoring of temperatures and spectral emissions.

Heat transfer can occur by means of any one, or a combination, of conduction, convection and radiation. In a solid heat transfer will be due to conduction alone. However, at a solid/fluid interface the resulting heat transfer will occur due to at least two of these three modes. Their relative contributions will be governed by, for example, the Reynolds number of the flow, a convective contribution, and the temperature of adjacent surfaces, a radiative contribution. Most heat flux sensors measure the total heat flux at the solid/fluid interface or surface. The measurement techniques available for determining heat flux can, for convenience, be broadly arranged into four categories:

1 *Differential temperature*. The heat flux across a spatial distance can be determined by means of measurement of temperatures at discrete locations and the heat flux related to this gradient and the material properties.

2 *Calorimetric methods.* A heat balance is applied to a particular region of the sensor and the temporal measurement of a temperature related to the heat flux.

3 *Energy supply or removal.* A heater or cooler is used to supply or remove energy from a system and a heat balance used to relate the temperature measured to the heat flux.

4 *Mass transfer analogy.* Measuring mass transfer in place of heat transfer and using an analogy to infer the heat transfer.

In reality practical implementation of a heat flux measurement may involve operation according to one of the above categories under one range of heat flux fluctuation and another category for a different range of heat flux fluctuation. For example, some differential temperature measuring sensors (category 1) must be analysed as a calorimeter (category 2) for high frequencies of heat flux variation. It is probably no surprise to the reader that because of the often conflicting requirements for sensitivity, robustness, size and cost, no one heat flux measurement technique is suitable for all applications. Indeed the trend is for increased diversity; in particular, recent developments in the use of deposition techniques and micromanufacture methods combined with high-speed data-acquisition systems have resulted in a number of new devices becoming available.

As described in Chapter 2, the inclusion of a sensor with different properties from that of the medium of interest will cause a disturbance to the temperature distribution within the medium. Section 12.2 introduces the issues of thermal disturbance with specific reference to the presence of a heat flux measurement device. The different methods of heat flux measurement are introduced in Sections 12.3, 12.4, 12.5, 12.6 and 12.7. Section 12.8 provides guidance for the selection of a particular type of measurement technique.

12.2 Thermal disruption

A goal of any measuring technique is to determine the desired quantity in a non-intrusive fashion without disturbing the quantity being measured. For convective heat flux measurement, this may entail matching the surface roughness and thermal radiation properties, and maintaining the surface heat flux and temperature field that would exist in the absence of any instrumentation. These restrictions are difficult to achieve in practice. A device applied to the surface will disrupt the geometric surface profile and an insert located within the surface may disrupt the thermal conditions because of a mismatch of thermal properties. If the heat flux sensor is manufactured as an insert with a different thermal conductivity from the component under investigation, the heat flux through the insert will be different from that through the geometry prior to installation of the instrumentation. These implications are illustrated in Figures 12.1 and 12.2 for a 0.5 mm thick thermopile heat flux sensor measuring the temperature differential across a polyimide wafer (k = 0.17 W/m·K) installed within a composite substrate layer ($k_{\text{along fibre}}$ = 11.3 W/m·K, $k_{\text{transverse fibre}}$ = 0.67 W/m·K). Here for the boundary conditions imposed the thermal disruption is significant with a 10°C temperature depression at the base of the 0.5 mm sensor in comparison to the undisturbed temperature. Of course, this could be compensated for by calibration or by the use of different

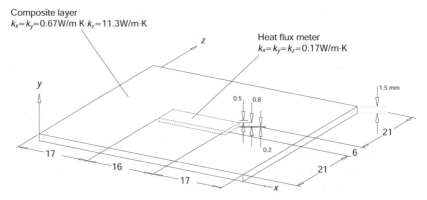

Figure 12.1 An example of the installation of a heat flux sensor manufactured on a polyimide wafer installed in a composite layer (Childs, 1991)

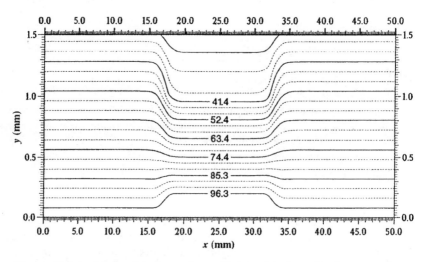

Figure 12.2 Isotherm plot, lines of constant temperature, for the geometry and boundary conditions of Figure 12.1 (Childs, 1991)

materials for the sensor which match the thermal conductivity of the substrate. The modification of thermal boundary conditions has been described by the terms hot or cold islands depending on the conditions of interest. With heat transfer from the surface to the convective flow, if the thermal conductivity of the material used for the heat flux sensor is below that of the component material the heat flow through the sensor will be less than that prior to the installation of the sensor and the temperature of the fluid in the immediate vicinity of the sensor will be reduced. Hence the term cold island (see Kim *et al.*, 1996; Dunn *et al.*, 1997 and Example 2.1). This, of course, can have implications for the whole system if there are small local disturbances to the thermal boundary layer.

A slightly different but potentially significant disruption caused by installation of heat flux measuring devices on a surface is illustrated in Figure 12.3 following an example given by Flanders (1985). The two potential problems that arise with the device are:

(a) disturbing the flow field and acting as a possible trip for a boundary layer causing transition from laminar to turbulent flow, and
(b) modification of the thermal boundary layer.

The first of these problems is encountered in the application of heat flux sensors to the measurement of heat fluxes on, for example, the wall of a building. Hot air from a central pool near the ceiling moves outwards towards and down the cool wall. The Nusselt number for natural convection

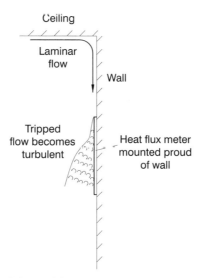

Figure 12.3 Fluid and thermal boundary layer disruption due to the installation of a heat flux sensor

of air with Pr = 0.72 over a vertical flat plate for laminar flow can be described, (Bejan, 1993) by

$$Nu = 0.387(Pr\ Gr_x)^{0.25} \tag{12.1}$$

where:

$$Nu = \frac{hx}{k} \tag{12.2}$$

$$Gr_x = \frac{\rho^2 g \beta (T - T_\infty)\ x^3}{\mu^2} \tag{12.3}$$

$$Pr = \frac{\mu c_p}{k} \tag{12.4}$$

where: Nu = the Nusselt number
Gr_x = the Grashof number
Pr = the Prandtl number
h = the heat transfer coefficient (W/m$^2\cdot$K)
x = local coordinate along the plate (m)
k = thermal conductivity of the fluid (W/m$\cdot$K)
ρ = density (kg/m^3)
g = acceleration due to gravity (m^2/s)
β = coefficient of volumetric expansion ((=1/T) for an ideal gas)
T = temperature (K)
T_∞ = free stream temperature (K)
μ = viscosity (Ns/m^2)
c_p = specific heat capacity (J/kg$\cdot$K).

Turbulent flow over a vertical flat plate is described by (Bayley, 1955):

$$Nu = 0.1(PrGr_x)^{1/3} \tag{12.5}$$

The flow can be assumed to be turbulent when the product $PrGr_x > 10^9$.

If the presence of the heat flux sensor causes the flow to trip locally from laminar to turbulent flow then the error indicated by the sensor owing to the change in flow regime can be determined by the ratio of equations (12.5) and (12.1). Assuming this to occur for a value of $PrGr_x = 10^9$ the error would be

$$\frac{0.1(10^9)^{1/3}}{0.387(10^9)^{0.25}} = 1.45$$

This represents a 45% overestimate. Of course, a heat flux installation of this type would measure the total heat flux due to convection and radiation. For typical wall temperatures in buildings the influence of the error owing to any change of flow regime would be less and typical values for the overall error would be of the order of 10%, (Flanders, 1985).

12.3 Differential temperature heat flux measurement techniques

The basis of these techniques is to monitor the difference in temperature between locations in a component and, with a knowledge of the thermal properties, to use a conduction analysis to determine the heat flux. Various devices have been developed which utilize this principle, including differential layer temperature sensors (see Section 12.3.1), planar differential temperature sensors (Section 12.3.2) and the Gardon gauge (Section 12.3.3).

12.3.1 Differential layer devices

A substantial proportion of commercial heat flux sensors in use are based on monitoring a temperature differential across a spatial distance within a medium. Fourier's one-dimensional law of conduction (equation (12.6) can then be used to determine the heat flux through the medium provided the thermal properties are known. The temperature difference can be determined by means of thermometers, thermistors, thermocouples and thermopiles or RTDs as well as thermal radiation-based devices and optical methods. An obvious limitation of this technique is that the heat flux must be one-dimensional across the region of concern,

$$q = -k \frac{dT}{dx} = -k \frac{T_2 - T_1}{\delta x} \tag{12.6}$$

where: q = heat flux (W/m^2)
   k = thermal conductivity of the material (W/m·K)
   x = local coordinate (m)
   T = temperature (K)
   δx = distance between the two locations of temperature (m).

The selection of a temperature-measuring technique for determining the difference in temperature depends on the need to provide reasonable sensitivity and signal output for the range of heat flux under consideration. A popular method is to use a thermopile formed around an electrically insulating layer as illustrated in Figure 12.4. The principle is that the thermocouples formed measure the temperature difference and the output signal is a simple multiple of the signal that would result from a single differential thermocouple

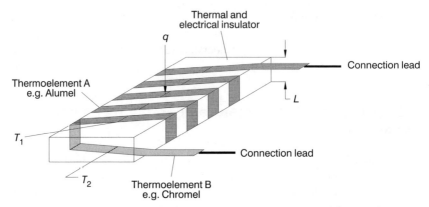

Figure 12.4 Differential thermopile heat flux sensor. Commercially available from the Rdf Corporation (Hudson, USA)

pair. This method was reported by Martinelli *et al.* (1942) and subsequently by Hartwig (1957) and Hartwig *et al.* (1957). In Hartwig's design, a sheet of glass of 0.16 mm^2 surface area and 0.18 mm thick was wrapped with fifty turns of 25 μm diameter constantan wire. One half of each loop was silver-plated by placing the edge of the sensor half-way into a silver-plating solution, forming a silver–constantan junction on each side of the flat sensor.

In Figure 12.4, a ten-junction chromel/alumel thermopile is illustrated formed on a polyimide insulating layer, which is similar to the sensors commercially available from the Rdf Corporation (Hudson, USA) and patented by Hines (1971) following the original design and patent of Hager (1967). Polyimide has a thermal conductivity of approximately 0.1 to 0.35 W/m·K, therefore giving a reasonable temperature differential and hence high sensitivity (V/(W·m^{-2})) for heat fluxes in the range 1 W/m^2 to 6 kW/m^2. The material can also be used at up to 200°C, which makes it particularly useful for heat transfer work. The sensitivity of a differential layer thermopile heat flux sensor is a function of the junction separation distance, δ, the Seebeck coefficient, S, for the thermocouple combination used, the number of junctions, N_J, and the thermal conductivity, k, of the differential thickness material:

$$\text{Sensitivity} = \frac{\text{output voltage}}{\text{heat flux}} = \frac{N_J \, S \delta}{k} \qquad (12.7)$$

where: N_J = number of thermocouple junctions
S = the Seebeck coefficient (V/K)
δ = thickness (m)
k = thermal conductivity (W/m·K).

The transient response is a function of the junction separation distance and the thermal diffusivity of the material. A one-dimensional analysis performed by Hager, (1965) gave the time required for a 98% response as

$$t = \frac{1.5\delta^2}{\alpha} \ (s) \tag{12.8}$$

where: t = time (s)
        δ = junction separation distance (m)
        α = thermal diffusivity (m²/s).

From equations (12.7) and (12.8) it can be seen that whilst sensitivity increases with the junction separation, δ, the time response also increases at a rate proportional to the square of the square of this distance. The design of such a heat flux sensor must consider this trade-off; that is, increasing the thickness raises the sensitivity but decreases the speed of measurement response. Combined with this consideration, the minimization of thermal disturbance when measuring heat flux at a convective boundary condition dictates that the ratio hδ/k should be kept low. Practically this necessitates the use of a small differential layer thickness, particularly when the heat transfer coefficient is high.

Of the various embodiments of the differential heat flux sensor, the device reported by Hager et al. (1989) and Holmberg and Diller (1995), and available from the Vatell Corporation (Virginia, USA), is particularly noteworthy. This consisted in one form, (Hager et al., 1994), of an eighty-junction thermopile formed around a staggered 1 μm silicon monoxide layer (see Figure 12.5). The device was formed by sputtering the thermocouple and differential layer

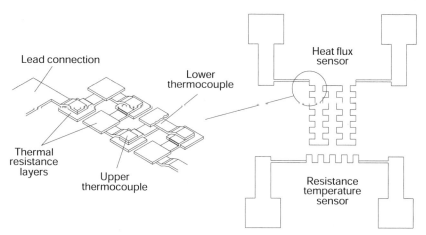

Figure 12.5 Multiple junction thermopile heat flux sensor (after Holmberg and Diller, 1995)

materials onto the substrate surface and using a masking and laser cutting process to form the thermocouple tracks. A PRT is used to monitor the surface temperature. As the device is physically and therefore thermally small it can be used for both steady and unsteady heat flux measurements. The response time has been estimated to be less than 10 μs. A similar device based on a circular geometry has been developed by Will (1992) and is illustrated in Figure 12.6.

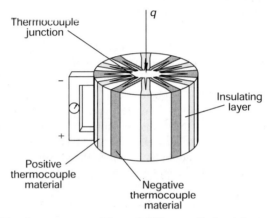

Figure 12.6 Circular geometry differential thermopile heat flux sensor (after Will, 1992)

Calibration of differential heat flux sensors can be achieved by any one or a combination of conduction, convection or radiation methods. One-dimensional conduction calibration can be carried out by mounting the sensor in a good thermal insulator with known heat addition and removal at each end. Problems with this method arise from the need to know the thermal conductivity of the different materials and also the interfacial contact resistances. A common laboratory method is the calibration of sensors against a well-defined convection correlation such as for jet impingement, (Martin, 1977). In this technique a sensor mounted flush with the surface is exposed to a fluid jet of known geometry, velocity and temperature; the expected value of the local flux is determined from an established correlation for this geometry and the given flow conditions. This is then used for the calibration against the electrical signal from the sensor. Alternatively, the sensor can be irradiated to a known intensity by, say, quartz lamps. Thermal radiation sources, including the use of a laser, are well suited for transient calibration where a mechanical shutter can be used to modulate the heat flux.

In the design of a differential heat flux sensor various options are open to the engineer. The differential layer material must be selected to be compatible with the requirements for minimal thermal disturbance and thermal response

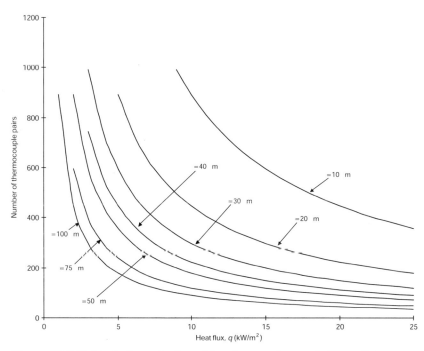

Figure 12.7 Guide to the selection of junction number and differential distance for differential thermopile heat flux sensors for a differential layer with thermal conductivity 0.2 W/m.K, thermocouple sensitivity 22.4 μV/°C and maximum thermopile output 10 mV

times. The thermocouple materials used to form the thermopile should be chosen according to manufacturing constraints and temperature limitations. The chart given in Figure 12.7 illustrates some of these criteria. A material with a thermal conductivity of 0.2 W/m·K, representative of polyamide, has been selected for the differential layer. The chart shows the number of thermocouple junctions, for a copper–nickel thermocouple, necessary to generate an emf of 10 mV as a function of heat flux. This variation is illustrated for a variety of differential distances. The figure of 10 mV for the maximum emf generated was selected to be compatible with the standard range of amplification available in data acquisition systems.

The microfabrication techniques used by Hager *et al.* (1990) involve a complex manufacturing process. The use of metal film sputtered sensors applied on either side of a polyimide sheet (see Figure 12.8) is reported by Epstein *et al.* (1985, 1986). This sheet was bonded to the surface of interest, in this case a turbine blade. The device could operate on the principle of a differential heat flux sensor for steady-state heat transfer or low frequencies of heat flux excitation. However, the physical phenomena under investigation may cause high-frequency fluctuation of the local heat flux, such as in the

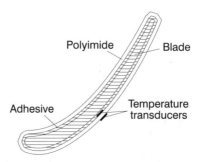

Figure 12.8 Differential PRT heat flux sensor

variation of heat flux caused by rows of turbine blades passing one another. In this case the device can then be analysed as a semi-infinite layer, assuming that the heat wave does not penetrate through the whole surface and that the temporal variation of the surface temperature can be related to the surface heat flux by a transient conduction analysis. Epstein *et al.* gave the time constant as $\tau = L^2/4\alpha$ for each case defining semi-infinite as $t \leq \tau$ and steady state as $t > 20\tau$. These methods are outlined in Section 12.4.

12.3.2 Planar thermal gradient methods

A variation on the differential thermopile heat flux sensor concept (where the temperature differential is measured across a distance in the direction of increasing depth into the surface) is to artificially create a thermal gradient on the surface plane of the component. This can be achieved, for example, by use of alternate layers of materials of different thermal conductivity or different thicknesses, or by exposing one half of the thermopile junctions to the convective boundary condition fluid as illustrated in Figure 12.9. Embodiments of this principle have been demonstrated by Godefroy *et al.* (1990), Bhatt and Fralick (1993), Cho *et al.* (1997a, b), Diller (1994) and Meyer and Keller (1996) all of which use thin-film thermocouples in a variety of configurations.

The use of this type of heat flux sensor can be illustrated by considering a device for measuring a heat flux of $20\,\text{kW/m}^2$ at 200°C. The sensor would require 400 thermocouple pairs of copper/nickel ($S_{200°C}=22.4\,\mu\text{V/°C}$) located across a $100\,\mu\text{m}$ differential layer manufactured by vacuum deposition from, for example, aluminium nitride ($k = 180\,\text{W/m·K}$) and silicon monoxide ($k = 0.2\,\text{W/m·K}$). This would provide a sensitivity (E/q) of $5\,\mu\text{V/(W/m}^2)$. The substantial difference in the thermal conductivity of the two thermal resistance layers generates a sufficiently large thermal gradient in the plane of the thermopile for sensitive measurements. The jet impingement method described in Section 12.3.1 could be used to calibrate this sensor.

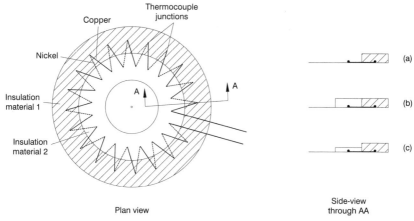

Figure 12.9 Planar differential heat flux sensors: (a) exposing just one junction directly to the convective boundary condition, (b) using different materials of the same thickness, (c) using different thicknesses of the same material

12.3.3 Gardon gauges

The Gardon gauge, or asymptotic calorimeter, named after Gardon (1953, 1960), comprises a thin disc connected to a heat sink at its periphery (Figure 12.10). As the surface is heated by a convective flow, heat is conducted radially to the edge of the disc. The temperature difference between the edge of the disc and the centre is proportional to the instantaneous heat flux. The temperature difference is measured by means of a thermocouple formed by the junction of a thermoelectrically dissimilar wire connected to the centre of the disc: a copper wire connected to a constantan disc in the bulk of such sensors.

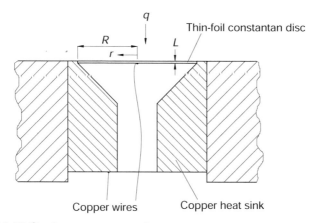

Figure 12.10 Gardon gauge geometry

Applying a heat balance to the geometry illustrated in Figure 12.10, with initial conditions $T = 0$ at $t = 0$ for $0<r<R$ and subsequently $T = 0$ at $0<t<\infty$ for $r = R$, upon simplification, gives

$$\frac{1}{\alpha} \frac{\partial T}{\partial T} = \frac{q}{Lk} + \frac{1}{r} \frac{\partial T}{\partial r} + \frac{\partial^2 T}{\partial r^2} \qquad (12.9)$$

where: α = thermal diffusivity (m²/s)
T = temperature (K)
t = time (s)
q = heat flux (W/m²)
L = thickness of the foil disc (m)
k = thermal conductivity of the foil (W/m·K)
r = local radial coordinate (m).

For steady-state conditions the $\partial T/\partial t$ term will be zero and equation (2.9) can be simplified to:

$$\frac{q}{Lk} + \frac{1}{r} \frac{\partial T}{\partial r} + \frac{\partial^2 T}{\partial r^2} = 0 \qquad (12.10)$$

Allowing for the variation of thermal conductivity with temperature of the thin-foil disc using

$$k = k_0 (1 + K_1 T) \qquad (12.11)$$

where: k_0 = the thermal conductivity of the material at a base temperature $T = 0$ (W/m·K)
K_1 is the fractional variation of conductivity per unit temperature rise (1/K),

gives

$$T \left(1 + \frac{K_1}{2} T \right) = q \frac{R^2 - r^2}{4 k_0 L} \qquad (12.12)$$

where R is the radius of the foil disc (m), and

$$\Delta T \left(1 + \frac{K_1}{2} \Delta T \right) = q \frac{R^2 - r^2}{4 k_0 L} \qquad (12.13)$$

where ΔT is the temperature difference between the centre ($r = 0$) and the periphery of the foil ($r = R$).

If the thermal conductivity can be assumed constant with temperature then

$$q = \frac{4 L k \Delta T}{R^2} \qquad (12.14)$$

If the thermal conductivity cannot be assumed constant, then, according to equation (12.13), there is a non-linear variation of heat flux with ΔT. Gardon's additional innovation was to select a copper–constantan thermocouple with the junction formed by soldering a copper wire to the centre of the constantan disc. The thermo-electric characteristic of the selected thermocouple is also non-linear and the temperature voltage characteristic of a copper–constantan thermocouple is such that it approximately cancels out the non-linearities inherent in the system heat balance. Taking the thermal conductivity of constantan as constant with $k = 21.8\ \text{W/m·K}$ and the thermo-electric characteristic of a copper–constantan thermocouple, neglecting third-order terms, as

$$E = 3.81 \times 10^{-5}\Delta T + 4.44 \times 10^{-8}\ \Delta T^2 \tag{12.15}$$

then, by substituting in equation (12.13) the sensitivity of the sensor is given by

$$\frac{E}{q} = 4.37 \times 10^{-9}\ \frac{R^2}{L} \tag{12.16}$$

Gardon originally obtained an approximate solution for the thermal response time of the sensor, given in equation (12.17), based on a one-dimensional analysis. An analytical analysis giving similar results was performed by Ash (1969):

$$t = \frac{R^2}{4\alpha} \tag{12.17}$$

In order to determine the initial dimensions for the foil diameter and thickness of a Gardon gauge, equations (12.16) and (12.17) can be used for a given sensitivity and time response compatible with the application. These equations have been plotted in Figure 12.11 following the original charts given in Gardon (1953).

Calibration of Gardon gauges has often been undertaken using a radiation heat source. Gardon gauges give accurate results for systems where radiation is the dominant mode of heat transfer. When convection is also significant, large errors can result due to non-uniformities in the foil temperature. This is confirmed in the analysis by Kuo and Kulkarni (1991), for the case when these sensors were used to determine the heat flux when convection is more significant than radiation in the contribution to the total heat flux. They defined a correction ratio given by

$$\frac{q_{\text{total}}}{q_{\text{radiation}}} = \frac{1 + (R\sqrt{h/k\delta}/2)^2}{1 + (R\sqrt{h/k\delta}/4)^2} \tag{12.18}$$

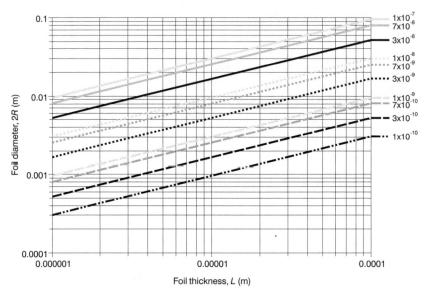

Figure 12.11 Guide to the specification of Gardon gauge geometry (after Gardon, 1953)

The error reduces as $R\sqrt{h/k\delta}$ reduces. Therefore a smaller-diameter foil is preferable but this reduces the sensitivity and a compromise is necessary. Alternatively, the sensitivity can be enhanced by use of a thermopile in place of the central thermocouple as reported by Trimmer *et al.* (1973). For applications involving continuous use a water-cooled gauge can be used with the water providing a continous heat sink enabling steady-state measurements. In general because of the thin-foil disc, Gardon gauges are not that robust and tend to be used in applications where radiation is the dominant mode of heat transfer.

12.4 Calorimetric heat flux measurement techniques

This method for determining heat transfer rates is realized through the measurement of the rate of change of temperature with time at a location near to or on the surface of interest. Analysis of these results with an appropriate form of the conduction equation or heat balance and accurate knowledge of material properties allows the heat flux to be quantified.

There are a number of heat flux measurement techniques based on this principle: including slug calorimeters, plug gauges, null point calorimeters, coaxial thermocouple gauges, thin-skin and thin-film sensors and surface temperature monitoring of the whole component.

12.4.1 Energy balance methods

The principle of a calorimeter sensor is to determine the instantaneous heat transfer rate to a surface by the measurement of the rate of change of thermal energy within an element at the surface. The thermal energy is determined by means of a temperature measurement on or near to the surface and the rate of change of this temperature with time can be used to determine the heat flux to or from the surface. In these methods, either an insert, commonly referred to as a slug, or part of the surface itself is utilized as the sensor.

A typical slug calorimeter comprising a quantity or 'slug' of material thermally insulated from its surroundings is illustrated in Figure 12.12. Typically, a single temperature measurement at the base of the slug is made and this is assumed to represent the temperature of the entire mass of the slug. This assumption is valid within a few per cent if the Biot number, Bi = hL/k (L is the depth of the slug), is less than 0.1. The slug is usually manufactured from metal, such as oxygen-free high-conductivity copper, but could in theory be manufactured using a material such as aluminium nitride, which can also have a relatively high thermally conductivity ($k \approx 180$ W/m.K).

The basis of this type of device is the lumped mass approximation of an element of material:

$$\text{Heat transfer in = heat stored} - \text{heat transferred out} \qquad (12.19)$$

i.e.

$$q_{in}A = \left(mc_p \frac{\partial T}{\partial t} + mT \frac{\partial c_p}{\partial t} \right) + Q_{loss} + Q_{out} \qquad (12.20)$$

where:
q_{in} = heat flux in (W/m^2)
A = the exposed surface area of the slug (m^2)
m = the mass of the slug (kg)
c_p = specific heat capacity (J/kg·K)
T = temperature (K)
t = time (s)
Q_{loss} = heat loss (W)
Q_{out} = heat transfer out (W).

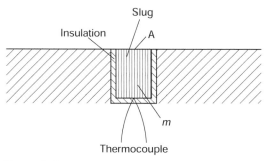

Figure 12.12 Slug calorimeter

Neglecting losses through the thermal insulation, and assuming uniform temperature and constant properties, the heat transferred to the calorimeter is equal to the energy stored and the heat flux is given by

$$q = \frac{mc_p}{A} \frac{dT}{dt} \tag{12.21}$$

Depending on the nature of the thermal boundary conditions, several solutions to equation (12.21) can be obtained. If the heat flux is constant, then

$$q = \frac{mc_p}{A} \frac{T_f - T_0}{t_f - t_0} \tag{12.22}$$

where: T_f = final temperature (K)
  T_0 = initial temperature (K)
  t_f = final time (s)
  t_0 = intial time (s).

Therefore, the heat flux can be determined from measurement of the transient temperature provided the thermal properties are known. If more accurate results are required the losses through the insulation layer should be modelled and accounted for by a correction term in equation (12.22):

$$q = \frac{mc_p}{A} \frac{dT}{dt} + q_{loss} = \frac{mc_p}{A} \frac{dT}{dt} + K_{loss} \Delta T \tag{12.23}$$

Here ΔT is the temperature difference between the slug calorimeter and the surrounding material. This can be taken as simply the rise in temperature of the slug assuming that the surrounding material temperature remains constant for the duration of the data acquisition. The constant K_{loss} for a given sensor can be determined by calibration and supplied by a laboratory or manufacturer.

If the sensor is exposed to a convective flow with a constant heat transfer coefficient then solution of equation (12.21) yields

$$\frac{T - T_\infty}{T_0 - T_\infty} = e^{-t/(mc_p/hA)} \tag{12.24}$$

The heat transfer coefficient appears in the exponential term and in practice can be evaluated from the gradient of the natural logarithm of the temperature ratio against time.

Slug calorimeters are usually used when the energy input to the surface is fairly constant. They can only produce useful data for short exposure; the duration of use is limited because T will approach T_∞ and they must be restored to an initial condition before re-use. Standard methods for the use of slug calorimeters and their design parameters are defined in ASTM E457-96 (1997). It is recommended that a period defined by equation (12.25) be allowed prior to logging of data:

$$t = L^2/2\alpha \qquad (12.25)$$

where: t = time period before logging (s)
         L = depth of the sensor (m)
         α = thermal diffusivity (m²/s).

Severe disadvantages of this form of device, particularly when measuring large heat fluxes, are the disruption to the thermal boundary layer owing to the thermal discontinuities introduced by the sensor material boundaries and the difficulty of quantifying the heat losses through the insulation layer. A recommendation in order to ensure minimization of non-uniform temperature effects is that the slug itself should be manufactured using the same material as the surroundings.

In the thin-skin method a section of surface of a model forms the slug calorimeter (Schultz and Jones, 1973). This method has been used to study heat transfer in, for example, re-entry vehicles (e.g. see Wood, 1968). In these studies part of the surface of the model or component is removed and replaced with a thin layer or 'skin' of a highly conductive material such as copper. The assumption in the analysis is that the rate of rise of the rear surface, which can be monitored with, say, thermocouples, is equal to the rate of rise of the mean temperature. Lateral conduction along the skin is ignored. Heat loss from the rear of the skin is assumed negligible so the rear of the skin must be insulated. Air, which has a relatively low thermal conductivity, can be used for this purpose as illustrated in Figure 12.13. If the thickness of the calorimeter skin is L and assuming there is no heat loss from the rear surface, the heat flux at the exposed surface is given by

$$q = \rho c_p L \frac{dT_{mean}}{dt} \approx \rho c_p L \frac{dT_{rear}}{dt} \qquad (12.26)$$

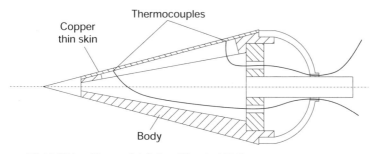

Figure 12.13 Thin-skin model (after Wood, 1968)

12.4.2 One-dimensional transient conduction analysis methods

The basic principle of these methods is that for the duration of the test the penetration of the thermal pulse into the surface is small in comparison to the relevant surrounding dimensions ($\alpha t/L^2 \ll 1$) and a one-dimensional transient analysis can be used. Typical applications of this form of measuring heat flux are in short-duration hypersonic facilities and blowdown turbomachinery experiments. For the surface shown in Figure 12.14 a one-dimensional analysis of the variation of temperature with time and depth into the surface is modelled by

$$\frac{\partial^2 T}{\partial x^2} = \frac{1}{\alpha} \frac{\partial T}{\partial t} \tag{12.27}$$

With the initial conditions of $T = T_0$ for $t = 0$ and $0 \leq x < \infty$, solutions can be developed to relate the temporal measurement of surface temperature to the heat flux through the surface (Schultz and Jones, 1973). For a step change in the surface temperature $T_1 - T_0$ at time $t = t_0$ the heat flux is given by

$$q(t) = \frac{\sqrt{\rho c_p k}}{\sqrt{\pi}} \frac{T_1 - T_0}{\sqrt{t - t_0}} \tag{12.28}$$

A general relation for any number of step changes of surface temperature can be determined by superposition:

$$q(t_n) = \frac{\sqrt{\rho c_p k}}{\sqrt{\pi}} \sum_{i=1}^{n} \frac{T_i - T_{i-1}}{\sqrt{t_n - t_{i-1}}} \tag{12.29}$$

This equation can be evaluated numerically from temperature measurements. However, as noted by Oldfield *et al.* (1978), since the heat flux is proportional to the change in temperature divided by the square root of time, the operation acts as a half derivative. It is therefore prone to introducing relatively large errors associated with small amounts of experimental noise. One scheme

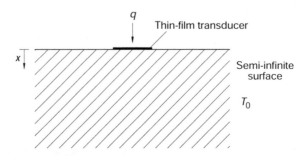

Figure 12.14 One-dimensional analysis for a semi-infinite geometry

reported by Cook *et al.* (1966, 1970), utilizes a piecewise linear function. For n uniform time steps, Δt,

$$q(t_n) = \frac{2\sqrt{\rho c_p k}}{\sqrt{\pi \Delta t}} \sum_{i=1}^{n} \frac{T_i - T_{i-1}}{\sqrt{n-i} + \sqrt{n+1-i}} \qquad (12.30)$$

Additional numerical schemes have been presented by Oldfield *et al.* (1978) and Diller (1996).

An alternative approach to the use of numerical techniques to evaluate the heat flux from the temporal measurements of temperature for these sensors is to use electrical analogue circuits (Schultz and Jones, 1973). Heat flow into a semi-infinite material is analogous to the flow of current in a resistor–capacitor, (RC), transmission line using the equivalencies of heat flux with current, temperature with voltage, thermal conductivity with the inverse of electrical resistance and the density and specific heat product with capacitance. Modern data-acquisition systems now have adequate resolution and speed so that this electrical analogue is not essential. If, however, the signal is transmitted from a rotating component as in a gas turbine rig, noise problems, from, for example, slip ring units, can swamp the high-frequency component of the heat flux signal. This can be moderated using an appropriate electrical circuit to improve the signal-to-noise ratio, in effect amplifying the useful information contained within the high-frequency output (Ainsworth *et al.*, 1989).

This analysis is the basis for a number of sensors including plug gauges, null point calorimeters, thick-film sensors and methods involving the monitoring of surface temperature by means of thin-film sensors, coaxial thermocouples, thermochromic liquid crystals and radiometers. The advantages of using this approach are the short duration of tests and fast response time of the transducers. The disadvantages are that the model cannot be used for steady-state applications and that multiple transducers are required in order to obtain a temperature history of the surface.

A development of the slug calorimeter is the plug gauge. This comprises a slug of material instrumented with multiple thermocouples to provide a temperature distribution within the slug and is illustrated in the form developed by Liebert (1994) and Liebert and Weikle (1989), in Figure 12.15. This kind of sensor can be manufactured directly on the surface of interest using electric discharge machining (EDM) techniques. The heat flux through the top surface can be determined by solution of

$$q_{x=0} = \int_0^L \rho c_p \frac{\partial T}{\partial t} \, dx + k \frac{\partial T}{\partial x} \bigg|_{x=L} \qquad (12.31)$$

where $\partial T/\partial t$ represents the temperature history and $\partial T/\partial x$ the temperature gradient which applies along the x-axis of the thermoplug and can be obtained from measured data.

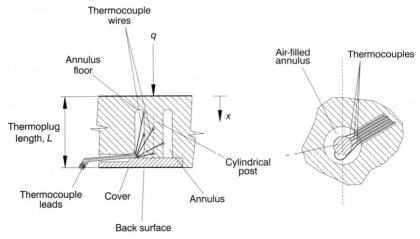

Figure 12.15 Plug gauge (after Liebert, 1994)

To improve the speed of response of the plug gauge the temperature-measuring location should be located closer to the surface. A method of achieving this is to drill a hole in the slug of material forming the calorimeter and to locate a thermocouple there as illustrated in Figure 12.16. The heat flux measurement is based on a one-dimensional transient conduction solution and is given, as for a semi-infinite geometry, by

$$q = \frac{2\sqrt{\rho c_p k}}{\sqrt{\pi}} \sum_{i=1}^{n} \frac{T_i - T_{i-1}}{\sqrt{t_n - t_i} + \sqrt{t_n - t_{i-1}}} \tag{12.32}$$

In order to eliminate the effects of the initial temperature transient on the indicated heat flux, a value of $a/b = 1.375$ (see Figure 12.16) is recommended by Kidd (1990a) for the ratio of hole diameter to thermocouple location from the surface. Measurements can be taken until the thermal transient reaches the back of the sensor at L, and the useful period of measurement can be bounded by

$$\frac{3b^2}{\alpha} < t < \frac{0.3L^2}{\alpha} \tag{12.33}$$

The term on the right-hand side of the above inequality is associated with an error in the semi-infinite approximation of less than 1% (Kidd, 1990a).

The principle of the coaxial gauge is to measure the variation in surface temperature directly as a function of time (Kidd, 1990b). A thermocouple is formed by the concentric arrangement illustrated in Figure 12.17. This consists of a central thermocouple material, which is electrically isolated from the outer thermocouple material by an insulating sheath. The thermocouple

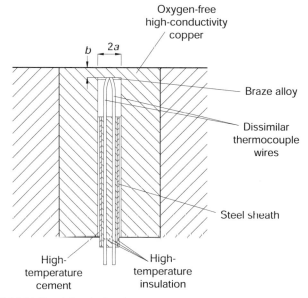

Figure 12.16 Null point calorimeter

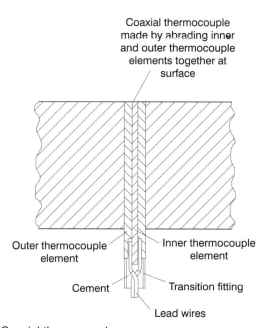

Figure 12.17 Coaxial thermocouple

circuit is completed at the surface by applying a thin layer of one of the thermocouple materials by, for example, vacuum deposition. Alternatively the thermocouple circuit can be completed by removing the insulation at the tip and soldering or merging the two metals together. Equation (12.32) can be used for determining the heat flux from the temporal temperature measurement. The average of the coaxial thermocouple material properties can be used for the material properties product, $\rho c_p k$.

The term 'thin-film sensor' is used here to refer to devices that are physically small, of the order of micrometres thick. They are applied to the surface of interest and due to their size, their speed of response to thermal conditions is fast, (Figure 12.18). The surface temperature is measured using a suitable method and the principles outlined in Section 12.4.2 used to determine the surface heat flux. In addition to surface temperature measurement (one or single-layer sensors), thin-film sensors can also be used to measure heat flux in layered or composite assemblies, although the analysis presented in Section 12.4.2 must be extended for these conditions.

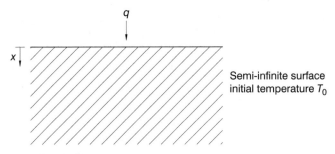

Figure 12.18 Semi-infinite geometry

For the single-layer sensor there are several options to measure the temperature. PRTs can be vacuum deposited or painted and sintered onto an electrically insulating surface such as Macor® (machineable glass ceramic). The painted PRT involves the use of metallo-organic paints, which comprise a fine suspension of a metal in a colloidal solution. Baking at elevated temperatures serves to sinter the metal onto the surface. Lead wires to the sensor can be produced using gold paints sintered onto the surface in a similar manner. The advantage of paints is in the simplicity of application. Alternatively, vacuum deposition techniques can be used to deposit both the sensor and the lead wires. Useful insight into vacuum deposition manufacture techniques applied to multi-layer thin film sensors is given in Lei *et al.* (1997). The advantages of the vacuum deposition techniques are the lower temperatures to which the substrate is exposed during manufacture, the purity of the deposited material and the potential inherent to the process for improved bonding. Calibration is a two-stage process requiring determination of the material properties in terms of the product of $\rho c_p k$ and also calibration of the temperature sensor. Procedures for obtaining values for the $\rho c_p k$ product are commonly based on electric discharge, radiant or pulsed laser heating and are described by Schultz and Jones (1973), Doorley (1985), and Lyons and Gai (1988).

It is not always practicable to manufacture components in an insulating material which will allow application of thin-film surface sensors for heat transfer studies. One technique to overcome this limitation involves the application of sensors to a polyimide film (Guo et al., 1995). This film can then be applied to the component of interest but a two-layer analysis must be employed with the base component that the film is bonded to assumed semi-infinite. An alternative development for use in aggressive environments such as turbomachinery research, for example, involves the application of vitreous enamel to a component to electrically isolate surface sensors, (Doorley, 1985, 1988). This form of sensor is illustrated in Figure 12.19 (see also Figure 12.20).

In the design and selection of a technique using this form of instrumentation it is important to establish the duration of an experiment. The period before an applied thermal boundary condition reaches the back surface of the substrate can be calculated using (Schultz and Jones, 1973)

$$t \le \frac{L^2}{16\alpha} \tag{12.34}$$

Some figures for the thermophysical properties in equation (12.34) will help to illustrate an important limitation. Steel and machineable glass ceramic

Figure 12.19 PRT transducer applied to a nozzle guide vane coated with vitreous enamel (Greenwood, 2000)

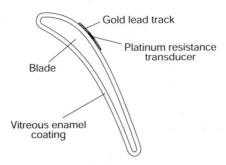

Figure 12.20 Schematic of the type of PRT transducer illustrated in Figure 12.19

(Macor®), are both commonly used materials and represent a reasonably good conductor and a thermal insulator. For 1 per cent carbon steel, $\alpha = 1.2 \times 10^{-5}$ m²/s, and the corresponding minimum thickness for a 0.3 s duration test is 7.6 mm. For Macor®, the respective values are $\alpha = 7.5 \times 10^{-7}$ m²/s, and 1.9 mm. These techniques are therefore limited by the availability of a facility to generate a rapid transient, such as a shock tube or blowdown facility (Schultz and Jones, 1973).

Thermochromic liquid crystals can be applied to a surface and a video recording of the surface temperature variation in response to a fluid temperature change can be used to determine the surface heat flux and corresponding surface heat transfer coefficient (see Ireland and Jones, 2000). A variety of techniques can be used to provide a fluid temperature change relative to the surface temperature. One method is to initially maintain the temperature for the wall at ambient conditions and raise the temperature of the flow by use of a valve switching system. Alternatively, the wall in, say, a wind tunnel can be preheated and the air flow then diverted through. The region of interest can be covered or shrouded providing insulation from the flow and the shroud removed thereby providing a step change in thermal conditions. A further method is to decrease the temperature of a steady flow suddenly by the introduction of, say, liquid nitrogen into an air flow. For semi-infinite conditions, the optically measured variation of temperature with time can be related to the heat transfer coefficient (assumed constant) by solution of

$$\frac{T - T_0}{T_\infty - T_0} = 1 - e^{\beta^2} \operatorname{erfc}(\beta) \tag{12.35}$$

where

$$\beta = h \sqrt{\frac{t}{\rho c_p k}} \tag{12.36}$$

and erfc is the complimentary error function which is tabulated in many mathematics texts and also in Carslaw and Jaeger (1959).

Any of the various forms of infrared thermometer can, in principle, be used to monitor the time-varying surface temperature provided the speed of response of the sensor is fast enough and the techniques given in Section 12.4.2 used to determine the heat flux.

12.5 Energy supply or removal heat flux measurement techniques

In these techniques a physical balance between incoming heat and heat loss is achieved by actively cooling or heating. Heating can be achieved by means of an electric circuit heater or by dissipation of a pulsed radiation source from a laser. Cooling can be achieved using convective passageways within the device or even the Peltier effect, (Shewan *et al.*, 1989). Owing to the response time of a typical active heat flux sensor arrangement, these methods are not recommended for either high heat fluxes or for use at high temperature.

A typical design of sensor for measuring the heat transfer coefficient in this manner for heat transfer into a surface is the use of heater strips which are exposed to the convective boundary condition but well insulated on the reverse and sides. Electric current is passed through the heater strip material until equilibrium, monitored by a surface temperature device such as a thermocouple, (Wilson and Pope, 1954) or RTD is achieved.

A sandwich construction can be applied to a surface comprising an insulating layer, heater, black background and a liquid crystal layer to obtain local steady-state heat transfer values. The heat transfer coefficient can then be determined from a heat balance:

$$h = \frac{(I^2 \Re / A) - \varepsilon \sigma (T_s^4 - T_\infty^4)}{T_\infty - T} \qquad (12.37)$$

where:
h	=	heat transfer coefficient (W/m²·K)
I	=	current (A)
$\Re$	=	resistance (Ω)
A	=	area (m²)
ε	=	total emissivity
σ	=	the Stefan–Boltzmann constant (5.67051×10^{-8} W/m²·K⁴ (Cohen and Taylor, 1999)
T_s	=	surface temperature (K)
T_∞	=	free stream temperature (K)
q_{loss}	=	a term to account for heat loss from the system (W/m²)
T	=	temperature.

Heaters used have included thermistors, thin layers of gold, Baughn *et al.* (1986), nichrome wires and carbon fibres. Figure 12.21 illustrates the use of a

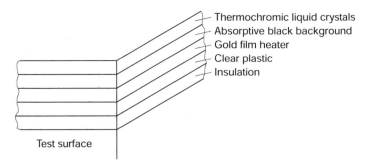

- Thermochromic liquid crystals
- Absorptive black background
- Gold film heater
- Clear plastic
- Insulation

Test surface

Figure 12.21 Sandwich construction heater method

gold layer sandwich construction. The difficulty in applying a uniform layer to a complex geometry can be overcome by means of producing a flexible sandwich, which can be stuck to the component of interest. One embodiment of this is to use carbon fibres as the heater in an epoxy resin. Thermochromic liquid crystals are applied to the surface to measure the temperature.

Energy supply methods can also be used in a transient mode by applying a known quantity of heat to a surface and measuring the thermal response. The use of a laser to apply a discrete local heat flux and an infrared camera system to determine the local surface temperature distribution due to heating is reported in Porro *et al.* (1991). A conduction solution using this measured temperature distribution can be used to calculate the heat transfer coefficient.

12.6 The mass transfer analogy

There is a mathematical similarity between the governing partial differential equations that describe heat transfer and the transport of mass. This can be practically exploited to make it possible to obtain heat transfer measurements from the results of mass transfer experiments. One of the best-known mass transfer methods is the naphthalene sublimation technique where the thickness of a surface coated with naphthalene is monitored. The advantage of mass transfer experiments is that they are generally much easier to carry out (especially on complex, or difficult geometries) than heat transfer experiments, are less prone to experimental error and uncertainty and it is much easier to model the equivalent of adiabatic and isothermal boundary conditions.

At atmospheric pressure and temperatures, naphthalene ($C_{10}H_8$) sublimes in air, i.e. there is a transition from solid to vapour without any intermediate liquid phase. This and the fact that naphthalene has relatively low toxicity, can be easily cast and machined and that the relevant thermophysical properties are well documented (Kudchadker *et al.*, 1978; Dean, 1987), make it an ideal choice for the method. The first stage in a mass transfer experiment is to apply naphthalene to the surface of interest. This can be done using casting, machining, spraying, dipping or a combination of these. Prior to beginning the experiment, the local mass distribution of naphthalene is measured by either weighing or using a surface profile gauge. For obtaining local rather than surface-average information, the surface of interest is divided into a number of individual segments, which are weighed separately. The experiment is run for enough time to create a reasonable loss of the solid naphthalene. The mass transferred is measured by either weighing or using a surface profile gauge at the completion of the experimental run.

Forced convection heat transfer can typically be modelled in the form

$$Nu = cRe^m Pr^n \qquad (12.38)$$

where Nu, Re and Pr are the Nusselt, Reynolds and Prandtl numbers, respectively. In a similar way, mass transfer data can be correlated using

$$Sh = cRe^m Sc^n \qquad (12.39)$$

where Sh is the Sherwood number (a dimensionless mass transfer coefficient), Sc the Schmidt number and c, n and m are constants. The translation from measurements of mass transfer to heat transfer coefficients is based on the equivalence of the Sherwood number to the Nusselt number, and the Schmidt number to the Prandtl number. This is expressed formally in equations (12.38) and (12.39) and, by rearranging, we find the desired relation between heat and mass transfer:

$$Nu = Sh \left(\frac{Pr}{Sc} \right)^n \qquad (12.40)$$

where, for example, for laminar flow from a flat plate, n takes the value of 1/3.

The Sherwood number, Sh, is defined as

$$Sh = h_m \frac{L}{D_{naph}} \qquad (12.41)$$

where: h_m = the local mass transfer coefficient (m/s)
         L = a characteristic length (m)
         D_{naph} = diffusion coefficient for naphthalene (m^2/s):

$$h_m = \frac{1}{\rho_v} \frac{dm}{dt} \qquad (12.42)$$

The differential dm/dt represents the amount of mass, m, of naphthalene transferred over a period, t. The density of naphthalene vapour, ρ_v, is calculated from the ideal gas equation using the vapour pressure of naphthalene and the wall temperature, L is a suitable characteristic length and D_{naph} is the diffusion coefficient for naphthalene in air. The Schmidt number is based solely on the properties of naphthalene, and

$$Sc = \frac{\upsilon}{D_{naph}} \approx 2.28 \text{ (at 1 bar and } 20°C) \tag{12.43}$$

where υ is the kinematic viscosity (m^2/s).

Using the relations given by equations (12.41), (12.42) and (12.43), in equation (12.40), it can be seen that it is possible to infer the Nusselt number and hence the convective heat transfer coefficient. A similar argument can be applied to buoyancy-driven flows (using the Rayleigh number); the resulting equivalence of Sherwood and Nusselt numbers is the same.

Although useful, the technique does have a number of drawbacks including:

- At low velocities, experimental times can be excessive and it may be difficult to control the ambient temperature over such long periods. The vapour pressure of naphthalene is very sensitive to ambient temperature (a 1°C change in temperature results in a 10% change in the vapour pressure). Therefore, a long experimental time can introduce significant experimental error.
- The sensitivity of vapour pressure to temperature is also of consequence at high velocities where viscous heating occurs.
- In high shear flows, naphthalene can be worn off the surface by erosion.
- A long experimental time may cause a significant change of shape in the mass transfer model.
- The results are, by definition, time average values.
- For wind tunnels, a naphthalene-free air stream should be maintained and variations in the air supply temperature minimized.
- The latent heat of vaporization of naphthalene will cause a drop in the actual surface temperature compared with that of the free stream. It is therefore advisable to measure the temperature of the naphthalene model by thermocouples on the surface.

Nonetheless, the technique has been used to provide accurate and valuable data for applications where conventional heat transfer instrumentation would be more difficult, in particular: external flows, internal flows, natural convection, fins and heat exchanger geometries, turbulence promoters, impinging jets, rotating surfaces and the cooling of electronics equipment. For further information on this technique the reader is referred to the review by Goldstein and Cho (1995) and a comprehensive survey of applications is given by Souza Mendez (1991).

12.7 Inverse conduction methods

Measured component temperatures can be used in conjunction with a conduction solution to determine the local heat flux and heat transfer coefficient. This method is routinely used for large-scale system analysis such as determining the heat transfer in the internal flow system of a gas turbine engine (Alexiou *et al.*, 2001). The method uses an identifiable step change in running conditions provided by, say, either an acceleration or a rapid change in the external flow temperature. The challenge facing the analyst is to match these measured temperatures with values calculated using a finite element-based conduction solution. The general procedure is as follows:

1 A basic model is set up comprising the geometry and area properties. An assumption of symmetry is usually made to reduce the complexity of the problem to be analysed.
2 Any time-varying parameters such as air pressure and temperature, mass flow rate and rotational speed are defined in a control database.
3 Thermal boundary conditions are imposed throughout the model.
4 Output points are defined at, say, thermocouple locations and a thermal analysis is run. The predicted temperatures are then compared with the measured values.
5 Based on the insight given by the comparison, the thermal boundary conditions are modified by, for example, introducing correction factors into the assumed heat transfer correlations or by changing the flow distribution. The model is then run again and the process repeated until a reasonable match between the measured and calculated data is obtained.

Manipulation of one boundary condition for one component influences the complete solution, and divergence away from conditions experienced is easily identified by the difference between the predicted and measured temperatures. Alexiou *et al.* (2001) note that even if a match is obtained it is possible that the modelling assumptions do not necessarily represent all the physical processes involved. Modelling using inverse heat transfer techniques is also described by Ozisik and Orlande (2000).

12.8 Conclusions

Heat flux sensors are used to measure the transfer of heat within a system. Heat flux meters can, for convenience, be divided into four categories: differential temperature, calorimetric, energy supply and those based on a mass transfer analogy. Not all types of heat flux sensor necessarily fall neatly into one of these categories and some may operate in one mode for steady conditions and in another under transient conditions. A summary of the characteristics of the techniques is presented in Table 12.1 and this can be used to assist in the selection of an appropriate method for a given application.

Table 12.1 Guide to heat flux measurement technique identification

Type	Steady state	Unsteady	Low heat flux < 10 kW/m²	High heat flux > 10 kW/m²	High temperature	Special manufacturing techniques	Commercially available	Fast response	Relative cost	Low thermal disturbance	High g capability	High signal
Differential thermocouple*	×A	✓C	×	✓E	✓E, F	×	×	✓	Low	×H	✓J	×
Differential PRT*	×A	✓C	×	✓E	✓E, F	×	×	✓	Low	×H	✓	✓ Bridged circuit
Differential thermopile	✓B	✓	✓B	✓E	✓E, F	✓G	✓	✓	Medium	Medium ×	×R	✓M
Transparent insert	✓Ng (1993, 1996)	✓	×	✓	✓	✓ Vacuum deposition	×	✓	High	×	×	× Subject to temperature difference
Planar thermopile	✓B	✓	✓B	✓E	✓E, F	✓ Vacuum deposition and lithographic technology	×	✓	High	× Subject to thickness and materials used	×R	✓M
Gardon gauge	✓ May require cooling	✓	×	✓ May require cooling	✓ Requires cooling	✓	✓	×	Medium	×	×	× Subject to heat flux
Interferometry	✓	×	✓	✓	×	×	×	×	High	✓	×	×
Slug calorimeters	×	✓	×	✓	✓F	×	×	✓	Medium	×K	×	×
Thin-skin sensors	×	✓D	×	✓	✓F	×	×	✓	Low	✓	✓N	×
Plug gauges	×	✓	×	✓	✓F	✓Q	×	✓	Medium	×K	×	×
Null point calorimeters	×	✓	×	✓	✓F	×	✓	✓	Medium	×K	×	×

Technique								Resolution			
Coaxial thermocouples	×	✓	×	✓	✓	✓F	✓ Wire extrusion	✓	Medium	×K	×	
One-layer thin-film sensors	×	✓D	✓D	×	✓E	✓F	×R	✓	Medium	✓	✓ Bridged circuit	
Multiple-layer thin-film sensors	× Unless the heat flux is high (Epstein 1985, 1986)	✓D	✓D	×	✓E	✓F	✓L	✓	High	✓	✓ Bridged circuit	
Thermochromic liquid crystals	×C	×C	✓D	×	✓E	×	×	✓	Medium	✓	× Crystals are subjected to strain	✓ Subject to colour resolution capability
Radiometry	×C	×C	✓D	×	✓	×	×	✓	High	✓	✓	
Heater or cooler methods	✓	✓	✓ Ideally suited to this application	×	×D	✓	×	×	High	High	×	✓
Mass transfer analogy	✓	×	✓ E,F	✓	×	×	×	×	Low	✓	×	×

* Conventional thermocouple wire or resistance thermometers.

A Unless the heat flux is high.
B Requires sufficient number of junctions.
C In conjunction with a 1D transient conduction analysis.
D 1D transient conduction analysis.
E Subject to thermal expansion considerations.
F Subject to material temperature limits.
G For miniature devices, requires vacuum deposition and lithographic technology.
H Unless applied directly across component.
J Subject to attachment method.
K Unless slug has similar thermal conductivity to surroundings.
L May require spray technology, lithography and vacuum deposition.
M Subject to number of junctions.
N Subject to attachment methods used for thermocouples.
P Time constants normally preclude their use.
Q May require electric discharge machining.
R Unless vacuum deposited to surface

References

Books and papers

Ainsworth, R.W., Allen J.L., Davies, M.R.D., Doorly, J.E., Forth, C.J.P., Hilditch, M.A., Oldfield, M.L.G. and Sheard, A.G. Developments in instrumentation and processing for transient heat transfer measurement in a full stage model turbine. *Trans. ASME, Journal of Turbomachinery*, **111**, 20–27, 1989.

Alexiou, A., Long, C.A., Turner, A.B. and Barnes, C.J. Thermal modelling of a rotating cavity rig to simulate the internal air system of a gas turbine H.P. compressor. 5th World Conference on Experimental Heat Transfer, Fluid Mechanics and Thermodynamics – ExHFT-5, Thessaloniki, Greece, 2001.

Ash, R.L. Response characteristics of thin foil heat flux sensors. *AIAA Journal*, **7**(12), 2332–2335, 1969.

Baughn, J.W., Cooper, D., Iacovides, H. and Jackson, D. Instruments for the measurement of heat flux from a surface with uniform temperature, *Review of Scientific Instruments*, **57** (5), 921–925, 1986.

Bayley, F.J. An analysis of turbulent free-convection heat transfer. *Proc. I. Mech. E.*, 169, 361–370, 1955.

Bejan, A. *Heat transfer*. Wiley, 1993.

Bhatt, H. and Fralick, G. Novel thin-film heat flux sensors: fabrication and calibration. *Proceedings of the Structural Testing Technology at High Temperature II*, Vol. **18**, pp. 88–97, 1993.

Carslaw, H.S. and Jaeger, J.C. *Conduction of Heat in Solids*, 2nd edition. Oxford University Press, 1959.

Childs, P.R.N. *Heat Transfer at the Surface of a Cylinder Rotating in an Annulus with a Stator Blade Row and Axial Throughflow*. DPhil thesis, University of Sussex, 1991.

Childs, P.R.N., Greenwood, J.R. and Long, C.A. Heat flux measurement techniques. *Proceedings of the Institution of Mechanical Engineers*, Part C, Vol. 213, pp. 655–677, 1999.

Cho, C.S.K., Fralick, G.C. and Bhatt, H.D. An experimental study of a radially arranged thin-film heat-flux gauge. *Measurement Science & Technology*, **8**(7), 721–727, 1997a.

Cho, C.S.K., Fralick, G.C. and Bhatt, H.D. Steady-state and frequency response of a thin-film heat flux gauge. *Journal of Spacecraft and Rockets*, **34**(6), 792–798, 1997b.

Cohen, E.R. and Taylor, B.N. The fundamental physical constants. *Physics Today*, BG5-BG9, 1999.

Cook, W.J. Determination of heat transfer rates from transient surface temperature measurements. *AIAA Journal*, **8**(7), 1366–1368, 1970.

Cook, W.J. and Felderman, E.J. Reduction of data from thin-film heat transfer gages: a concise numerical technique. *AIAA Journal*, 561–562, 1966.

Dean, J. A. *Handbook of Organic Chemistry*. McGraw-Hill, 1987.

Diller, T.E. Heat flux instrumentation for Hyflite thermal protection system. *NASA CR-197715*, 1994.

Diller, T.E. Methods of determining heat flux from temperature measurements. *ISA*, pp. 251–262, 1996.

Doorly, J.E. *The Development of a Heat Transfer Measurement Technique for Application to Rotating Turbine Blades*. DPhil thesis, Somerville College, Oxford University, 1985.

Doorly, J.E. Procedures for determining surface heat flux using thin film gages on a coated metal model in a transient test facility. *Trans. ASME, Journal of Turbomachinery*, **110**, 242–250, 1988.

Dunn, M.G., Kim, J. and Rae, W.J. Investigation of the heat-island effect for heat-flux measurements in short-duration facilities. *Trans. ASME, Journal of Turbomachinery*, **119**(4), 753–760, 1997.

Epstein, A.H., Guenette, G.R., Norton, R.J.G. and Yuzhang, C. High frequency response heat flux gauge for metal blading. *AGARD CP 390*, pp. 30(1)–30(16), 1985.

Epstein, A.H., Guenette, G.R., Norton, R.J.G. and Yuzhang, C. High frequency response heat flux gauge. *Review of Scientific Instruments*. **57**, 4, 639–649, 1986.

Flanders, S.N. Heat flow sensors on walls – what can we learn? In Bales, E., Bomberg, M. and Courville, G.E. (Editors), *Building Applications of Heat Flux Transducers*. ASTM STP 885, pp. 140–159, 1985.

Gardon, R. An instrument for the direct measurement of intense thermal radiation. *Review of Scientific Instruments*, **24**(5), 366–370, 1953.

Gardon, R. A transducer for the measurement of heat-flow rate. *Journal of Heat Transfer*, 396–398, 1960.

Godefroy, J.C., Clery, M., Gageant, C., Francois, D. and Servouze, Y. Thin film temperature heat fluxmeters. *Thin Solid Films*, **193**(1–2), 924–934, 1990.

Goldstein, R.J. and Cho, H.H. A review of mass transfer measurements using naphthalene sublimation. *Experimental Thermal and Fluid Science*, **10**, 416–434, 1995.

Greenwood, J.R. *The Development of Robust Heat Transfer Instrumentation for Rotating Turbomachinery*. DPhil thesis, University of Sussex, 2000.

Guo, S.M., Spencer, M.C., Lock, G.D., Jones, T.V. and Harvey, N.W. The application of thin film gauges on flexible plastic substrates to the gas turbine situation. ASME Paper 95-GT-357, 1995.

Hager, J.M., Onishi, S., Langley, L.W. and Diller, T.E. Heat flux microsensors. In: *Heat Transfer Measurements, Analysis, and Flow Visualisation*, The National Heat Transfer Conference. ASME HTD-112, pp. 1–8, 1989.

Hager, J.M., Simmons, S., Smith, D., Onishi, S., Langley, L.W. and Diller, T.E. Experimental performance of a heat flux microsensor. ASME Paper 90-GT-256, 1990.

Hager J.M., Terrell J.P., Sivertson, E. and Diller, T.E. *In-situ* calibration of a heat flux microsensor using surface temperature measurements. Vatell Corporation, Christiansburg, VA, USA, Paper No. 0227-7576/94, 261–269, 1994.

Hager, N.E. Jr. Thin foil heat meter. *Review of Scientific Instruments*, **36**(11), 1564–1570, 1965.

Hager, N.E. Jr. Temperature sensing probe. Patent number US 3 354 720, USA, 1967.

Hartwig, F.W. Development and application of a technique for steady state aerodynamic heat transfer measurements. Hypersonic Research Project No 37, Guggenheim Aeronautical Laboratory California Institute of Technology, OOR Project No. 1600-PE, 1957.

Hartwig, F.W., Bartsch, C.A. and McDonald, H. Miniaturised heat meter for steady state aerodynamic heat transfer measurements. *Journal of the Aeronautical Sciences*, 239, 1957.

Hines, F. Thermal apparatus. Patent number US 3 607 445, USA, 1971.

Holmberg, D.G. and Diller, T.E. High frequency heat flux sensor calibration and modelling. *Journal of Fluids Engineering*, **117**, 659–664, 1995.

Ireland, P.T. and Jones, T.V. Liquid crystal measurements of heat transfer and surface shear stress. *Meas. Sci. Technol.*, **11**, 969–986, 2000.

Kidd, C.T. Recent developments in high heat flux measurement techniques at the AEDC. *Proceedings of the ISA Aerospace Instrumentation Symposium*, Vol. 36, pp. 477–492, 1990a.

Kidd, C.T. Coaxial surface thermocouples: Analytical and experimental considerations for aerothermal heat flux measurement applications. *Proc 36th Int. Instrumentation Symposium*, 1990b.

Kim, J., Ross, R.A. and Dunn, M.G. Numerical investigation of the heat-island effect for button-type, transient, heat-flux gage measurements. *Proceedings 31st ASME National Heat Transfer Conference*, Vol. 327(5), pp. 33–39, 1996.

Kudchadker, A.P., Kudchadker, S.A. and Wilhoit, R.C. *Naphthalene*. API Monograph Ser. 707, American Petroleum Institute, Washington, DC, 1978.

Kuo, C.H. and Kulkarni, A.K. Analysis of heat flux measurement by circular foil gages in a mixed convection/radiation environment. *Journal of Heat Transfer*, **113**, 1037–1040, 1991.

Lei, J.F., Martin, L.C. and Will, H.A. Advances in thin film sensor technologies for engine applications. ASME Paper 97-GT-458, 1997.

Liebert, C.H. Miniature convection cooled plug-type heat flux gauges. *Instrumentation in the Aerospace Industry – 40th Proceedings*, pp. 289–302, 1994.

Liebert, C.H. and Weikle, D.H. Heat flux measurements. ASME Paper 89-GT-107, 1989.

Long, C.A. *Essential Heat Transfer*. Longman, 1999.

Lyons, P.R.A. and Gai, S.L. A method for the accurate determination of the thermal product ($\rho c_p k$) for thin film heat transfer or surface thermocouple gauges. *Journal of Physics E*, **21**, 445–448, 1988.

Martin, H. Heat and mass transfer between impinging gas jets and solid surfaces. In *Advances in Heat Transfer*, Vol. 13, pp. 1–59, Academic Press, 1977.

Martinelli, R.C., Morrin, E.H. and Boelter, L.M.K. An investigation of aircraft heaters V – Theory and use of heat meters for the measurement of rates of heat transfer which are independent of time. NACA, 1942.

Meyer, V.M. and Keller, B. A new heat flux sensor: from microvolts to millivolts. *Sensors and Materials*, **8**, 6, 345–356, 1996.

Oldfield, M.L.G., Jones, T.V. and Shultz, D.L. On-line computer for transient turbine cascade instrumentation. *IEEE Transactions on Aerospace and Electronic Systems*, **AES-14**, No. 5, 738–749, 1978.

Ozisik, M.N. and Orlande, H.R.B. *Inverse Heat Transfer*. Taylor and Francis, 2000.

Porro, A.R., Keith, T.G. Jr and Hingst, W.R. A laser-induced heat flux technique for convective heat transfer measurements in high speed flows. *ICIASF Record*, International Congress on Instrumentation in Aerospace Simulation Facilities, pp. 146–155, 1991.

Schultz, D. and Jones, T.V. Heat transfer measurements in short duration hypersonic facilities. AGARD 165, 1973.

Shewan, E.C., Hollands, K.G.T., and Raithby, G.D. The measurement of surface heat flux using the Peltier effect. *Journal of Heat Transfer*, **111**, 798–803, 1989.

Souza Mendes, P.R. The naphthalene sublimation technique. *Experimental Thermal and Fluid Science*, **4**, 510–523, 1991.

Trimmer, L.L., Matthews, R.K. and Buchanan, T.D. Measurement of aerodynamic heat rates at the Von Karman facility. IEEE Congress on Instrumentation in International Congress on Instrumentation in Aerospace Simulation Facilities, 1973.

Will, H.A. Fabrication of thin film heat flux sensors. NASA Langley Measurement Technology Conference, 93N13667, pp. 97–106, 1992.

Wilson, D.G. and Pope, J.A. Convective heat transfer to gas turbine blade surfaces. *Proc. I. Mech E.*, **168**, 861–874, 1954.

Wood, N.B. Hypersonic heat transfer and boundary layer transition on sharp and blunted cones. Royal Armam. R&D Est. Memo 40/68, 1968.

Standards

ASTM. Standard test method for E457–96 measuring heat transfer rate using a thermal capacitance (slug) calorimeter. American Society for Testing and Materials, 1997.

Websites

At the time of going to press the world wide web contained useful information relating to this chapter at the following sites.

http://cyclops-mac.larc.nasa.gov/AFCwww/AFC.html
http://members.aol.com/desmondint/Captec.html
http://www.bph.hbt.ethz.ch/
http://www.branom.com/LITERATURE/thermo-reg.html
http://www.cels.corning.com/
http://www.chemeng.ed.ac.uk/people/henry/masstran/cdg/masstind.html
http://www.dexterresearch.com/products.htm
http://www.electriciti.com/~thermo/htt.html
http://www.eng.ox.ac.uk/lc/research/introf.html
http://www.g3.net/vatell/
http://www.hukseflux.com/heat%20flux/heatflux.htm
http://www.hull.ac.uk/php/chpsmt/lc/history.html
http://www.inframetrics.com/indexn.htm
http://www.jbme.com/INDEXGB.HTM
http://www.landinst.com/infr/
http://www.lci.kent.edu/lc.html#Description
http://www.lerc.nasa.gov/WWW/RT1996/2000/2510f2.htm
http://www.minco.com/
http://www.omega.com/pdf/temperature/Z/zsection.asp
http://www.rdfcorp.com/
http://www.sensorsmag.com/articles/0199/flu0199/main.shtml
http://www.thiesclima.com/radiation/heat_flux.htm
http://www.tpd.tno.nl:80/TPD/smartsite101.html
http://www.tpd.tno.nl:80/TPD/smartsite101.html
http://www.wuntronic.de/heat_flux/cocept.htm

Nomenclature

A	=	area (m^2)
Bi	=	Biot number
c_p	=	specific heat capacity (J/kg·K)
D_{naph}	=	diffusion coefficient for naphthalene (m^2/s)
E	=	electromotive force (V)
g	=	acceleration due to gravity (m^2/s)
h	=	heat transfer coefficient (W/m^2·K)
h_m	–	mass transfer coefficient (m/s)
Gr_x	=	Grashof number
I	=	current (A)
k	=	thermal conductivity (W/m·K)
k_o	=	thermal conductivity of the material at a base temperature $T = 0$ (W/m·K)
K_1	=	the fractional variation of conductivity per unit temperature rise (1/K)
L	=	length (m)
m	=	the mass of the slug (kg)
n	=	normal coordinate (m)
N_J	=	number of thermocouple junctions
Nu	=	Nusselt number
Pr	=	Prandtl number
q	=	heat flux (W/m^2)
Q_{loss}	=	heat loss (W)
Q_{out}	=	heat transfer out (W)
r	=	local radial coordinate (m)
R	=	radius (m)
$\Re$	=	resistance (Ω)
Re	=	Reynolds number
S	=	Seebeck coefficient (V/K)
Sc	=	Schmidt number
Sh	=	Sherwood number
t	=	time (s)
t_f	=	final time (s)
t_0	=	intial time (s)
T	=	temperature (K)
T_f	=	final temperature (K)
T_s	=	surface temperature (K)
T_0	=	initial temperature (K)
T_∞	=	*free stream temperature (K)*
x	=	local coordinate (m)
α	=	thermal diffusivity (m^2/s)
β	=	coefficient of volumetric expansion (($=1/T$) for an ideal gas)
δ	=	thickness (m)

δx = differential distance (m)

ε = total emissivity

μ = viscosity (Ns/m^2)

ρ = density (kg/m^3)

ρ_v = vapour density (kg/m^3)

σ = the Stefan–Boltzmann constant (W/m^2·K^4)

τ = time constant (s)

υ = kinematic viscosity (m^2/s)

ΔT = temperature difference (K)

Index _____